Heavy Metals in the Environment - Contamination, Risk, and Remediation

Edited by Mitsuo Yoshida

Published in London, United Kingdom

Heavy Metals in the Environment - Contamination, Risk, and Remediation
http://dx.doi.org/10.5772/intechopen.1003438
Edited by Mitsuo Yoshida

Contributors
Abdul Majeed Khan, Angelica A. Chacon, Babalwa Gqomfa, Benett Siyabonga Madonsela, Bibi Shazia, Boredi Silas Chidi, Carlos R. Cabrera, Chuanpin Chen, Dilgam Tagiyev, Femi Peter Adesina, Fisseha Itanna, Getabu Albert, Hamid Ullah, Ibrahim Shaw, Jozef Hančuľák, Karabo Concelia Malakane, Ketiyala Ao, Latonglila Jamir, Lemzila Rudithongru, Mahdi Banaee, Mitsuo Yoshida, Motsamai Tjakata, Muhammad Asif, Muhammad Imtiaz, Nizami Zeynalov, Nyabaro Obed, Omondi Argwings, Omondi Reuben, Orina Paul, Rayori Douglas Mosoti, Robert Birundu Onyancha, Sadeeq Ullah, Sevda Fatullayeva, Shamil Tagiyev, Takalani Terry Phungela, Thabang Maphanga, Tomáš Faragó, Tsenbeni N Lotha, Tímea Brázová, Xolisiwe Sinalo Grangxabe, Yang Liu, Yimer Seid Ali

First published in London, United Kingdom, 2025 by IntechOpen
IntechOpen is the global imprint of INTECHOPEN LIMITED, registered in England and Wales, registration number: 11086078, 167-169 Great Portland Street, London, W1W 5PF, United Kingdom

For EU product safety concerns: IN TECH d.o.o., Prolaz Marije Krucifikse Kozulić 3, 51000 Rijeka, Croatia, info@intechopen.com or visit our website at intechopen.com.

British Library Cataloguing-in-Publication Data
A catalogue record for this book is available from the British Library

Heavy Metals in the Environment - Contamination, Risk, and Remediation
Edited by Mitsuo Yoshida
p. cm.
Print ISBN 978-0-85014-847-3
Online ISBN 978-0-85014-846-6
eBook (PDF) ISBN 978-0-85014-848-0

If disposing of this product, please recycle the paper responsibly.

Meet the editor

Mitsuo Yoshida serves as the Senior Advisor on Environmental Management and Climate Change at the Global Environment Department, Japan International Cooperation Agency (JICA) in Tokyo, Japan. He is also the Director and CEO of the International Network for Environmental & Humanitarian Cooperation (iNehc), a nonprofit organization based in Tokyo. Dr. Yoshida has held notable positions in academia at leading universities in Japan. He served as a Visiting Professor of International Environmental Cooperation in the Department of International Studies, Division of Environment, Graduate School of Frontier Sciences, at the University of Tokyo, from 2012 to 2017. Before that, he was an Environmental Science and Technology Visiting Professor at the Multidisciplinary Graduate School of Science and Engineering, Tokyo Institute of Technology, from 2008 to 2012. Yoshida's expertise spans Environmental Management Engineering, Waste Management, Soil and Groundwater Contamination, and Environmental Geology. He holds a Ph.D. from the Graduate School of Science at Hokkaido University, Sapporo, Japan. He has authored a number of research papers in the fields of environmental science and international technical cooperation. These publications (in English) are accessible for download via his profile on the following platform: https://mitsuoyoshida.academia.edu/.

Contents

Preface

Heavy metals are not consistently defined across scientific and engineering disciplines; their definitions vary depending on the specific field of study and intended application [1]. Generally, the term refers to metallic elements with a density exceeding 5 g/cm^3, including elements such as chromium (Cr), manganese (Mn), iron (Fe), cobalt (Co), nickel (Ni), copper (Cu), zinc (Zn), molybdenum (Mo), cadmium (Cd), mercury (Hg), and lead (Pb). Many of these elements are integral to industrial processes and are also essential to human life as micronutrient metals (e.g., Cr, Mn, Fe, Co, Mo, Ni, and Zn). However, when exposure surpasses safe thresholds, these elements can exhibit toxic effects, posing significant risks to human health and ecosystem in the environment. Environmental contamination caused by heavy metals, often due to mining and industrial activities, necessitates effective countermeasures and remediation strategies. The issue is particularly pronounced in economically developing countries, where rapid industrialization and mining activities are central to economic advancement but frequently lead to severe environmental challenges.

The book "*Heavy Metals in the Environment – Contamination, Risk, and Remediation*" presents a comprehensive synthesis of contemporary research on heavy metal contamination, associated risks, and remediation strategies. This volume is a valuable resource for experts and researchers across diverse fields, including environmental science, chemistry, engineering, ecology, and public health.

The book is structured into three parts comprising 13 chapters:

Section 1: Contamination of Heavy Metals

The first section focuses on heavy metals' migration, contamination, and transformation within various environmental compartments. It elucidates the chemical, physical, and biological processes governing the behavior of heavy metals in natural systems. Heavy metals migrate through multiple media, including air, water, and soil, undergoing complex transformations influenced by specific environmental conditions. This section critically reviews existing research while presenting case studies to provide a detailed understanding of these mechanisms.

- Chapter 1: Takalani Terry Phungela et al. examine the relationship between metal concentrations in river water and climate change, using the Kaapmuiden River as a case study. The authors discuss that climate variability, such as fluctuations in precipitation patterns and temperature, can impact metal concentration levels in water resources.

- Chapter 2: Sevda Fatullayeva et al. review current knowledge on the distribution, toxicity, and removal of heavy metals from the environment and living organisms, focusing on novel sorbents.

- Chapter 3: Mitsuo Yoshida investigates mercury methylation and its increasing toxicity based on studies of mercury-contaminated river wetlands in Uruguay.

Section 2: Risks of Heavy Metals

The second section explores the risks posed by heavy metals to public health and ecosystems. Emphasis is placed on the toxicological effects of heavy metals, particularly in economically developing regions and areas with high industrial activity. The section also discusses bioaccumulation processes and their impacts on ecosystem services from multiple perspectives.

- Chapter 4: Mahdi Banaee investigates heavy metals' sources, pathways, and toxicological impacts in aquatic ecosystems based on literature review.

- Chapter 5: Tímea Brázová et al. demonstrate the utility of fish parasites as bioindicators of heavy metal contamination in aquatic environments.

- Chapter 6: Femi Peter Adesina assesses the ecological risks and toxicity of heavy metals in various aquatic organisms.

- Chapter 7: Rayori Douglas Mosoti et al. discussed that phytoplankton could contribute to wastewater polishing by converting nutrients into their biomass and removing heavy metals from the wastewater through bioaccumulation, based on the analysis of phytoplankton community structure in a wastewater treatment plant in Kenya.

- Chapter 8: Motsamai Tjakata and Fisseha Itanna study the risks associated with sorghum and pearl millet cultivated in soils contaminated by heavy metals near diamond mining sites.

- Chapter 9: Hamid Ullah et al. (Pakistan, Saudi Arabia) discuss heavy metal contamination in food, its associated health risks, and policy recommendations for public safety.

- Chapter 10: Angelica A. Chacon and Carlos R. Cabrera examine uranium's dual role as a valuable resource and a potential environmental hazard, emphasizing the need for balanced management strategies.

Section 3: Remediation of Heavy Metals Pollution

The final section addresses remediation technologies and countermeasures for heavy metal contamination. It highlights innovative approaches integrating physical, chemical, and biological techniques while promoting sustainable solutions through resource recovery and policy frameworks.

- Chapter 11: Tsenbeni N. Lotha et al. review the development of activated carbon derived from biowaste for treating heavy metal-contaminated wastewater.

- Chapter 12: Robert Birundu Onyancha discusses the efficacy of agricultural waste adsorbents as cost-effective, environmentally safe solutions for heavy metal decontamination.

- Chapter 13: Yimer Seid Ali et al. provide an overview of advanced nanotechnology-based heavy metal remediation and analysis approaches.

The chapters included in this volume were carefully selected following a rigorous peer review process. Out of 50 initial proposals, 13 chapters authored by researchers and practitioners from 15 countries—Azerbaijan, China, Ethiopia, India, Iran, Japan, Kenya, Lesotho, Nigeria, Pakistan, Saudi Arabia, Slovakia, South Africa, Tunisia, and the United States (in alphabetical order of country names)—were chosen for publication.

The editor extends our deepest gratitude to the contributing authors, reviewers, and others who supported the editing and publication processes. I hope this book will serve as a valuable reference for researchers, policymakers, and practitioners engaged in addressing heavy metal pollution and will foster further discourse and innovative action toward environmental conservation and sustainable development.

Mitsuo Yoshida
General Director/CEO of International Network for International Network for Environmental and Humanitarian Cooperation,
Nonprofit Inc.,
Tokyo, Japan

Senior Advisor,
Global Environmental Department,
Japan International Cooperation Agency (JICA),
Tokyo, Japan

References

[1] Duffus JH. "Heavy Metals"—A Meaningless Term? (International Union Of Pure And Applied Chemistry (IUPAC) Technical Report). Pure and Applied Chemistry. 2002;**74**(5):793-807. DOI: 10.1351/pac200274050793

Section 1

Contamination of Heavy Metals

Chapter 1

Implication of Climate Variability and Seasonality on Metal Concentrations in Water Resources

Takalani Terry Phungela, Babalwa Gqomfa, Karabo Concelia Malakane, Xolisiwe Sinalo Grangxabe, Benett Siyabonga Madonsela, Boredi Silas Chidi and Thabang Maphanga

Abstract

The exploitation of resources and industrialized and societal activities has resulted in global freshwater contamination, which is a serious environmental concern. Climate variability exerts a significant influence on the metal concentrations in water sources. The monitoring of metal concentrations in surface water systems amidst climate change and variability has received limited attention, especially in southern Africa. Therefore, the objective of this study is to evaluate temporal variations and seasonal patterns that impact metal levels in the Kaapmuiden River. The results show that the river observed higher iron in the autumn season of 2021 and 2016, recording 0.90 mg/l and 0.64 mg/l, respectively. Manganese was high in two different seasons: spring (2016) recording 0.060 mg/l and winter (2021) recording 0.053 mg/l. The result of this study shows that iron concentration was within the acceptable level of approximately 0.5–1 mg/l set for the ecological health of the river. Subsequently, results show high levels of aluminium in autumn for 2017 and 2021 with a concentration of 0.59 mg/l and 1.10 mg/l, respectively. High concentrations of manganese and aluminium in water resources have an ecotoxicological effect on aquatic biota; hence, adequate measures should be placed to minimize the introduction of these metals from anthropogenic activities.

Keywords: climate change, rainfall, surface water, spatio-temporal, seasonal variations

1. Introduction

Climate change poses a serious threat to water resource availability and health, and long-term changes in temperature and rainfall patterns present shifts in stress that can affect water resource availability and quality [1–3]. The interaction between surface water resource quality and climate change is well documented [4–6], investigating the prevalence of certain constituents in water resources and their relationship with climate variability and seasonality. Most of the research has focused mainly on conventional parameters, such as the impact of seasonality on ammonia, nitrate, phosphate, and pH in rivers. The long-term shifts in average temperatures and weather patterns that influence micro- and macroclimates on Earth, coupled with anthropogenic activities, play a major role in the process of metal concentration enrichment in rivers. Moreover, this directly or indirectly influences the chemical composition of water because of the complex multi-stage progressions in streams, reservoirs, and rivers. Furthermore, climate change and variability in the effects of contaminant dynamics involve higher nutrient and sediment runoff to surface water bodies during the rainy season.

Heavy metals are present in trace amounts in aquatic environments; however, their concentrations have increased significantly due to anthropogenic activities, such as industrial, rapid urbanisation, and mining activities [7]. The increase in heavy metals due to anthropogenic activities affects the physiochemical characteristics of water resources, thus affecting aquatic fauna and flora [8]. The presence of heavy metals in potable water also has a detrimental impact on human health. Hence, a study is needed to understand the influence of seasonality and climate variability on the occurrence of heavy metals in water resources [9]. It is anticipated that climate change will lead to an increase in the occurrence of extreme weather events, thereby exerting direct and indirect impacts on the dynamics of pollutants within aquatic ecosystems [10]. An illustration may be observed in the alterations occurring within the physical and geochemical characteristics of the soil-contaminant water system [4]. The influence of climate change on river water quality has been investigated by several researchers [4–6, 8, 9]. Several studies have examined the correlation between temperature and water quality, whereas others have focused on the effects of surface runoff [11, 12]. These studies provide evidence that climate change exerts both direct and indirect impacts on surface water quality [5, 13]. As a consequence of climate change, the magnitude of the effects associated with elevated temperatures has increased [14, 15].

A multitude of scholarly publications in the scientific realm have explored heavy metal contamination, toxicity, and methods for remediation. For instance, Maphanga et al. [16] examined the potential impact of long-term shifts in average temperatures and weather patterns, as well as human activities, on the accumulation of metal contaminants in rivers. Their study aimed to determine the extent to which these factors contribute to the formation of such contaminants in both micro- and macroclimates on Earth. However, Maphanga et al. [16] investigated the relationship between physicochemical parameters (pH and EC) and a broader range of metals, including iron, aluminium, manganese, and arsenic. The current study examines the temporal variations and seasonal patterns that impact the levels of specific metals in the Kaapmuiden River. This research is crucial for understanding the relationship and patterns between climate change and the presence of heavy metals in South African water resources, an area that has not been extensively studied. In addition, this study specifically examined three metals: iron, manganese, and aluminium.

The study conducted by Schiedek et al. [17] investigated the qualitative effects of climate change on heavy metal contamination of marine ecosystems. Climate change has the potential to elevate the likelihood of flood occurrences, thereby amplifying the probability of toxins being reactivated within floodwater and then contaminating sediment and water, which ultimately makes their way into both freshwater and marine ecosystems [18]. Iron, as a crucial transition metal with redox activity, can regulate the geochemical cycle of trace elements [19]. Moreover, this nutrient plays a vital role in marine ecosystems by restricting primary productivity in many oceanic regions and intensifying issues related to cyanobacterial blooms. Comprehension of the temporal fluctuations and factors influencing manganese (Mn) levels in runoff from upland catchments facilitates the capacity to forecast undesirable Mn concentrations in water sources as well as the recognition of water supply catchments that are vulnerable to elevated Mn levels [11]. Although manganese (Mn) is abundant in the natural environment, it has received limited attention in comprehensive hydrological studies of mountainous catchments. The presence of iron in water is predominantly in the form of ferrous iron, which exhibits solubility, or ferric iron, which demonstrates insolubility. Because of the complete degradation of iron, water containing ferrous iron exhibits clarity and lacks discernible colour. Therefore, the objective of this study was to evaluate the temporal variations and seasonal patterns affecting the levels of certain metals in the Kaapmuiden River, which is of utmost importance in order to understand the relationship and patterns between changing climate and the presence of heavy metals in South African water resources where studies of such nature have not been undertaken extensively.

2. Materials and method

2.1 Study area

Kaapmuiden is a small farming town situated in the Lowveld region, approximately 40 km east of Mbombela (previously known as Nelspruit) at the confluence of the Kaap River and Crocodile River in Mpumalanga, South Africa. The area is characterised by sub-tropical climate conditions and fertile farmlands used for cultivating sugarcane and other subtropical fruits and vegetables. The Kaap River is one of the largest tributaries of the Crocodile River system, with a catchment of a total surface area of 1640 km^2, flowing and draining past areas dominated by mining activities (active mines, defunct mines, and decanting mines) and agricultural activities before joining the Crocodile River at Kaapmuiden. After its confluence with Kaap River, Crocodile River water quality is generally characterised by high metal contamination, specifically manganese and arsenic, as well as high sulphates arising from mining activities in the Kaap River sub-catchment [16]. The sub-catchment plays a pivotal role in the region mainly because it fosters socioeconomic development and, furthermore, forms part of water resources of international importance, flowing into Mozambique. The riverbanks are densely covered with marginal vegetation (reeds) and the river is broad and braided. On each bank of the river intensive citrus is grown. The weir acts as a barrier to fish migration under normal flows as the height of the weir is six metres.

The water quality around this vicinity should indicate the effects of the intensive sugarcane farming in the lower reaches of the Crocodile River as well as the sum of all the developments within the Crocodile River catchment upstream.

2.2 Field data collection and analysis

Water samples were collected and analysed for metals such as iron, manganese, and aluminium to evaluate their concentrations in the Crocodile River relative to seasonality and climate in a study area where prevalent land use activities are dominantly farmlands dedicated to sugarcane, subtropical fruit, and vegetables, and a tributary to the river confluence at the area flows through historic mine workings. Moreover, the study forms part of a continuous project within the Crocodile River whereby some research has been published such as Maphanga et al.'s study [16]. To determine the influence of seasonal and temporal variations on heavy metal concentrations, three sampling stations were selected from the Kaap River f (see **Figure 1**) based on their proximity to activities such as agriculture, runoff from old mine workings and other anthropogenic activities with potential to impact the water resource. Sampling was conducted between 2016 and 2022 throughout all seasons (wet and dry seasons) at three strategic points downstream, midstream and upstream of the confluence of the Crocodile and Kaap Rivers using the grab sampling technique. Sampling was conducted during clear weather conditions with minimal or no precipitation, or turbulence in order to minimise the sampling of highly turbid river water, and acid-rinsed polyethylene bottles were used for sample collection, and the sampling dates and times were marked with a marker. The samples were taken in duplicates and then placed in a cooler and transported to a laboratory accredited in accordance with the South African National Accreditation System (SANAS) for the determination of Al, Iron and Mn. Atomic Absorption Spectrometer was utilised to analyse metals (Fe, Al and Mn), and the analytical approach is on the Beer Lambert's law which asserts that the absorbance of light emitted by a light to a detector is directly proportional to the concentration. A sample containing a metal to be analysed is exposed to the light at the characteristic wavelength of that metal, and its atoms will absorb the light. A series of standard solutions of studied metals (Fe, Al and Mn) from commercially available metal standards were prepared, diluting these standards to appropriate

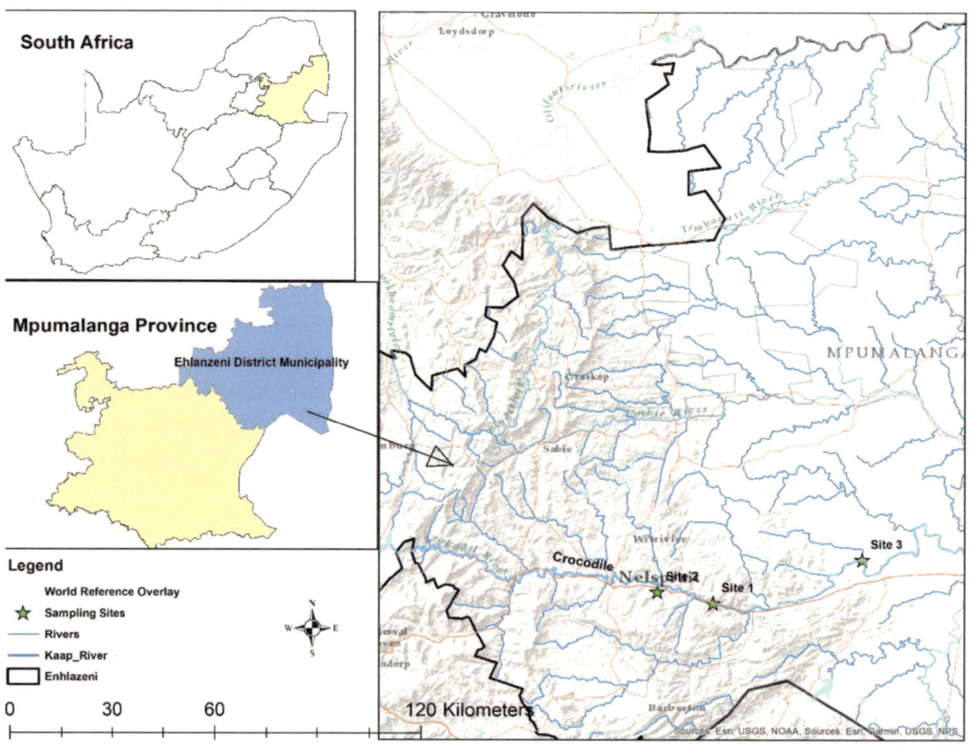

Figure 1.
The sampling site in the Crocodile River tributary.

concentrations using deionised water and nitric acid solvents. The calibration curve was then generated by running the standard solutions though the AAS for each metal, and the curve was used to relate the absorbance to the metal concentration in the sample. Samples were then run in the AAS instrument, with appropriate wavelength set for each metal, recording the absorbance reading. Reagents and standards that were utilised were handled with care, stored, and disposed of as per the material safety data sheet information. Running reagent blanks, analysing duplicate samples, and adding known quantities of the study metals and then analysing them to check recovery rates were undertaken as an effort for quality control and validation.

2.3 Statistical analysis

Since this current study dealt with large data set, the Seaborn software was used to do the statistical analysis using a python that is a popular programming language that offers numerous data analysis and visualisation tools and libraries. Water quality data sets were analysed using Seaborn (Python), including plotting and analysing the distribution and correlation of variables with Matplotlib libraries since correlation aids in measuring and identifying relationships between variables. To better understand the data and spot trends and patterns, heatmaps are frequently used in data analysis.

3. Results and discussion

Figure 2a–c depicts a time series analysis for iron (Fe), manganese (Mn) and aluminium (Al) concentration within the Kaapmuiden sampling site, indicating a spatio-temporal variation of both metals in the Crocodile River. An average concentration of 0.14 mg/l was observed for iron over the study period (2016–2022). Contrary to this study, a study carried out in Moradabad district, India, where the inductively coupled plasma atomic emission spectroscopy (ICP-OES) technique has been used to evaluate

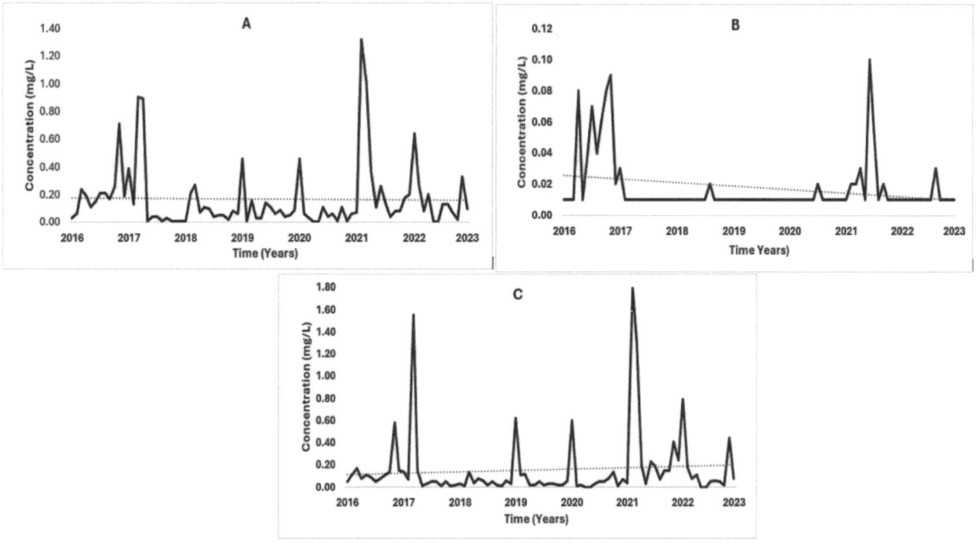

Figure 2.
Iron (a), manganese (b), and aluminium (c) time series analysis.

the metal concentration in surface and groundwater water quality, showed that over 50% of the samples had iron concentrations that were over the maximum permitted level. The maximum amount of iron found in surface water and groundwater samples was 6294 and 3820 ppb, respectively, when the allowable limit is 300 ppb [20]. This, when converted, is way higher than this study. The concentrations of iron in water resources can vary depending on the intensity and pattern of precipitation. High precipitation can cause an increase in runoff, which can mobilise sediments rich in iron, thus increasing the iron concentrations in water bodies [21]. On the other hand, less precipitation can result in less runoff, which can lower the concentration of iron. Temperature also plays a significant role in influencing the iron concentrations in water resources. Higher temperatures can increase the solubility of iron, resulting in increased quantities of iron in water [22]. This is because iron dissolves more readily in warmer water due to its increased solubility [23]. On the other hand, cooler temperatures can result in lower iron concentrations.

The results shown in **Figure 2** demonstrates the fluctuation of Al concentrations between the years 2016 and 2022. Notably, the greatest levels of Al concentrations were seen in 2017 and 2021, with recorded values of 1.15 and 1.75 mg/l, respectively. The highest peak of 0.9 mg/l was, however, noted during 2021, which can be attributed to land use activities correlating with the area. However, this study showed that the iron concentration was within an acceptable level of approximately 0.5–1.0 mg/l set for the ecological health of the river. An average of 0.01 mg/l was observed during the study period for Manganese, with major peaks observed during 2016 (0.06 mg/l) and 2021 (0.053 mg/l). Senze et al. [24] found that Nysa Szalona and Strzegomka rivers had the largest concentrations of aluminium (Al), while the Bystrzyca River had the lowest concentrations. During a four-year cycle, Senze et al. [24] concluded that aluminium concentrations were influenced by modernisation efforts undertaken in riverbeds rather than environmental changes. This finding corroborates the observation documented in the aforementioned study, which highlights a significant fluctuation in the concentrations of manganese (Mn) and aluminium (Al) throughout the period spanning from 2016 to 2022. The issue with the detrimental impact of aluminium on living creatures has been apparent due to the intensified human activities mostly linked with industrial and agricultural growth [24–26]. Aluminium (Al) has been widely acknowledged as a hazardous substance for freshwater creatures in significant quantities. This assertion has particular validity in cases when industrial point sources release process water containing substantial amounts of aluminium.

Figure 3a–c depicts the seasonal and temporal analysis of manganese and aluminium, showing the highest concentration of manganese noted (0.06 mg/l) during the spring season of 2016. The second highest concentration noted (0.053 mg/l) was during the winter of 2021. Figure shows the highest aluminium concentration recorded (1.10 mg/l) during autumn in 2021, which was not consistent with the overall average Al concentration within the studied site over time (2016–2022). This phenomenon may suggest a potential anthropogenic activity taking place within the catchment and high aluminium concentrations are introduced through stormwater runoff during rainy seasons, causing a rapid spike in aluminium concentration. Literature has shown that metal and metalloids present bioavailability, potential toxic effects and environmental persistence. The concentrations of Mn exhibited an upward trend in the following sequence: winter, fall, spring and summer. In the year 2016, the manganese concentration in river water exhibited a significantly elevated level, hence posing possible risks to human health and necessitating immediate environmental remediation measures. Significant fluctuations in the amounts of aluminium (Al) were observed

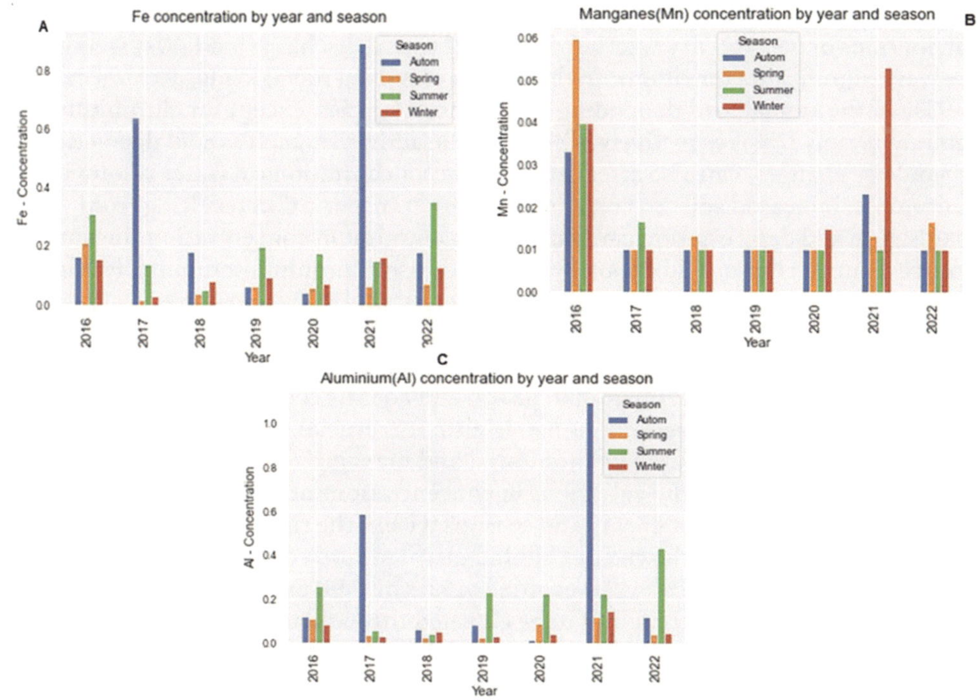

Figure 3.
Seasonal and temporal analysis of Fe (a), Mn (b) and Al (c).

across all four seasons and during the research period spanning from 2016 to 2022. The concentrations of aluminium exhibited a significant increase in all four seasons of the year 2021, as compared to the corresponding periods of the previous year. The presence of aluminium was observed in all samples evaluated over the seven-year study period (2016–2022), owing to the significant abundance of this element in the Earth's highest crustal layer. In comparison with the above results of 2016–2022 which focused on all seasons, another study on seasonal variation of heavy metals and their potential ecological risk in Nzhelele River, South Africa, revealed that the rainy season in the river had higher quantities of Al (10.565 mg/l) than the dry season (2.963 mg/l). For both seasons, the mean difference in Al in the water varied significantly ($p < 0.05$) [27]. Climate variability, including variations in precipitation patterns and temperature, can affect the concentrations of aluminium in water resources. For example, rising temperatures can cause aluminium to become more soluble, leading to higher concentrations in water sources [28]. Furthermore, variations in precipitation patterns may affect the transportation of aluminium from soil to water bodies, resulting in variations in the aluminium concentrations. Seasonality can also affect the concentrations of aluminium in water resources. For instance, heavy rains may transport aluminium-rich soil particles into water bodies, thus increasing the concentration of aluminium. Conversely, reduced runoff and sediment transport during periods of drought may result in lower aluminium concentrations [16, 29]. As part of a monitoring program in a similar study of a watershed in Korea, 11 heavy metal species were measured during dry and wet weather conditions in the YS River, where Gwangju City (GJ), a subcatchment of the YS River, was further monitored to clarify the responsibility of different metal species discharged into the mainstream. The most abundant metal species discharged from GJ were manganese, aluminium and iron with different contributions of wet and dry weather flows to the total discharge load. Wet weather flow was a significant contributor to the annual dissolved metal loads, accounting for 44–93% of the annual load depending on the metal species, except for chromium and cadmium (9 and 27%, respectively) [30]. Human activities, such as acid deposition and land use changes, can also affect aluminium concentrations in water resources. For example, increased soil erosion and sediment transportation resulting from urbanisation and deforestation can increase the aluminium concentrations in water resources. Furthermore, acid deposition can increase aluminium solubility, leading to increased aluminium concentrations in water resources [31].

Climate variability and seasonality on aluminium concentrations in water resources can significantly impact water quality. Increased aluminium concentrations can have detrimental effects on human health and aquatic ecosystems. Consequently, to protect aquatic ecosystems and human health, it is crucial that the aluminium concentrations in water sources be monitored and managed [32].

Figure 3a–c illustrates the variations in concentrations of manganese, and aluminium throughout different years and seasons within the time frame of 2016–2022. The findings indicate that the river exhibited elevated iron concentrations during the fall seasons of 2021 and 2016, with recorded values of 0.90 and 0.64 mg/l, respectively. Manganese levels were found to be elevated throughout two distinct seasons: spring (2016) with a recorded concentration of 0.060 mg/l and winter (2021) with a recorded concentration of 0.053 mg/l. The subsequent findings indicate elevated amounts of aluminium throughout the fall seasons of 2017 and 2021, with concentrations of 0.59 and 1.10 mg/l, respectively. The long-term shifts in average temperatures and weather patterns that influence micro- and macroclimates on Earth coupled with anthropogenic activities play a major role in in the process of metal contaminants in rivers. Moreover, this directly or indirectly influences the chemical composition of water because of complex multi-stage progressions transpiring in streams, reservoirs and rivers or catchment areas. It is also plausible that the precipitation throughout the summer season facilitated the transportation of metallic substances from the neighbouring agricultural lands and mining sites into the river, so augmenting the river's elevated levels of metal concentrations. As the water's acidity level rises, the release of aluminium (Al) into the water occurs, reaching concentrations of up to 5 mg/l. Kaap River is a significant tributary within the Crocodile River system, characterised by a catchment region primarily influenced by mining operations (both active and inactive) as well as intense agricultural practices. These activities have perhaps had a role in the introduction of manganese (Mn) pollution into the river.

According to the study conducted by Nhantumbo et al. [33], the high concentration of Fe and Mn can be attributed to the mining operations that are taking place there. Several studies have shown that mining activities have significant negative environmental and water resource effects worldwide. To extract certain minerals from ore, mining firms commonly use chemical substances, such as sulphuric acid or cyanide. Nhantumbo et al. [33] describe contamination as a result of leaching, leaks, spills and infiltration of aforementioned chemicals into nearby water bodies. Moreover, mining operations have been found to significantly degrade the quality of water through sand extraction [34]. Manganese (Mn) is considered to be an important element for human health. However, the occurrence of manganese shortage is deemed rare due to the presence of reasonable amounts of this element in the lithosphere, soil and various dietary sources [35, 36]. On the other hand, heightened amounts of manganese have been linked to harmful effects [37, 38]. According to Matveeva et al. [39], Russia has set standards for soil stability and mobility specifically for the chemical element Mn. These recommendations are based on variations in humus content, acidity and precipitation. When the concentration of water exceeds the established threshold of 0.3 mg/l, it results in changes in turbidity, colour and taste [40]. The occurrence of Mn might potentially be ascribed to the development of metal complexes, which serve to augment the solubility of metals [41, 42]. Several variables, including air and water temperatures, water levels, wind directions, and speeds, can affect seasonal manganese contamination in rivers. Peng et al. [43] assessed reservoir contamination using numerical models. In contrast to low air temperature, low water level and high wind speed, which alleviate pollution, hot air temperature, high water level and low wind speed intensify it.

4. Conclusion and recommendations

Fluctuation of heavy metal concentrations were noted through different season during the period of the study. However, in 2016, manganese concentration was above the average concentration throughout all season, which indicates a possible land use activity significantly altering the concentration in the water resource. Spring, however, had the highest peak which can be attributed to high surface runoff associated with the rainy season. Major peaks were also noted for Mn and Al during 2021 study period. High concentrations of manganese and aluminium in water resources have an ecotoxicological effect since they are highly bioaccumulated, affecting both humans and aquatic biota. As a result, persisting deterioration of surface water quality in the region will decrease its potential usability for socioeconomic development and will require immediate management and control measures. In addition, the number of heavy metals exceeding the threshold values set by both the World Health Organisation (WHO) and the United States Environmental Protection Agency (USEPA) standards varied among the five continents: two in North America, six in Europe, seven in South America, nine in Asia and ten in Africa. Climate variability, such as fluctuations in precipitation patterns and temperature, can impact the levels of aluminium in water resources. Increasing temperatures can result in greater solubility of aluminium, resulting in elevated concentrations in water sources. Moreover, changes in precipitation patterns can impact the movement of aluminium from soil to water bodies, leading to fluctuations in aluminium concentrations. Fluctuations in the levels of aluminium in water resources can also be influenced by seasonal changes. For example, intense rainfall can carry soil particles containing high levels of aluminium into bodies of water, thereby raising the concentration of aluminium. Among the recommendations are that sediments on the benthic zone should also be tested, in addition to water, when doing sampling. Investigations on the relationship between PH and metal precipitates are also necessary. Metals can linger in sediments for extended periods of time, so technology that could help remove the metals from the sediments has to be investigated.

Acknowledgements

Our sincere gratitude is extended the department of water and sanitation in Mpumalanga province.

Funding

This study received no internal or external funding.

Conflict of interests

The authors declare no conflict of interest.

Data availability

All data generated or analysed during this study are included in this published article.

Consent for publication

We certify that the manuscript you have received has not been published elsewhere or is under consideration by any other journals.

Ethics approval

We are solely responsible for the manuscript's veracity, validity and originality, and we are the sole owners of its contents. Additionally, we certify that this manuscript is our original work and has not been plagiarised.

Author details

Takalani Terry Phungela[1,2], Babalwa Gqomfa[1], Karabo Concelia Malakane[3], Xolisiwe Sinalo Grangxabe[1], Benett Siyabonga Madonsela[1], Boredi Silas Chidi[4]* and Thabang Maphanga[1]

1 Department of Environmental and Occupational Studies, Faculty of Applied Sciences, Cape Peninsula University of Technology, Zonnebloem, Cape Town, South Africa

2 Department of Water and Sanitation, Mbombela, South Africa

3 Department of Biodiversity, Faculty of Science and Agriculture, University of Limpopo, Polokwane, South Africa

4 Bioresource Engineering Research Group (BioERG), Cape Peninsula University of Technology, Cape Town, South Africa

*Address all correspondence to: chidib@cput.ac.za

References

[1] Nijhawan A, Howard G. Associations between climate variables and water quality in low- and middle-income countries: A scoping review. Water Research. 2022;**210**(4):117996

[2] Döll P, Fritsche M, Eicker A, Müller Schmied H. Seasonal water storage variations as impacted by water abstractions: Comparing the output of a global hydrological model with GRACE and GPS observations. Surveys in Geophysics. 2014;**35**:1311-1331

[3] IPCC. Climate Change 2021: The Physical Science Basis. 2021. Available from: https://www.ipcc.ch/report/ar6/wg1/

[4] Fan M, Shibata H. Simulation of watershed hydrology and stream water quality under land use and climate change scenarios in Teshio River watershed, Northern Japan. Ecological Indicators. 2015;**50**:79-89

[5] Rostami S, He J, Hassan QK. Riverine water quality response to precipitation and its change. Environments. 2018;**5**(1):8

[6] Xia XH, Wu Q, Mou XL, Lai YJ. Potential impacts of climate change on the water quality of different water bodies. Journal of Environmental Informatics. 2015;**25**(2):85-98

[7] Zeitoun MM, Mehana E, E. Impact of water pollution with heavy metals on fish health: Overview and updates. Global Veterinaria. 2014;**12**(2):219-231

[8] Tu J. Combined impact of climate and land use changes on streamflow and water quality in eastern Massachusetts, USA. Journal of Hydrology. 2009;**379**(3-4):268-283

[9] Bae S, Hwang YS, Lee YJ, Lee SK. Effects of water chemistry on aggregation and soil adsorption of silver nanoparticles. Environmental Health and Toxicology. 2013;**28**:e2013006

[10] Okamoto S, Eltis LD. The biological occurrence and trafficking of cobalt. Metallomics. 2011;**3**(10):963-970

[11] Horner TJ, Little SH, Conway TM, Farmer JR, Hertzberg JE, Janssen DJ, et al. Bioactive trace metals and their isotopes as paleoproductivity proxies: An assessment using GEOTRACES-era data. Global Biogeochemical Cycles. 2021;**35**(11):e2020GB006814

[12] Horner RM, Harto CB, Murphy DJ, Clark CE, Shupryt JP. Water Use and Mangement in the Bakken Shale Oil Play. Argonne, IL, United States: Argonne National Lab (ANL); 2014

[13] Moyo NAG, Rapatsa MM, Mbokane EM. Evaluation of Heavy Metals and Microbiological Contamination and Assessment of the Suitability of the Sand River Water for Irrigation in the Limpopo Province. WRC Report No. TT 835/20. 2020

[14] Döll P, Jiménez-Cisneros B, Oki T, Arnell NW, Benito G, Cogley JG, et al. Integrating risks of climate change into water management. Hydrological Sciences Journal. 2014;**60**(1):4-13. DOI: 10.1080/02626667.2014.967250

[15] Luo Y, Rao J, Jia Q. Heavy metal pollution and environmental risks in the water of Rongna River caused by natural AMD around Tiegelongnan copper deposit, Northern Tibet, China. PLoS One. 2022;**17**(14):e0266700

[16] Maphanga T, Chidi BS, Phungela TT, et al. The interplay between temporal

and seasonal distribution of heavy metals and physiochemical properties in Kaap River. International Journal of Environmental Science and Technology. 2024;**21**:6053-6064. DOI: 10.1007/s13762-023-05401-x

[17] Schiedek D, Sundelin B, Readman JW, Macdonald RW. Interactions between climate change and contaminants. Marine Pollution Bulletin. 2007;**54**(12):1845-1856

[18] Lønborg C, Carreira C, Jickells T, Álvarez-Salgado XA. Impacts of global change on ocean dissolved organic carbon (DOC) cycling. Frontiers in Marine Science. 2020;**7**:466

[19] Pagano T, Bida M, Kenny JE. Trends in levels of allochthonous dissolved organic carbon in natural water: A review of potential mechanisms under a changing climate. Water. 2014;**6**(10):2862-2897

[20] Kumar V, Bharti PK, Talwar M, Tyagi AK, Kumar P. Studies on high iron content in water resources of Moradabad district (UP), India. Water Science. 2017;**31**(1):44-51

[21] Björnerås C, Persson P, Weyhenmeyer GA, Hammarlund D, Kritzberg ES. The lake as an iron sink - new insights on the role of iron speciation. Chemical Geology. 2021;**584**:1-11

[22] Félix-Bermúdez A, Delgadillo-Hinojosa F, Torres-Delgado EV, Muñoz-Barbosa A. Does seasurface temperature affect solubility of iron in mineral dust? The Gulf of California as a case study. Journal of Geophysical Research: Oceans. 2020;**125**:1-18

[23] Kritzberg ES, Ekström SM. Increasing iron concentrations in surface waters—A factor behind brownification? Biogeosciences. 2012;**9**:1465-1478

[24] Senze M, Kowalska-Góralska M, Czyż K. Availability of Aluminium in river water supplying dam reservoirs in lower Silesia considering the hydrochemical conditions. Environmental Nanotechnology, Monitoring & Management. 2021;**16**:100535

[25] Botté A, Zaidi M, Guery J, Fichet D, Leignel V. Aluminium in aquatic environments: Abundance and ecotoxicological impacts. Aquatic Ecology. 2022;**56**(3):751-773

[26] Rosseland BO, Eldhuset TD, Staurnes M. Environmental effects of aluminium. Environmental Geochemistry and Health. 1990;**12**(1-2):17-27. DOI: 10.1007/BF01734045

[27] Edokpayi JN, Odiyo JO, Popoola EOM, Msagati TAM. Evaluation of temporary seasonal variation of heavy metals and their potential ecological risk in Nzhelele River, South Africa. Open Chemistry. 2017;**15**(1):272-282

[28] Trueman BF, Bleasdale-Pollowy A, Locsin JA, Bennett JL, Krkošek WH, Gagnon GA. Seasonal lead release into drinking water and the effect of aluminium. ACS ES T Water. 2022;**2**(5):710-720

[29] Zhou L, Tan Y, Huang L, Fortin C, Campbell PGC. Aluminium effects on marine phytoplankton: Implications for a revised iron hypothesis (iron–Aluminium hypothesis). Biogeochemistry. 2018;**139**(2):123-137

[30] Kang JH, Lee YS, Ki SJ, Lee YG, Cha SM, Cho KH, et al. Characteristics of wet and dry weather heavy metal discharges in the Yeongsan Watershed,

Korea. Science of the Total Environment. 2009;**407**(11):3482-3493

[31] Munasinghe M. Sustainable development: Basic concepts and application to energy. In: Cleveland CJ, editor. Encyclopedia of Energy. Elsevier; 2004. pp. 789-808. DOI: 10.1016/B0-12-176480-X/00441-1. Available from: https://www.sciencedirect.com/science/article/pii/B012176480X004411. ISBN 9780121764807

[32] Alasfar RH, Isaifan RJ. Aluminium environmental pollution: The silent killer. Environmental Science and Pollution Research. 2021;**28**:44587-44597

[33] Nhantumbo C, Pondja E, Juízo D, Cumbane A, Matsinhe N, Paqueleque B, et al. Effect of mining to water quality in Chua and Revué Rivers, Mozambique. In: International Mine Water Association Congress New Zealand 2020. Christchurch, New Zealand: International Mine Water Association Mine Water Solutions; 2020. p. 57. ISBN 978-3-00-067297-2

[34] Rentier ES, Cammeraat LH. The environmental impacts of river sand mining. Science of the Total Environment. 2022;**838**:155877

[35] Goss MJ, Carvalho MJGPR, Cosimini V, Fearnhead ML. An approach to the identification of potentially toxic concentrations of manganese in soils. Soil Use and Management. 1992;**8**(1):40-43

[36] Semenkov IN, Koroleva TV. Guideline values for the content of chemical elements in soils of urban functional zones: A review. Eurasian Soil Science. 2022;**55**(1):81-89

[37] Crossgrove J, Zheng W. Manganese toxicity upon overexposure. NMR in Biomedicine: An International Journal Devoted to the Development and Application of Magnetic Resonance In Vivo. 2004;**17**(8):544-553

[38] Luo X, Ren B, Hursthouse AS, Thacker JR, Wang Z. Soil from an abandoned manganese mining area (Hunan, China): Significance of health risk from potentially toxic element pollution and its spatial context. International Journal of Environmental Research and Public Health. 2020;**17**(18):6554

[39] Matveeva VA, Alekseenko AV, Karthe D, Puzanov AV. Manganese pollution in mining-influenced rivers and lakes: Current state and forecast under climate change in the Russian Arctic. Water. 2022;**14**(7):1091

[40] WHO. Guidelines for Drinking-Water Quality. 4th ed. Vol. 1. Geneva, Switzerland: World Health Organization; 2011

[41] Adeyeye O, Xiao C, Zhang Z, Liang X. State, source and triggering mechanism of iron and manganese pollution in groundwater of Changchun, Northeastern China. Environmental Monitoring and Assessment. 2020;**192**:1-15

[42] Gerringa LJA, Slagter HA, Bown J, van Haren H, Laan P, De Baar HJW, et al. Dissolved Fe and Fe-binding organic ligands in the Mediterranean Sea– GEOTRACES G04. Marine Chemistry. 2017;**194**:100-113

[43] Peng H, Zheng X, Chen L, et al. Analysis of numerical simulations and influencing factors of seasonal manganese pollution in reservoirs. Environmental Science and Pollution Research. 2016;**23**:14362-14372

Chapter 2

Perspective Chapter: Heavy Metals – Sources of Releasing into Ecosystems, Biological Importance, Toxicity, and Sorption Methods

Sevda Fatullayeva, Dilgam Tagiyev, Nizami Zeynalov and Shamil Tagiyev

Abstract

Environmental pollution with heavy metals is one of the most important environmental problems of this century. Bioaccumulating in the food chain, heavy metals exhibit high toxicity toward living organisms. To prevent serious risks to human health, eliminating the toxicity of these metals in air, soil, and water is of great importance. The chapter reviews modern data on the distribution of heavy metals in the environment, living organisms, their toxicity, and the development of new effective sorbents for their removal from the environment and living organisms. A large number of studies are devoted to the synthesis, modification, and properties of new effective sorption materials toward heavy metals. New information is presented in the field of application of the enterosorption method as one of the promising areas for removing heavy metals from the living organism, as well as in the prophylaxis and treatment of various diseases.

Keywords: heavy metal, environmental pollution, sorption, living organism, toxicity, concentration, disease

1. Introduction

At the present time, the environmental situation in the world is one of the most important problems of modern society. The release of huge amounts of toxic substances into nature due to anthropogenic impact leads to negative consequences affecting the health of all living beings on the planet. Environmental pollution leads to ozone depletion, emissions of greenhouse gases, the formation of acid rain, the death of flora and fauna, and an increase in natural disasters, which have dire consequences for all living things. The impact of soil, water, and air pollution on human health is very great, causing serious diseases and pathological processes [1]. Thus, the problem of environmental protection is becoming increasingly serious, requiring urgent action in order to reduce environmental pollution and decrease the influence of adverse conditions on living organisms [2].

Heavy metals are considered persistent environmental pollutants that can lead to irreversible processes and represent a threat to living organisms [3]. Despite the fact that heavy metals are necessary in certain quantities for the normal biochemical and physiological functioning of living organisms, they can be dangerous at accumulating in the body [4]. Deviations in their concentrations from the norm can lead to some changes in the organism and certain diseases. Minimization and removal of excessive amount of heavy metal ions (in these concentrations the metal ion passes into the category of toxic substances) from the organism is an important area of modern research [5].

To avoid serious problems for human health, heavy metal ions should be removed from the environment. It is necessary to find effective methods for removing heavy metals and eliminating the toxicity of these metals in air, soil, and water and to develop eco-friendly and low cost-effective removal techniques for sorption of heavy metal contaminants [6].

2. Heavy metals

Heavy metals (mostly transition elements) are a group of metals with relatively high density (above 5 g/cm^3), atomic weights in the range of 63.5–200.6 g/mol and showing toxicity at low concentrations (even at trace ppb level) [7]. Heavy metals, due to bioaccumulation in the food chain, cause destabilization of the ecosystem [8]. There are essential (with known biological functions: Mn, Fe, Co, Ni, Cu, Zn, Mo, etc.) and non-essential (Cd, Hg, Pb) heavy metals.

We focused on information about the sources of heavy metals, their distribution in nature and in living organisms, indicated the toxic effects of these metals on human health when accumulated in excess, and reviewed some international regulations regarding their maximum permissible concentrations. We searched for review and research articles published mainly between 2019 and 2024 and extracted data from 86 articles through literature review and laboratory experiments.

2.1 Distribution of heavy metals in nature

Distribution of heavy metals in the environment is possible in different ways. Heavy metals are deposited in the soil and spread through air and water. The presence of heavy metal pollutants in the environment is mainly found in various natural (e.g. rocks, soils, and water) and anthropogenic (sewage discharges, industrial wastes, mining, and agricultural activities) sources, which have a great influence on their accumulation in nature and contribute to a negative impact on the ecosystem and the health of living organisms [9–11]. The content of heavy metals, which are considered environmental pollutants and exhibit toxic effects, resistance, non-biodegradability, and bioaccumulation, must be constantly monitored [12, 13]. The circulation of heavy metals in the environment occurs as a result of human activity and biogeochemical processes. Depending on the flow power, particle size and physicochemical parameters of the aquatic environment, flowing water plays a significant role in the distribution of heavy metals in the water-soil–plant–animal-human system [14].

In the earth's crust, heavy metals are found in the form of ores and minerals. Thus, some of them (Fe, Zn, Co, Ni, Pb) occur in most ores in sulfide forms, others

(Mn, Al, Cr) are found in the form of oxides. In addition, industrial products (batteries, coatings, cosmetics, and paints) consist of heavy metals, which are sources of heavy metal exposure to living organisms. Heavy metals are ultimately absorbed by plants, animals, and assimilated by humans through food consumption [15, 16]. Taking into account the health risks associated with the spread of heavy metals in water resources, it is very important to control their presence and study their impact on crops through irrigation (**Table 1**) [21].

It has been studied the bioaccumulation of heavy metals (Cu, Cr, Ni, Hg, Cd, As, Pb, and Zn) in soil-crop ecosystems and used machine learning algorithms to identify the transfer process and controlling factors [17]. Increased uptake of heavy metals by aquatic medium affects food quality and safety, causing a health threat to humans and animals in the ecosystem as a whole [22]. Heavy metal contamination can pose a potential threat to the food chain and human health. Environmental pollution by heavy metals is being investigated, and the impact of natural conditions, various types of industries, and the development of road traffic in various regions on human health is being studied [23]. Thus, rural, suburban, and urban areas were studied according to the main physical–chemical parameters and the content of heavy metals in soils. Soil samples were evaluated using a variance analysis (ANOVA) (**Table 2**) [24].

The results of the research presented in **Tables 1** and **2** showed that in most cases the average content of heavy metals (e.g., copper, chromium, cadmium, and cobalt) exceeds the maximum permissible levels, recommended by the Food and Agricultural Organization (FAO) and the World Health Organization (WHO) (FAO/WHO), which

Crop type	Heavy metals	Concentrations
Barley	Cd, Co, Cr, Cu, Fe, Mn, Mo, Ni, Pb, Zn	0.1–1 mg Cd/kg, 0.1–18 mg Co/kg, 3–85 mg Cr/kg, 2–48 mg Cu/kg, 121–11,165 mg Fe/kg, 0.4–538 mg Mn/kg, 1–5 mg Mo/kg, 0.3–42 mg Ni/kg, 0.1–5 mg Pb/kg, 6–194 mg Zn/kg
Cauliflower	Cd, Cr, Cu, Fe, Mn, Zn	0.001–1.290 mg Cd/kg, 0.104–12.580 mg Cr/kg, 0.313–11.450 mg Cu/kg, 2.649–22.840 mg Fe/kg, 2.073–26.750 mg Mn/kg, 1.640–25.820 mg Zn/kg
Grass, Nettle	Cd, Cu, Pb, Zn	11. 5 mg Cd/kg, 60 mg Cu/kg, 130 mg Pb/kg, 475 mg Zn/kg in nettle; 10 mg Cd/kg, 217 mg Pb/kg, 633 for mg Zn/kg in grass
Rice	Cd	0.021–0.33 mg Cd/kg
Sunflowers	Cd, Ni, Pb, Zn	79.54 mg Cd/kg, 33.89 mg Ni/kg, 141.63 mg Pb/kg, 324.65 mg Zn/kg
Water lettuce	Cd, Cu, Fe, Pb, Mn, Zn	0.204 mg Cd/kg, 0.788 mg Cu/kg, 14.83 mg Fe/kg, 0.297 mg Pb/kg, 10.351 mg Mn/kg, 4.181 mg Zn/kg
Wheat	Cd, Cr, Cu, Hg, Mn, Ni, Pb, Zn,	0.07–0.55 mg Cd/kg, 6.55–26.57 mg Cr/kg, 6.21–26.2 mg Cu/kg, 0.57–4.64 mg Hg/kg, 164.0–1087.0 mg Mn/kg, 5.24–16.7 mg Ni/kg, 13.27–47.1 mg Pb/kg, 21.00–100.70 mg Zn/kg

Table 1.
Estimation of heavy metal concentrations in agricultural crops [17–20].

Elements	**Studied areas**	**(average ± standard deviation, mg/kg)**	
	Rural	**Suburban**	**Urban**
Cd	1.6 ± 0.1	1.5 ± 0.1	1.4 ± 0.1
Co	7.7 ± 0.7	7.4 ± 0.3	5.5 ± 0.2
Cr	19 ± 1	19 ± 1	15 ± 1
Cu	25 ± 3	39 ± 3	49 ± 2
Mn	623 ± 72	553 ± 22	464 ± 19
Ni	28 ± 2	26 ± 1	20 ± 1
Pb	25 ± 4	36 ± 3	79 ± 8
Zn	78 ± 7	88 ± 3	130 ± 5
Parameters	*Rural*	*Suburban*	*Urban*
pH	7.1 ± 0.1	7.8 ± 0.1	7.5 ± 0.1
Electrical conductivity, μS/cm^2	372 ± 91	931 ± 73	516 ± 67
Moisture content, %	35 ± 3	28 ± 4	22 ± 3
Organic matter, %	12.7 ± 0.8	12.3 ± 0.5	10.0 ± 2.6
Calcium carbonate, %	6.7 ± 2.3	7.8 ± 1.3	6.0 ± 1.5
Plasticity index	36 ± 4	43 ± 4	41 ± 3
Soil texture	Sandy loam	Clay loam	Loam

Table 2.
Concentration of some heavy metals (mg/kg) in the soil samples and physical–chemical characteristics of soil along the urbanization gradient.

poses a risk to human health. Thus, the recommended maximum concentration for copper is 10 mg/kg in plants, 20 mg/kg in soil, and 0.55 mg/kg in water; for chromium, it is 1.3 mg/kg in plants, 100 mg/kg in soil, and 0.017 mg/kg in water; for cadmium, it is 0.02 mg/kg in plants, 0.03 mg/kg in soil, and 0.02 mg/kg in water; for cobalt, it is 50 mg/kg in plants, 8 mg/kg in soil, and 0.05 mg/kg in water [25–27]. It should also be noted that maximum permissible levels vary depending on the kind of food, water (drinking and irrigation), and soil (commercial, residential, and agricultural) [28]. The influence of heavy metals on a living organism and the toxicity of these metals will be discussed below (see Section 2.2.).

The distribution of various chemical forms of heavy metals (determining their possible toxic effects) in suspended atmospheric substances is different and may vary depending on the territory (**Table 3**). Thus, in rural areas, the majority of heavy metal pollutants are Cd, Co, Mn, and Ni, possibly resulting from soil changes. High levels of Pb pollution have been identified in urban areas as a result of incomplete combustion in automobile engines and from industrial branches.

(Information on permissible levels of heavy metals can be found in the documents: Economic Commission for Europe Executive Body for the Convention on Long-range Transboundary Air Pollution 1998 Protocol on Heavy Metals, as amended on December 13, 2012; Directive 2008/50/EC; Directive 2004/107/EC; Commission Directive (EU) 2015/1480; 2011/850/EU Commission Implementing Decisions of the European Parliament and of the Council as regards the reporting on ambient air quality)

City	Pb	V	As	Mn	Ni	Cr	Cd
China, average of major cities	304.2	18.7	46.6	151.7	33.7	97.4	12.9
Tokyo, Japan	125	0.06	—	40.1	5.63	6.09	—
Agra, India	1100	21	—	900	200	300	—
Lahore, Pakistan	4400	1. 4	22	300	18	30	77
Islamabad, Pakistan	144	—	—	79	24	10	4
Ankara, Turkey	71	3.9	1.5	4.9	3.1	3.2	0.11
Vienna, Austria	11	1.4	0.9	13	5.7	5.5	0.5
Bern, Switzerland	49	2.9	0.8	25	3	—	0.26
Budapest, Hungary	24	—	1. 4	30	3.2	8.9	—
Seoul, South Korea	96.4	—	—	39	19.6	13.7	—
California, USA	39	33	6	17	12	15	43
Toulouse, France	9.7	—	—	7.5	35.22	35.97	0.25
New York, USA	7.9	9.2	—	—	10	2.7	0.34
Ile-Ife, Nigeria	2762.9	—	—	1123.7	323.6	33	731.8
Durban, South Africa	77	—	—	49	—	—	—

Table 3.
Heavy metal concentrations in aerosols in some world cities (ng/m^3) [29].

To assess the state of pollution, regular monitoring of the content of heavy metals in ecosystems is necessary, since in concentrations above the limit norm they are dangerous to human health. Oral ingestion, inhalation, and skin contact are the main routes of entry of heavy metals into the human body. Circulating in the blood and accumulating in various organs and tissues, they can cause mutations, lead to various carcinogenic and other chronic diseases, regardless of age and gender [8].

2.2 Influence of heavy metals on the vital activity of organisms and toxicity

Some heavy metals are related to “essential in traces” bioelements that occur in living organisms in very low concentrations. In a certain amount, heavy metals such as iron, zinc, copper, manganese, and cobalt participate in various physiological, metabolic processes, enzymatic reactions, and transport of substances [30].

Thus, iron is indispensable in the processes of hematopoiesis and intracellular metabolism. It plays an important role in the processes of energy release and respiration, in enzymatic and redox reactions, in ensuring immune functions, and in cholesterol metabolism. The daily requirement of an adult healthy person for iron is 10–20 mg and is replenished with a regular balanced diet.

The adult human organism contains about 100 mg of copper (liver, bones, brain), which increases the activity of a number of enzymes. It is involved in hematopoiesis, enzymatic oxidation, tissue respiration, and immune processes. The human requirement for copper is 2–3 mg per day. The main function of copper is catalytic. It is known that copper ions, compared to other ions, react more actively with amino acids and proteins; therefore, copper forms the most stable complex compounds with biologically active substances.

The human organism weighing 70 kg contains about 14 mg of cobalt. The main places where cobalt accumulates in the human body are liver, blood, spleen, bone tissue, thyroid gland, and pituitary gland. The biological activity of cobalt is determined by its participation in the construction of the vitamin B_{12} molecule. It is necessary for the activity of a number of enzymes, affects the synthesis of proteins and nucleic acids, the metabolism of carbohydrates and fats, and redox reactions. Cobalt is a powerful activator of hematopoiesis, actively participates in the formation of thyroid hormones, promotes the excretion of water by the kidneys, and increases the absorption of iron and hemoglobin synthesis. It has been established that the process of hematopoiesis in animals and humans can only occur with the normal interaction of cobalt, copper, and iron ions [31].

Molybdenum accumulates mainly in the liver, kidneys, endocrine glands, and skin. It is evenly distributed in the blood between the formed elements and plasma. It is part of the enzymes involved in purine metabolism, participates in the synthesis of vitamins B_{12} and E, and various redox processes. Molybdenum affects the activity of the enzyme xanthine oxidase, catalyzing the oxidation of xanthine by oxygen into uric acid, and prevents the development of dental caries.

Zinc is found in all human organs and tissues. The human body (weighing 70 kg) contains up to 3 g of zinc. It is part of the most important enzymes, participates in protein metabolism, in the synthesis of mRNA molecules in the corresponding sections of DNA (transcription), and in stabilizing the structures of ribosomes and biopolymers (RNA, DNA, and some proteins). Zinc is an essential part of blood enzymes. It is necessary for maintaining normal skin, hair, and nail growth and also for wound healing, since it is involved in protein synthesis. This metal is part of insulin, a pancreatic hormone that regulates blood sugar levels. Zinc activates the biosynthesis of vitamins C and B.

Manganese is found in many human tissues and organs (liver, skeleton, and thyroid gland are richest in manganese). It actively affects the metabolism of proteins, carbohydrates, and fats. It is a catalyst for metabolism and participates in the formation of bone tissue. It is also necessary for the functioning of enzyme systems and the regulation of vitamin metabolism, and maintains a certain level of cholesterol in the blood. Furthermore, Mn^{2+} ions stabilize the structure of nucleic acids, affect hematopoietic processes, accelerate the formation of antibodies, act on the central nervous system, affect the ability to reproduce, and strengthen the immune system. The daily human need is about 3–8 milligrams of manganese.

Chromium (III) is involved in carbohydrate and lipid metabolism and in the proper functioning of the immune system [32]. Nickel is involved in the metabolism of glucose, hormones, lipids, adrenaline, and cell membranes and is also responsible for the absorption of iron and the synthesis of red blood cells [33].

As can be seen from the abovementioned, heavy metals in a certain amount are necessary for the normal functioning of the organism [34]. However, literature data show that any heavy metals present above safe levels can lead to various physiological abnormalities and genetic mutations (according to Paracelsus, "the dose makes the poison") [35]. Their concentrations being higher than normal are toxic because heavy metals do not decompose but accumulate and cause pathological processes in the body (liver, kidney diseases, respiratory and vision problems, decreased energy levels, failure in brain processes, etc.) [36]. Gastrointestinal and immune system dysfunction, nervous system disorders, vascular lesions, congenital malformations, and cancer are examples of complications of heavy metal toxicity. Heavy metals destroy cell membranes, increase the permeability of biological barriers, bind to proteins, changing secondary and tertiary structures, block many enzyme systems, which ultimately leads to changes and disorders in the organism.

Each heavy metal has its own mechanism of toxicity, such as inhibiting enzymatic activity, weakening antioxidant defenses, altering the functioning of the central nervous system, and changing important organ functions, which leads to various diseases. As a result of simultaneous exposure to several metals, the toxic effect may increase [37].

Mercury, lead, and cadmium (non-essential heavy metals) are the most dangerous to human health, having a toxic effect on various tissues and organs [38]. These three metals are related to global environmental pollution.

It should be noted that any heavy metal quickly enters the body's bloodstream and then passes into internal tissues and organs. It was found that the places of maximum accumulation of metals in the body are the skeletal system, liver, and kidneys. Heavy metals entering the human body with food are absorbed through the gastrointestinal tract (for example, lead is more efficiently absorbed on an empty stomach). Distributed throughout the body, heavy metals have a toxic effect on certain organs and tissues.

Pathologies caused by the toxic effects of heavy metals on the human body are presented in **Figure 1**, obtained from literature data on the study of excess content of some heavy metals in the organism [2, 39].

Cadmium and hexavalent chromium enter the body into the bloodstream through the lungs, inhibit enzyme activity, damage DNA, and cause genetic abnormalities. Cadmium enhances its toxic effects when it competes with zinc for binding to a protein responsible for regulating zinc homeostasis. Chromium (VI) compounds are very toxic (capable of easily penetrating biological membranes and causing damage to cell membranes); they cause lung cancer and various allergic diseases.

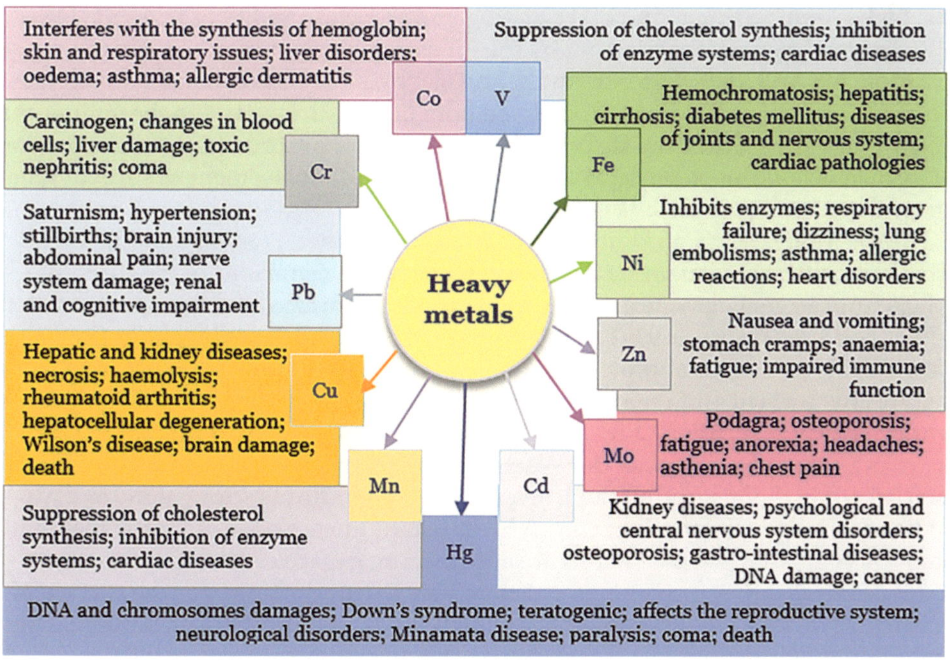

Figure 1.
Various pathologies of the organisms connected with excess amounts of some heavy metals.

Mercury and lead can penetrate the skin barrier and enter the bloodstream. Mercury causes neurological toxicity, affecting the brain, and central nervous system. Lead accumulates in bones and can persist for years, displacing calcium ions, and impairing neurotransmission and muscle function.

Molybdenum excess leads to anemia, gout, endemic goiter (with simultaneous iodine deficiency), joint diseases, and intestinal dysfunction.

Some heavy metals lead to the formation of excess amounts of reactive oxygen species, thereby suppressing the body's antioxidant defenses.

As can be seen from the literature data, in-depth knowledge of heavy metals is essential in order to provide protection against excessive exposure [40, 41]. Therefore, more in-depth research in this area can be aimed at developing new effective and environment-friendly methods and sorbents for the removal of heavy metals [42, 43].

3. Sorption of heavy metals

In order to prevent serious risks to human health, the development of effective methods for the removal of heavy metals and to eliminate the toxicity of these metals in air, soil, and water is of great importance. Among the known methods (chemical precipitation, complexation, ion exchange, electrodialysis, reverse osmosis, electrocoagulation, solvent extraction, ultrafiltration, photocatalysis, and others) for removing metal ions, sorption is considered the most promising due to its easy applicability, high removal efficiency over a wide pH range and low cost, and therefore, it is preferable to develop highly effective, practical, economical, and environment-friendly sorbents [44]. It should be noted that the sorption method proceeds without introducing any additional substances into the system to be purified [45]. The main difficulties of sorption methods are the simultaneous removal of different kinds of ions, high retention times, and regenerating and recycling ability of sorbents [46].

Sorption of metals is carried out by the following mechanisms: coordination, precipitation, ion exchange, Van der Waals, and electrostatic interactions. The interaction of the sorbent and sorbate is observed both on the surface and in the volume due to the porous structure of the sorbent. Accumulation of adsorbed substances around the sorbent occurs due to physical and chemical forces, replacement of sorbent ions with sorbate ions takes place (**Figure 2**).

The sorption process, in which the polluting heavy metal ion is bound to a solid sorbent, includes three main stages: transport of heavy metal ions to the sorbent surface from an aqueous solution, sorption on a solid surface, and transport inside the sorbent particle (absorption). The presence of pores and cavities increases the surface area available for sorption, providing more opportunities for the interaction of metal ions with the sorbent and their retention by it. There is physical sorption and chemisorption. Physical sorption (a reversible process) between the sorbate and the sorbent occurs as a result of a weak Van der Waals force independent of the nature of the material, while chemisorption (an irreversible process) forms strong ionic or covalent bonds through chemical reactions. Electrostatic attraction leads to the fact that positively charged heavy metal ions are adsorbed on negatively charged functional groups of the adsorbent, since heavy metals have a strong affinity for the surface of the sorbent. In addition, hydrophobic interactions as well as hydrogen bonds can be involved in the process of metal sorption on the surface of sorbents [43].

Precipitation is considered one of the most effective methods of removing heavy metal ions, converting dissolved metal ions into solid insoluble particles, and

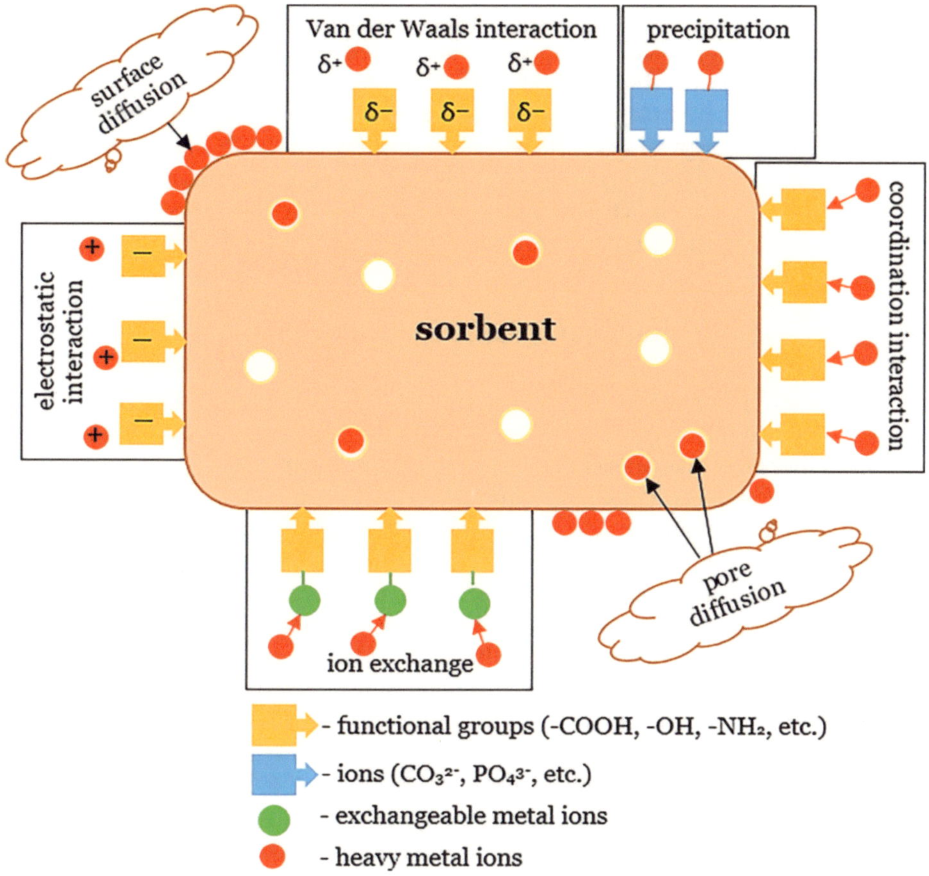

Figure 2.
A schematic representation of possible sorption mechanisms of heavy metals.

precipitating them by changing pH, electrooxidation potential, etc. The ion exchange method is a reversible chemical reaction used to replace an undesirable heavy metal ion with a harmless and environmentally friendly one. Coordination interaction describes the ability of heavy metal ions (Lewis acids) to interact or enter into a coordination bond with ligands (Lewis bases) and shows that this depends on the availability of their outer electrons and empty molecular orbitals.

The sorption process depends on many factors, such as the physicochemical properties of the sorbent and heavy metals, as well as the process conditions (temperature, solution pH, contact time, initial concentration of metal ions, sorbent dosage, ionic strength, the type of active centers on the surface of the sorbent, the point of zero charge of the sorbent, etc.) [47].

Information about the reactive centers of the sorbent participating in the binding of heavy metal ions can be obtained using physicochemical analytical methods, including nuclear magnetic resonance spectroscopy (NMR) for determining the molecular structure and chemical composition of the sorbent; atomic absorption spectroscopy (AAS) and inductively coupled plasma (ICP) for determining the metal concentration in an aqueous solution; Fourier-transform infrared spectroscopy (FTIR) for identifying the functional groups of the sorbent; UV–visible spectrophotometry for determining the metal concentration by measuring the intensity of its color in the aqueous medium; energy-dispersive X-ray spectroscopy (EDS) for elemental analysis; X-ray diffraction spectroscopy (XRD) for determining of crystal structure, phase, structural parameters and chemical composition of the sorbent; X-ray photoelectron spectroscopy (XPS) for determining the oxidation state of the metal bound to the sorbent; scanning electron microscopy (SEM) and transmission electron microscopy (TEM) for analysis of the surface and internal morphology of the sorbent, respectively; thermogravimetry (TGA), differential scanning calorimetry (DSC) for evaluating the correlation between chemical structure, thermal transitions and stability of sorbent and others.

3.1 Types of sorbents for removal of heavy metals

Nowadays, different types of sorbents for removal of heavy metal ions have been developed: polymer materials, natural zeolites, clays, bio-adsorbents, activated carbon, organic, and inorganic mesoporous silica [48]. Some of them are presented below.

Carbon-based sorbents. These sorbents have properties, such as high sorption capacity, ability to surface modification, distinctive chemical properties, thermal stability as well. Graphene and carbon nanotubes have found greater use in sorption methods due to simple preparation and functionalization techniques compared to other carbon sorbents, such as activated carbon, fullerenes, carbon nanofibers, and graphitic carbon nitride. Numerous carbon-containing materials, such as wood, peat, coal, coconut shells, and biomass waste from pulp production, are used to develop activated carbons, which are considered effective sorbents for the removal of heavy metals (Pb^{2+}, Cu^{2+}, Co^{2+}, Ni^{2+}, and Cd^{2+}) due to the presence of transport and adsorption pores for restoring natural infrastructure [49]. The porous structure of carbon-based sorbents, small particle size, large surface area, surface functional groups, and the ability to adsorb various substances make them a valuable material for the effective removal of heavy metals. The absorption capacity of these sorbents is limited by several factors, including loss of adsorption efficiency during regeneration and the possibility of secondary contamination. The modification process can lead to the emergence of new sources of pollution (it is necessary to assess their impact on the environment), create problems with cost, yield, and operational practicality.

Polymer-based sorbents. Recently, hybrid polymer sorbents consisting of organic–inorganic components have gained popularity. Due to its high selectivity, high thermal stability and complete insolubility in water, the hybrid polymer has found use as an effective sorbent for Cd^{2+}, Pb^{2+}, Co^{2+}, Ni^{2+}, and Hg^{2+} ions. The thermostable inorganic component and the presence of organic fragments with various functional groups determine the high sorption capacity of the sorbent and increase the probability of binding certain heavy metal ions [50]. Molecularly imprinted polymers obtained as a result of polymerization reactions between the target matrix and monomers with reactive functional groups were used. Due to their stability, selectivity, resistance to acids and bases, low cost, and simple preparation method, these polymer materials have proven themselves to be highly effective sorbents for drinking and wastewater purification [51]. A new type with improved mechanical and morphological properties of submicrometric hyperbranched poly(β-cyclodextrin) fibers forming molecular inclusion complexes with various organic and inorganic compounds was prepared and characterized as effective sorbents of heavy metals (Pb^{2+}, Ni^{2+}, and Cd^{2+}) [52]. Despite its unique properties, chitosan is difficult to use in powder or flake form due to its low porosity, small surface area, and resistance to mass transfer. Therefore, structural and chemical modifications (cross-linking, grafting) have been proposed to overcome these drawbacks. It has been proposed to combine chitosan with other adsorption materials to improve the adsorption capacity for heavy metal ions, mechanical strength and thermal stability.

Mineral sorbents (zeolite, silica, and clay). Diatomite and its modified (sulfuric acid) form have been used as adsorbents to remove heavy metal ions from aqueous solutions. Diatomite treated with sulfuric acid (which led to an increase in specific surface area by 23%) showed higher sorption capacity results as an adsorbent of heavy metals [53]. Natural clay minerals including bentonite, volcanic ash, and red soil were used to remove copper, zinc, and nickel ions from acid mine drainage. The experimental results showed that the percentage of removal of heavy metals by bentonite was 89–99.9%. The volcanic ash mineral was characterized by simplicity of operation, low sediment formation, and suitability for low concentrations [54]. Natural and synthetic aluminosilicate minerals, in particular zeolites, due to their unique physical and chemical properties, are considered to be highly effective, inexpensive and environmentally friendly sorbents in the purification processes of water, contaminated with heavy metals, various environmental pollutants as well [55]. Mineral sorbents are considered effective materials for the sorption of different kinds of pollutants. Modification methods (calcination, impregnation) have been proposed to increase the versatility and efficiency of removing heavy metals using these sorbents after several cycles of application. Although, as the literature data show, these modifications entailed additional costs for the technological process and the release of new chemicals into the environment.

Magnetic sorbents. Magnetic sorbents are produced by incorporating magnetic nanoparticles into adsorbent materials. A method of magnetic solid-phase microextraction has been proposed, in which magnetic nanoparticles coated or modified with various materials (carbon, organometallic, and silica) are used as sorbents for the extraction of heavy metals from food samples. The presence of high surface area, adsorption capacity, thermal stability, and other qualities increases the efficiency and sensitivity of the process. After contact with food samples, the sorbents are separated by a magnetic field and analyzed. This makes them ideal for continuous water treatment, as they can be recovered and reused multiple times [56, 57]. Magnetic sorbents have shown low cost and simplicity of synthesis. However, their sorption properties strongly depend on the structure of the sorbent, the amount of active hydroxyl groups on the surface of Fe_3O_4 particles and reactive functional groups. The adsorbents obtained did not have sufficient cyclic stability, which is an obstacle to industrial use.

Biosorbents (fungi, algae, and bacteria). Biosorbents obtained from plant, animal and marine origin wastes, and modified to enhance adsorption properties, have proven to be highly effective sorbents in relation to various pollutants, including heavy metals [58]. Fruit peel, leaves, bark, roots, husk, sawdust, biofertilizers, peat, bone gelatin mass, and their modifications, due to the presence of organic compounds bearing polar functional groups (hydroxyl, carbonyl, carboxyl, amino, etc.) show a high complex capacity toward ions of heavy metals [59]. These agricultural waste materials, considered as natural organic sorbents due to their wide availability, environmental friendliness, and easy processing into biosorbents using simple preparation methods, such as washing, drying, and grinding, are an efficient candidate for removing heavy metal pollutants [60]. Despite the fact that biosorbents are readily available, environment-friendly, and can be regenerated, their sorption capacity and selectivity toward heavy metal ions may vary depending on the production and processing methods.

Nanomaterials. Nanomaterials are promising sorbents for efficient removal of heavy metals from waste water due to their special properties such as small size, large specific surface area and morphological structure. It was found that the sorption capacity of these nanomaterials increases due to certain reactions carried out on their surface (oxidation, modification, and others) [61]. Materials based on carbon, metals, metal oxides, dendrimers (particle sizes ranging from less than 10 nm to several hundred nanometers), and nanocomposites have been studied. Studies have been conducted depending on the type of nanoparticles, kind and concentration of the metal in the aqueous solution, and it has been shown that carbon-based nanoparticles are less effective than other sorbents applicable for the removal of heavy metal ions [62]. Nanomaterials made of metal oxides (iron, manganese, etc.) as effective sorbents have attracted wide attention due to their ease of production, a wider range of raw materials and less environmental impact. The redox process, complex precipitation, electrostatic attraction, and ion exchange have been proposed as mechanisms for removing heavy metals from an aqueous solution. Improved production methods have been developed to increase the number of surface functional groups, the specific surface area and the stability of the nanosorbent for the removal of heavy metals [63]. Nanomaterials have the greatest advantages for the sorption of heavy metals due to their small size, large surface area and high reactivity. An improvement in their adsorption capacity, the possibility of recycling and reuse have been shown due to the modifications carried out (introduction of magnetic particles, various functional groups). However, issues regarding environmental impact and maintaining stability in various media require more in-depth study.

Metal–organic framework sorbents. Metal–organic frameworks are porous, extensive crystalline structures with molecular-sized channels and voids with metal-containing nodes and organic fragments. Having a large surface area and capable of being modified by various functional (amine, sulfate, hydroxyl, and alkyl) groups, these structures exhibit high adsorption activity toward heavy metal ions (Cu^{2+}, Pb^{2+}, and Zn^{2+}) [64]. Metal–organic frameworks are promising sorbents due to their controlled properties, large surface area, and sustainable performance compared to traditional materials such as activated carbon and zeolites. The features of these materials that influence their sorption characteristics, such as pore size, surface area, the presence of functional groups, and the possibility of recycling and regeneration, structural diversity, and simple functionalization, are considered. The important role of ligands and metal nodes in tailoring the properties of metal–organic frameworks for heavy metals removal is indicated [65, 66]. Due to the introduction of some functional groups into the structure, metal–organic frameworks are promising materials for the sorption of heavy metals. However, despite the good ability to remove heavy metal ions, they have micropores that are inaccessible to some target metals. It has been observed that most of the organic ligands used to form many metal–organic frameworks are very expensive and toxic. Most of them have low stability in water. Therefore, further research is still needed to adjust, improve the stability of the metal–organic framework structure, and implement these materials in industrial heavy metal sorption processes.

In addition, the development of effective decisions on desorption is crucial to ensure complete utilization and reuse of sorbents. Also, additional research should be directed to the development of effective sorbents for the simultaneous removal of several heavy metal ions.

3.2 Sorption of heavy metals by polymer-based materials

In connection with growing problems of ecosystems and, as a consequence, its negative impact on the health of living organisms, intensive scientific research is required in the field of creating new generations of sorbents with improved properties. A change in porosity, the introduction of reactive groups into the structure of the sorbent, a change and increase in reaction sites contribute to an increase in adsorbing and complexing activity toward heavy metals. Among the variety of sorbents, polymer-based materials and their derivatives are of particular interest [47].

It has been proposed to introduce certain functional groups into the polymer structure in order to increase the surface area of the polymer available for the sorption of heavy metal ions and provide a stronger interaction with pollutants. Polymer/polymer and polymer/inorganic hybrids (copolymer composites, polymer/carbon composites, polymer/clay composites, mineral composites, magnetic polymer composites, and polymer/metal composites) have been synthesized using various polymers and also by introducing inorganic nanoparticles inside the polymer skeleton. The hybrid polymer materials produced by this method, in addition to retaining their properties, also showed improved properties such as higher processability and high chemical stability [67]. Research has shown that the properties of these materials depend both on the physicochemical properties of the individual fragments and the structure of the polymer.

Hydrogels, cross-linked polymers that form a three-dimensional network structure, have been proposed for water purification. Due to properties such as biodegradability, swelling, and modification, they have proven to be excellent sorbents for the removal of heavy metals [68]. Polysaccharide-based hydrogels have been prepared by crosslinking polymer chains with either metal ions, metal oxides, or covalent bonding and then used as effective adsorbents for numerous metal ions (Cr^{3+}, Zn^{2+}, and Mn^{2+}). Recent advances in the development of sorbent materials based on (cellulose, chitosan, agar, and hyaluronic acid) hydrogels used for wastewater treatment, in particular the mechanisms of synthesis and adsorption, the study of properties, as well as their classification, have been studied in detail [69]. The research results showed that polysaccharide-based hydrogels exhibit chemical and physical stability, polymer community flexibility, the possibility of reuse, and multifunctionality [70]. Methods involving the use of molecularly imprinted polymers for the detection of heavy metals were discussed. Adsorption properties, sensitivity, low cost, and simplicity of preparation enhance their applicability for the removal of many environmental pollutants, indicating their potential future as excellent sorbents for removal of heavy metals [51].

The new Schiff base polymer poly-2,2′-hexamethylenebis(oxybenzaldehyde)-4,4′-diiminophenylsulfone (PoHOBD) was obtained by the polycondensation reaction of 2,2′-hexamethylenebis(oxybenzaldehyde) and diamine. PoHOBD was used as a sorbent for the removal of Co^{2+}(78%), Cu^{2+}(99%), and Au^{3+}(98%) ions from contaminated wastewater. The results indicated the chemisorption/ion exchange nature of adsorption [71].

Owing to high ability to reduce the concentration of transition metal ions, biopolymers (agar, pectin, starch, cellulose, chitosan, alginic acid, carrageenan, xanthan, dextran, gelatin, guar gum, etc.), being renewable and biodegradable, have attracted the attention of many scientists. Among biopolymers and a variety of sorbents, special attention is paid to chitosan and its derivatives, as the most promising group of substances, which, due to their unique properties, such as biocompatibility, stability in the natural environment, non-toxicity, high biological activity, economic availability, chelation of metal ions, have found wide application in various fields of science and industries [72].

Chitosan is a natural polymer possessing sorption properties and affinity toward pollutants due to the presence of reactive amino ($-NH_2$) and hydroxyl ($-OH$) groups. Due to the large number of binding sites, this polymer can chelate with metal ions and is able to adsorb metal ions (Pb^{2+}, Cu^{2+} and Cd^{2+}) with a maximum removal capacity of 176.50 mg/g, which proves its excellent effect on the adsorption of heavy metals [73]. A series of modifications of chitosan such as cross-linking, grafting, combining with other adsorption materials (carbon nanotubes, minerals, polymers, etc.) were carried out, which led to the formation of structures with much higher adsorption properties and hydrogels, nanoparticles and nanofibers were obtained, which significantly improved its adsorption properties toward heavy metal ions [74]. New synthesized sodium tripolyphosphate (TPP) and vanillin (V)-modified chitosan-based magnetic nanosorbents (TPP-CMN and V-CMN) nanosorbents were used for treatment of wastewater, containing Cd(II), Co(II), Cu(II) and Pb(II) ions both in laboratory and industrial conditions. The obtained results show high adsorption rate (>80%) of metals due to the large number of active sites on the sorbent surface and which was maintained even after fourfold recycling of these sorbents. The synthesized nanosorbents also have a high ability to remove other metal ions (Cr (II), Zn (II), Mn (II), As (II), etc.) [75].

The methods for the preparation of magnetic chitosan nanoparticles (M-Ch-NPs) using magnetite to increase the mechanical strength, reusability opportunities, and minimize the difficulty of separating chitosan nanoparticles from aqueous medium are described. The results of the study indicate that M-Ch-NPs is an effective biosorbent for the removal of Cr(VI) from industrial wastewater. Improved physicomechanical and thermal parameters facilitate the possibility of recycling and reuse of M-Ch-NPs as biosorbents for the removal of Cr(VI), which will significantly reduce the cost of industrial wastewater treatment [76]. Based on the publications of recent years, in which chitosan and chitosan-based magnetic nanocomposites are used as sorbents for removal of heavy metal ions, it can be concluded that chitosan and its derivatives can be used as effective sorbents for environmental purposes in order to adsorb Cr(III), Co(II), Ni(II), Cu(II), and Pb(II) ions from aqueous media [77].

The properties of chitosan and its numerous advantages, including adsorption properties, availability, ease of modification, and relatively low costs (of acquisition and use), have led to an increase in the interest of scientists in polymer-based adsorption materials and, consequently, the development of new, sustainable, environmentally friendly effective sorption methods intended to meet global environmental goals and ensure the long-term viability of biopolymer sorbents for removal of heavy metal ions from the environment [78–80].

3.3 Enterosorption as a method of removal of toxic metals from organisms

Environmental pollution with heavy metals affects the health of living organisms, leading to various pathologies and the occurrence of various diseases. As a result of the accumulation of heavy metals in the human body, intoxication of the organism is observed, causing disturbances in its functioning. There is a need for research aimed at developing methods to protect and treat their adverse effects, including through enterosorption method. Enterosorption is an effective method of binding and removing from the body various exogenous (heavy metal ions, bacterial toxins, chemical warfare agents, drugs, etc.) and endogenous (metabolism products) toxic substances using enterosorbents (oral sorbents) [81]. These sorbents are drugs ("smart sponge") that can absorb various toxic substances in the lumen of the gastrointestinal tract and are used in the prevention and treatment of certain diseases, poisoning and correction of pathological conditions associated with endo- and exotoxicosis [82]. Currently, enterosorbents are used in toxicology, allergology, dermatology, gastroenterology, narcology, and oncology to prevent toxic-allergic reactions, intoxication, reduce the metabolic load on the excretory and detoxifying organs. The World Health Organization included enterosorbents in the anatomical-therapeutic-chemical classification as "Intestinal adsorbents" of group A07B. It should be noted that enterosorbents must have a large active surface area, high sorption capacity in relation to heavy metal ions, absence of toxic and allergic properties, neutral taste, selectivity of action with minimal loss of essential micronutrients, and appropriate pharmacokinetics [2]. The study of the effectiveness of enterosorbents of various chemical structures that can be used to remove excessive amounts of heavy metals from the organism, and their selectivity in relation to a specific xenobiotic, seems to be a relevant issue [83]. A large number of enterosorbents, based on carbon sorbents, polymers, zeolites, and various composites, have been proposed to remove heavy metals and radionuclides from the organism. Compounds based on polysaccharides are of great interest among enterosorbents aimed at cleaning the internal environment of the organism and binding heavy metals. These compounds (alginates, pectin, chitin, and their derivatives) are used as enterosorbents and, in addition, have a wide range of pharmacological properties [82].

The mechanisms of action of enterosorbents are still a subject of discussion. The method of sorption detoxification of the organism is based on scientifically substantiated mechanisms for reducing the concentration of toxic substances. The presence of ionized polar groups that bind heavy metal ions by ion exchange; functional groups (with high affinity for specific heavy metal) capable of complex formation with heavy metal ions; and molecular adsorption of species containing heavy metal determine the binding capacity of enterosorbents toward heavy metal ions [84].

Enterosorbents based on natural polymers or inorganic minerals contain various functional groups. Thus, polysaccharides have -OH and -COOH, chitosan has -OH and $-NH_2$, lignin has phenolic groups, inorganic minerals contain hydroxyl groups bound with Si, Al, or Mg. In natural minerals, partial replacement of hydrogen by Na and Ca ions is observed. Inorganic enterosorbents (based on zeolites and clays) have ion-exchange groups capable of binding a number of heavy metal ions. Literature data have been obtained describing the removal of lead and the prevention of cadmium accumulation using zeolite-based enterosorbents. Thus, the use of zeolite normalized the mineral balance in children, a decrease in the concentration of Cr, As, and Ni and an increase in the amount of Mg and Ca in tissues, bringing them closer to normal concentration, were observed. However, it was suggested that enterosorbents obtained on the basis of natural minerals should be subjected to additional purification before oral administration due to the possible presence of traces of heavy metals as natural impurities. The experiments carried out showed that the retention of heavy metals predominates and prevents their release into the organism [85]. "Enterosgel" is used as a synthetic sorbent based on highly dispersed silica. The presence of surface hydroxyl groups in the structure capable of binding heavy metal ions contributed to a decrease in the level of lead, cadmium and some other heavy metals in the blood and tissues of animals, normalized their biochemical metabolism. Oral administration of enterosorbents based on natural biopolymers, such as alginates and pectins, for 2–4 weeks can reduce the amount of Pb in the blood by an average of 161%, and lead excretion in the urine will increase by an average of 132%. Experimental data on the use of citrus pectin in patients with various diseases were presented, where after treatment a decrease in amount of heavy metals by 74% was observed. Modified and composite sorbents based on activated carbons have demonstrated selectivity for heavy metals due to the presence of a porous structure and functional groups on the surface of the material and have shown high efficiency in the treatment of patients with chronic intoxication with cadmium, mercury and lead [86]. Chitosan, compared to other enterosorbents, is superior in the efficiency of binding heavy metal ions. Furthermore, chitosan has an antioxidant property, fixing Cu^{2+}, Zn^{2+}, Ni^{2+}, and Fe^{3+} ions, which are catalysts for oxidative processes. Chitosan, in addition to reducing the level of lead by 64.4%, nickel by 68.55%, and iron by 35.3%, also improves the hematological indices of the organism.

Thus, the research results show that the prophylactic use of enterosorption effectively prevents the accumulation of heavy metals and radioactive elements in the organism as a result of industrial activities or man-made and natural disasters, thereby reducing the negative impact of heavy metals on human health.

4. Conclusion

Heavy metals are found in the environment and very important for our lives, but they can become dangerous when they accumulate in living organisms. Environmental pollution with heavy metals is one of the most important environmental problems of this century and poses a serious threat to living organisms for normal functioning. In order to prevent serious risks to human health, the development of effective methods for the removal of heavy metals and to eliminate the toxicity of these metals in air, soil, and water is of great importance.

The chapter reviewed the literature data of recent years in the field of using new sorbents for removing heavy metals, and also provides information on heavy metals, their distribution in the environment and living organisms, as well as their toxicity. A large number of studies have been devoted to the synthesis, modification, and study of the properties of new effective sorption materials toward heavy metals. The synthesized materials contribute to the development of the field of sorption processes, studying the mechanisms of this process. As can be seen from the literature data, the possibilities of the enterosorption method are far from exhausted. Its further development is considered as one of the promising directions in the removal of heavy metals from the living organism, as well as in the treatment and prevention of diseases of various etiologies.

Moreover, the growth of industrial production, the introduction of nanotechnology in the production process, and the deterioration of the environmental situation in the world make it relevant to create new materials and methods for the prevention and treatment of chronic intoxications caused by heavy metals. The literature data of recent years show that a large number of enterosorbents approved for use in clinical practice have been developed on the basis of various both natural and synthetic materials. However, there are a number of issues that require further study and should be considered in future research. These issues include the study of enterosorbent selectivity to a certain metal to be eliminated from the human organism; dose and frequency of usage; the presence or absence of adverse effects with prolonged use of enterosorbent, and others. In general, further research is also needed to reduce the overall cost of pretreatment or to develop new effective and cheap methods for removing heavy metals.

Conflict of interest

The authors declare no conflict of interest.

Author details

Sevda Fatullayeva[1*], Dilgam Tagiyev[1], Nizami Zeynalov[1] and Shamil Tagiyev[2]

1 Institute of Catalysis and Inorganic Chemistry Named After Academician M.F. Nagiyev of Ministry of Science and Education of the Republic of Azerbaijan, Baku, Azerbaijan

2 Azerbaijan Medical University, Baku, Azerbaijan

*Address all correspondence to: sevafatullayeva@hotmail.com

References

[1] Shichkin VP, Kurchenko OV, Okhotnikova EN, Chopyak VV, Delfino DV. Enterosorbents in complex therapy of food allergies: A focus on digestive disorders and systemic toxicity in children. Frontier Immunology. Nutritional Immunology. 2023;**14**:1210481. DOI: 10.3389/fimmu.2023.1210481

[2] Fatullayeva S, Tagiyev D, Zeynalov N. A review on enterosorbents and their application in clinical practice: Removal of toxic metals. Colloid and Interface Science Communications. 2021;**45**:100545. DOI: 10.1016/j.colcom.2021. 100545

[3] Liu Z, Xu Z, Xu L, Buyong F, Chay T, Li Z, et al. Modified biochar: Synthesis and mechanism for removal of environmental heavy metals. Carbon Research. 2022;**1**:8. DOI: 10.1007/s44246-022-00007-3

[4] Mitra S, Chakraborty A, Tareq A, Emran T, Nainu F, Khusro A, et al. Impact of heavy metals on the environment and human health: Novel therapeutic insights to counter the toxicity. Journal of King Saud University Science. 2022;**34**:101865. DOI: 10.1016/j.jksus.2022.101865

[5] Fatullayeva S, Tagiyev D, Zeynalov N, Mammadova S, Aliyeva E. Recent advances of chitosan-based polymers in biomedical applications and environmental protection. Journal of Polymer Research. 2022;**29**:259. DOI: 10.1007/s10965-022-03121-3

[6] Ali S, Abbas Z, Rizwan M, Zaheer I, Yava I, Ünay A, et al. Application of floating aquatic plants in phytoremediation of heavy metals polluted water: A review. Sustainability. 2020;**12**:1927. DOI: 10.3390/su12051927

[7] Eskandari E, Kosari M, Farahani M, Khiavi N, Saeedikhani M, Katal R, et al. A review on polyaniline-based materials applications in heavy metals removal and catalytic processes. Separation and Purification Technology. 2020;**231**(16):115901. DOI: 10.1016/j.seppur.2019.115901

[8] Witkowska D, Słowik J, Chilicka K. Heavy metals and human health: Possible exposure pathways and the competition for protein binding sites. Molecules. 2021;**26**:6060. DOI: 10.3390/molecules26196060

[9] Perumal K, Antony J, Muthuramalingam S. Heavy metal pollutants and their spatial distribution in surface sediments from Thondi coast, Palk Bay, South India. Environmental Sciences Europe. 2021;**33**:63. DOI: 10.1186/s12302-021-00501-2

[10] Xiao J, Wang Y, Fu S, Yang S, Liu G, Luo X, et al. Distribution of heavy metals and the exploration of potential indicators and hyperaccumulators in Jiang'an river, Chengdu, PR China. Ecological Indicators. 2022;**145**:109665. DOI: 10.1016/j.ecolind.2022.109665

[11] Liu X, Zhang J, Huang X, Zhang L, Yang C, Li Z. Heavy metal distribution and bioaccumulation combined with ecological and human health risk evaluation in a typical urban plateau Lake, Southwest China. Frontiers in Environmental Science. 2022;**10**:814678. DOI: 10.3389/fenvs.2022.814678

[12] Yuan Y, Liu B, Liu H. Spatial distribution and source identification for heavy metals in surface sediments of east Dongting Lake, China. Scientific Reports. 2022;**12**:7940. DOI: 10.1038/s41598-022-12148-x

[13] Alengebawy A, Abdelkhalek S, Qureshi S, Wang M-Q. Heavy metals and pesticides toxicity in agricultural soil and plants: Ecological risks and human health implications. Toxics. 2021;**9**:42. DOI: 10.3390/toxics9030042

[14] Rezapour S, Asadzade F, Nouri A, Khodaverdiloo H, Heidari M. Distribution, source apportionment, and risk analysis of heavy metals in river sediments of the Urmia Lake basin. Scientific Reports. 2022;**12**:17455. DOI: 10.1038/s41598-022-21752-w

[15] Ohiagu F, Chikezie P, Ahaneku C, Chikezie C. Human exposure to heavy metals: Toxicity mechanisms and health implications. Material Science and Engineering International Journal. 2022;**6**(2):78-87. DOI: 10.15406/mseij.2022.06.00183

[16] Ibrahim E, Selim E. Pollution and health risk assessment of trace metal in vegetable field soils in the eastern Nile Delta, Egypt. Environmental Monitoring and Assessment. 2022;**194**:540. DOI: 10.1007/s10661-022-10199-1

[17] Montazeri A, Chahkandi B, Gheibi M, Eftekhari M, Wacławek S, Behzadian K, et al. A novel AI-based approach for modelling the fate, transportation and prediction of chromium in rivers and agricultural crops: A case study in Iran. Ecotoxicology and Environmental Safety. 2023;**263**:115269. DOI: 10.1016/j.ecoenv.2023.115269

[18] FAO/WHO. Report of the 11th session of the codex committee on contaminants in foods. (CICG). Joint FAO/WHO food standards programme codex alimentarius commission 40th session. 2017;1-69

[19] EPA. U. S. Guidelines for human exposure assessment. (EPA/100/B-19/001). Washington, D.C: Risk Assessment Forum, U.S. EPA. U.S. EPA (U.S. Environmental Protection Agency); 2019

[20] Ejaz U, Khan SM, Khalid N, Ahmad Z, Jehangir S, Fatima Rizvi Z, et al. A detoxifying the heavy metals: A multipronged study of tolerance strategies against heavy metals toxicity in plants. Frontiers in Plant Science. 2023;**14**:1154571. DOI: 10.3389/fpls.2023.1154571

[21] Li X, Liu N, Meng W, He J, Wu P. Accumulation and health risk assessment of heavy metal(loid)s in soil-crop systems from Central Guizhou, Southwest China. Agriculture. 2022;**12**:981. DOI: 10.3390/agriculture12070981

[22] Ndur SA, Nyarko SY, Quaicoe I, Osei LB. Heavy metal loading in surface sediments along the Kawere stream, Tarkwa, Ghana. Ghana Mining Journal. 2020;**20**(2):77-85. DOI: 10.4314/gm.v20i2.9

[23] Gong C, Lu H, Zhang Z, Wang L, Xia X, Wang L, et al. Spatial distribution characteristics of heavy metal(loid)s health risk in soil at scale on town level. Scientific Reports. 2022;**12**:19195. DOI: 10.1038/s41598-022-20867-4

[24] Bibi D, Tőzsér D, Tóthmérész B, Simon E. Heavy metal pollution of soil in Vienna, Austria. Water, Air, and Soil Pollution. 2023;**234**:232. DOI: 10.1007/s11270-023-06244-5

[25] FAO/WHO. Codex Alimentarius. General standard for contaminants and toxins in food and feed. Shedule 1 maximum and Guidline levels for contaminants and toxins in food. In: Reference CX/FAC 02/16. Joint FAO/WHO Food Standards Programme. Rotterdam, The Nertherlands: Codex Committe; 2002

[26] FAO/WHO. Joint AO/WHO Food Standards Program Codex Committe on Contaminants in Food, Food CF/5INF/1. The Hague, The Netherlands: Fifth Session; 2011

[27] Akhtar S, Shafqat A, Rani N, Younas I, Afzal A. Comparatively study and analysis of heavy metal in the cultivated wetland of Baien stream, Shakargarh, Pakistan. Plantation Health. 2022;**01**(01):09-16. DOI: 10.33687/planthealth.01.01.4534

[28] Monchanina C, Devaud J-M, Barron A, Lihoreau M. Current permissible levels of metal pollutants harm terrestrial invertebrates. Science of the Total Environment. 2021;**779**:146398. DOI: 10.1016/j.scitotenv.2021.146398

[29] Jin M, Yuan H, Liu B, Peng J, Xu L, Yang D. Review of the distribution and detection methods of heavy metals in environment. Analytical Methods. 2020;**12**:5747-5766. DOI: 10.1039/D0AY01577F

[30] Ibrahimov R. The effects of heavy metals on biochemical processes in the human body (review). Journal of Life Sciences and Biomedicine. 2020;**2**(75):38-46. DOI: 10.29228/jlsb.40

[31] Oves M, Saghir K, Huda Q, Nadeen F, Almeelbi T. Heavy metals: Biological importance and detoxification strategies. Journal of Bioremediation & Biodegradation. 2016;**7**(2):1000334. DOI: 10.4172/2155-6199.1000334

[32] Lewicki S, Zdanowski R, Krzyżowska M, Lewicka A, Dębski B, Niemcewicz M, et al. The role of chromium III in the organism and its possible use in diabetes and obesity treatment. Annals of Agricultural and Environmental Medicine. 2014;**21**(2):331-335. DOI: 10.5604/1232-1966.1108599

[33] Baj J, Bargieł J, Cabaj J, Skierkowski B, Hunek G, Portincasa P, et al. Trace elements levels in major depressive disorder—Evaluation of potential threats and possible therapeutic approaches. International Journal of Molecular Sciences. 2023;**24**:15071. DOI: 10.3390/ijms242015071

[34] Jadaa W, Mohammed H. Heavy metals – Definition, natural and anthropogenic sources of releasing into ecosystems, toxicity, and removal methods – An overview study. Journal of Ecological Engineering. 2023;**24**(6):249-271. DOI: 10.12911/22998993/162955

[35] Coradduzza D, Congiargiu A, Azara E, Mammani I, Miglio M, Zinellu A, et al. Heavy metals in biological samples of cancer patients: A systematic literature review. Biometals. 2024;**37**:803-817. DOI: 10.1007/s10534-024-00583-4

[36] Živković B, Đokić J. Distribution of heavy metals from tailings ponds and landfills into underground and surface waters. Metallurgical and Materials Engineering. 2022;**28**(4):685-696. DOI: 10.56801/MME859

[37] Balali-Mood M, Naseri K, Tahergorabi Z, Khazdair MR, Sadeghi M. Toxic mechanisms of five heavy metals: Mercury, lead, chromium, cadmium, and arsenic. Frontiers in Pharmacology. 2021;**12**:643972. DOI: 10.3389/fphar.2021.643972

[38] Abd Elnabi MK, Elkaliny NE, Elyazied MM, Azab SH, Elkhalifa SA, Elmasry S, et al. Toxicity of heavy metals and recent advances in their removal: A review. Toxics. 2023;**11**:580. DOI: 10.3390/toxics11070580

[39] Sall ML, Diaw AKD, Gningue-Sall D, et al. Toxic heavy metals: Impact on the environment and human health,

and treatment with conducting organic polymers, a review. Environmental Science and Pollution Research. 2020;**27**:29927-29942. DOI: 10.1007/s11356-020-09354-3

[40] Peng H. The toxic mechanism of heavy metals and their effects on human health, highlights in science. Engineering and Technology. 2023;**74**:1759-1764. DOI: 10.54097/my35g210

[41] Dahiya V. Heavy metal toxicity of drinking water: A silent killer. GSC Biological and Pharmaceutical Sciences. 2022;**19**(01):020-025. DOI: 10.30574/gscbps.2022.19.1.0107

[42] Smyrnov V, Mitryasova O, Salamon I, Smyrnova S, Chvyr V, Mats A. The distribution of heavy metals mobile forms in the industrial urban agglomeration soil. Journal of Ecological Engineering. 2023;**24**(11):303-313. DOI: 10.12911/22998993/171774

[43] Raji Z, Karim A, Karam A, Khalloufi S. Adsorption of heavy metals: Mechanisms, kinetics, and applications of various adsorbents in wastewater remediation—A review. Waste. 2023;**1**:775-805. DOI: 10.3390/waste1030046

[44] Ozdemir S, Turp S, Oz N. Simultaneous dry-sorption of heavy metals by porous adsorbents during sludge composting. Environmental Engineering Research. 2020;**25**(2):258-265. DOI: 10.4491/eer.2019.071

[45] Kharlyamov D, Mavrin G, Sippel I. About the possibility of sorption concentration of heavy metals using magnetite. Life Science Journal. 2014;**11**(8):607-610. Available from: http://www.lifesciencesite.com.89

[46] Naef AAQ, Ramy HM, Dahiru UL. Removal of heavy metal ions from wastewater: A comprehensive and critical review. Clean Water. 2021;**4**:36. DOI: 10.1038/s41545-021-00127-0

[47] Nastasovic A, Markovic B, Surucic L, Onjia A. Methacrylate-based polymeric sorbents for recovery of metals from aqueous solutions. Metals. 2022;**12**:814. DOI: 10.3390/met12050814

[48] Flieger J, Kawka J, Płaziński W, Panek R, Madej J. Sorption of heavy metal ions of chromium, manganese, selenium, nickel, cobalt, iron from aqueous acidic solutions in batch and dynamic conditions on natural and synthetic aluminosilicate sorbents. Materials. 2020;**13**:5271. DOI: 10.3390/ma13225271

[49] Hashemi B, Rezania S. Carbon-based sorbents and their nanocomposites for the enrichment of heavy metal ions: A review. Microchimica Acta. 2019;**186**:578. DOI: 10.1007/s00604-019-3668-2

[50] Aldawsari AM, Alsohaimi IH, Hassan HMA, Al-Abduly AJ, Hassan A, Saleh EAM. Multiuse silicon hybrid polyurea-based polymer for highly effective removal of heavy metal ions from aqueous solution. International Journal of Environmental Science and Technology. 2020;**19**:20. DOI: 10.1007/s13762-021-03355-6

[51] Metwally MG, Benhawy AH, Khalifa RM, El Nashar RM, Trojanowicz M. Application of molecularly imprinted polymers in the analysis of waters and wastewaters. Molecules. 2021;**26**:6515. DOI: 10.3390/molecules26216515

[52] Anceschi A, Caldera F, Bertasa M, Trotta F, Bracco P, Cecone C, et al. New poly(β-cyclodextrin)/poly (vinyl alcohol) electrospun sub-micrometric fibers and their potential application for wastewater treatments. Nanomaterials.

2020;**10**:482. DOI: 10.3390/nano10030482

[53] Muntean SG, Nistor MA, Nastas R, Petuhov O. Dyes and heavy metals removal from aqueous solutions using raw and modified diatomite. PRO. 2023;**11**:2245. DOI: 10.3390/pr11082245

[54] Esmaeili A, Mobini M, Eslami H. Removal of heavy metals from acid mine drainage by native natural clay minerals, batch and continuous studies. Applied Water Science. 2019;**9**:97. DOI: 10.1007/s13201-019-0977-x

[55] Kozera-Sucharda B, Gworek B, Kondzielski I. The simultaneous removal of zinc and cadmium from multicomponent aqueous solutions by their sorption onto selected natural and synthetic zeolites. Minerals. 2020;**10**:343. DOI: 10.3390/min10040343

[56] Lashkarian E, Ahmadi S, Beigmohammadi F. Recent application of nanomaterials-based magnetic solid phase microextraction for heavy metals food toxicity. International Journal of New Chemistry. 2024;**11**(4):216-286

[57] Soares SF, Fernandes T, Trindade T, et al. Recent advances on magnetic biosorbents and their applications for water treatment. Environmental Chemistry Letters. 2020;**18**:151-164. DOI: 10.1007/s10311-019-00931-8

[58] Osman A, El-Monaem E, Elgarahy A, Aniagor C, Hosny M, Farghali M, et al. Methods to prepare biosorbents and magnetic sorbents for water treatment: A review. Environmental Chemistry Letters. 2023;**21**:2337-2398. DOI: 10.1007/s10311-023-01603-4

[59] Balintova M, Demcak S, Pagacova B. A study of sorption heavy metals by natural organic sorbents. International Journal of Energy and Environment. 2016;**10**:189-194

[60] Rahman ZD, Vijayaraghavan J, Thivya J. A comprehensive review on zinc(II) sequestration from wastewater using various natural/modified low-cost agro-waste sorbents. Biomass Conversion and Biorefinery. 2021;**13**(2):1-31. DOI: 10.1007/s13399-021-01822-1

[61] Kumar R, Chawla J, Kaur I. Removal of cadmium ion from wastewater by carbon-based nanosorbents: A review. Journal of Water and Health. 2015;**13**(1):18-33. DOI: 10.2166/wh.2014.024

[62] Adabi S, Yazdanbakhsh A, Shahsavani A, Hadi M. Removal of heavy metals from the aqueous solution by nanomaterials: A review with analyzing and categorizing the studies. Journal of Environmental Health Science and Engineering. 2023;**21**:305-318. DOI: 10.1007/s40201-023-00863-0

[63] Li M, Kuang S, Kang Y, Ma H, Dong J, Guo Z. Recent advances in application of iron-manganese oxide nanomaterials for removal of heavy metals in the aquatic environment. Science of The Total Environment. 2022;**819**:153157. DOI: 10.1016/j.scitotenv.2022.153157

[64] Eden N, Zadehahmadi F, Konstas K, Mardel J, Shaibani M, Mahdavi H. Removal of metals from water using MOF-based composite adsorbents. Environmental Science: Water Research & Technology. 2023;**9**:1305. DOI: 10.1039/d2ew00941b

[65] Essalmi S, Lotfi S, BaQais A, Saadi M, Arab M, AitAhsaine H. Design and application of metal organic frameworks for heavy metals adsorption in water: A review. RSC Advances. 2024;**14**(13):9365-9390. DOI: 10.1039/d3ra08815d

[66] Wang H, Wang S, Wang S, Fu L, Zhang L. Efficient metal-organic

framework adsorbents for removal of harmful heavy metal Pb(II) from solution: Activation energy and interaction mechanism. Journal of Environmental Chemical Engineering. 2023;**11**(2):109335. DOI: 10.1016/j.jece.2023.109335

[67] Wang Q, Zhu S, Xi C, Zhang F. A review: Adsorption and removal of heavy metals based on polyamide-amines composites. Frontiers in Chemistry. 2022;**10**:814643. DOI: 10.3389/fchem.2022.814643

[68] Darban Z, Shahabuddin S, Gaur R, Ahmad I, Sridewi N. Hydrogel-based adsorbent material for the effective removal of heavy metals from wastewater: A comprehensive review. Gels. 2022;**8**:263. DOI: 10.3390/gels8050263

[69] Jadoun S, Fuentes J, Urbano B, Yáñez J. A review on adsorption of heavy metals from wastewater using conducting polymer-based materials. Journal of Environmental Chemical Engineering. 2023;**11**(1):109226. DOI: 10.1016/j.jece.2022.109226

[70] Levy-Ontman O, Yanay C, Alfi Y, Paz-Tal O, Wolfson A. Selective sorption of heavy metals by renewable polysaccharides. Polymers. 2023;**15**:4457. DOI: 10.3390/polym15224457

[71] Qureshi F, Memon SQ, Khuhawar MY, et al. Removal of Co^{2+}, Cu^{2+} and Au^{3+} ions from contaminated wastewater by using new fluorescent and antibacterial polymer as sorbent. Polymer Bulletin. 2021;**78**:1505-1533. DOI: 10.1007/s00289-020-03170-y

[72] Borsagli F, Borsagli A. Chemically modified chitosan bio-sorbents for the competitive complexation of heavy metals ions: A potential model for the treatment of wastewaters and industrial spills. Journal of Polymers and the Environment. 2019;**27**:1542-1556. DOI: 10.1007/s10924-019-01449-4

[73] Zhao C, Liu G, Tan Q, Gao M, Chen G, Huang X, et al. Polysaccharide-based biopolymer hydrogels for heavy metal detection and adsorption. Journal of Advanced Research. 2023;**44**:53-70. DOI: 10.1016/j.jare.2022.04.005

[74] Upadhyay U, Sreedhar I, Singh S, Patel C, Anitha K. Recent advances in heavy metal removal by chitosan based adsorbents. Carbohydrate Polymers. 2021;**251**:117000. DOI: 10.1016/j.carbpol.2020.117000

[75] Chandra D, Molla T, Bashar A, Islam S, Ahsan S. Chitosan-based nano-sorbents: Synthesis, surface modification, characterisation and application in Cd (II), Co (II), Cu (II) and Pb (II) ions removal from wastewater. Scientific Reports. 2023;**13**:6050. DOI: 10.1038/s41598-023-32847-3

[76] Muddin N, Badsha M, Merican Z, Arafath A, Hossain S. Magnetic chitosan nanoparticles as a potential bio-sorbent for the removal of Cr(VI) from wastewater: Synthesis, environmental impact and challenges. Desalination and Water Treatment. 2024;**319**:100449. DOI: 10.1016/j.dwt.2024.100449

[77] Rostami M, Khodaei M. Recent advances in chitosan-based nanocomposites for adsorption and removal of heavy metal ions. International Journal of Biological Macromolecules. 2024;**270**(2):132386. DOI: 10.1016/j.ijbiomac.2024.132386

[78] Kaczorowska MA, Bozejewicz D. The application of chitosan-based adsorbents for the removal of hazardous pollutants from aqueous solutions—A review. Sustainability. 2024;**16**:2615. DOI: 10.3390/su16072615

[79] Annu MM, Tripathi S, Shin DK. Biopolymeric nanocomposites for wastewater remediation: An overview on recent progress and challenges. Polymers. 2024;**16**:294. DOI: 10.3390/polym16020294

[80] Russo T, Fucile P, Giacometti R, Sannino F. Sustainable removal of contaminants by biopolymers: A novel approach for wastewater treatment. Current state and future perspectives. Processes. 2021;**9**:719. DOI: 10.3390/pr9040719

[81] Kormishina AE, Mizina PG, Kormishin VA. Composite enterosorbent for multipurpose use. Patent [in Russian]. RU 2682623 C1. 2019;8

[82] Andrusenko SF, Anfinogenova OI, Denisova EV. The assessment of the sorption capacity of enterosorbents at the risk of heavy metal poisoning. Entomology and Applied Science Letters. 2020;**7**(1):10-13

[83] Tunakova Y, Novikova S, Ragimov A, Faizullin R, Valiev V. A method for assessing the retention of trace elements in human body using neural network technology. Journal of Healthcare Engineering. 2017;7:3471616. DOI: 10.1155/2017/3471616

[84] Mikhalovsky S, Voytko O, Demchenko V, Demchenko P. Enterosorption in the treatment of heavy metal poisoning. Chemistry Journal of Moldova. 2021;**16**(2):9-27. DOI: 10.19261/cjm.2021.852

[85] Wang M, Bera G, Mitra K, Wade T, Knap A, Phillips T. Tight sorption of arsenic, cadmium, mercury, and lead by edible activated carbon and acid-processed montmorillonite clay. Environmental Science and Pollution Research International. 2021;**28**(6):6758-6770. DOI: 10.1007/s11356-020-10973-z

[86] Jandosov J, Alavijeh M, Sultakhan S, Baimenov A, Bernardo M, Sakipova Z, et al. Activated carbon/pectin composite enterosorbent for human protection from intoxication with xenobiotics Pb(II) and sodium diclofenac. Molecules. 2022;**27**:2296. DOI: 10.3390/molecules27072296

Chapter 3

Microbial Methylation of Mercury in Wetlands and Its Environmental Risks

Mitsuo Yoshida

Abstract

Global mercury contamination presents a substantial threat to ecosystems and human health, primarily due to the conversion of inorganic mercury from anthropogenic sources into methylmercury by microorganisms in the environment. Wetlands are especially prone to this transformation due to their anaerobic conditions, abundant organic matter, and the prevalence of metal-reducing bacteria. This chapter explores the mechanisms of inorganic mercury methylation and the subsequent diffusion and accumulation of methylmercury in organisms and ecosystems, using a case study of mercury contamination in the Santa Lucia River wetlands in Uruguay. In this case, high concentrations of mercury were detected in the sediments over a 15,000 square meter area, exceeding regulated limits. The contamination was linked to the discharge of untreated wastewater containing inorganic mercury from a chlor-alkali plant. Although a wastewater treatment system was later installed, significant mercury contamination and hotspots persisted in the wetlands. These areas contained methylmercury in both the sediments and the biota, including reeds, shellfish, earthworms, and fish. Microbial analysis revealed the presence of *Geobacter sulfurreducens* and *Geobacter metallireducens*, bacteria known to methylate inorganic mercury. The existence of these microbial communities in areas of high mercury contamination poses serious risks at four levels to public health and ecosystems due to ongoing *in situ* methylation.

Keywords: chlor-alkali plant, mercury contamination, wetlands, microbial methylation, *Geobacter*

1. Introduction

Mercury, a naturally occurring heavy metal, becomes a significant environmental pollutant when its concentration is increased by anthropogenic activities, particularly mining and industrial processes. The rapid industrialization has led to a substantial increase in the release of harmful heavy metals, including copper, lead, nickel, cadmium, chromium, selenium, vanadium, zinc, silver, and mercury, into the environment. These discharges have raised considerable concerns regarding their adverse effects on the natural environment and ecosystems, as well as on human health [1]. Mercury is released into the environment from both natural sources, such as volcanic activity, and anthropogenic emissions. It is distributed across various environmental compartments, including the hydrosphere, atmosphere, and terrestrial soils and sediments. Mercury is transported through the water cycle, leading to contamination of rivers and oceans, while gaseous mercury disperses through the atmosphere, contributing to global environmental pollution [2–4].

Figure 1 illustrates the global mercury cycle based on estimates from the United Nations Environment Programme (UNEP) [5, 6]. According to these estimates, global anthropogenic emissions of mercury range between 2000 and 3000 tons per year into the atmosphere, with an additional 600 tons per year emitted onto terrestrial areas, including rivers and lakes. Mercury released into the atmosphere, whether from anthropogenic or from natural sources such as volcanic eruptions, is dispersed globally. However, mercury emitted onto land, primarily in the form of solid waste or industrial wastewater, results in localized contamination of rivers, lakes, and soils. It is imperative to control these anthropogenic emissions of mercury to prevent further environmental degradation.

Mercury poses a severe toxic threat to human health, particularly to the development of the fetus and young children. It exists in various forms, each with different levels of toxicity: elemental mercury (metallic), inorganic mercury (e.g., mercuric chloride), and organic mercury (e.g., methylmercury). These forms of mercury present distinct health risks and require different preventive measures [7]. The World Health Organization (WHO) has established guideline values for mercury pollution, including a limit of 6 μg/liter for inorganic mercury in water [8], 1 $\mu g/m^3$ (annual average) for mercury in air [9], a tolerable concentration of 0.2 $\mu g/m^3$ for long-term inhalation of elemental mercury vapor, and a tolerable intake of 2 μg/kg body weight per day for total mercury [10]. While most mercury is discharged in inorganic forms, it can be transformed into the highly toxic methylmercury (MeHg) upon entering aquatic environments. Methylmercury is notorious for causing Minamata disease, a severe neurological disorder resulting from the consumption of fish and shellfish contaminated with MeHg [11–13].

To address the global mercury pollution issue, the Minamata Convention on Mercury was adopted in 2013 and came into force in 2017 [14]. The Convention

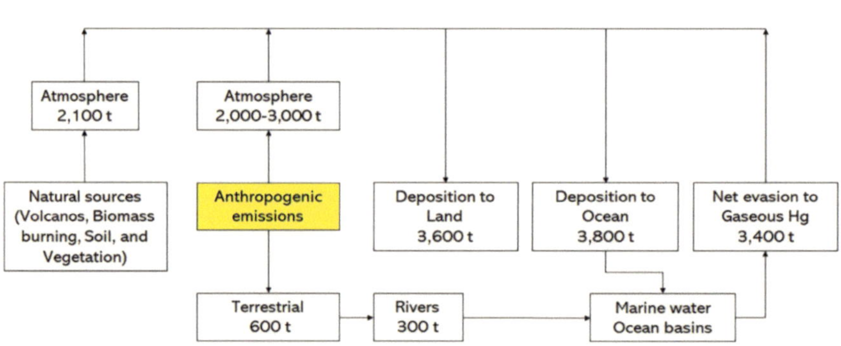

Figure 1.
Outline of the global mercury cycle (estimated amounts of mercury were based on UNEP Global Mercury Report 2018) [5, 6].

outlines three immediate actions: (1) phasing out mercury-using plants and promoting mercury-free industrial processes through cleaner production methods, (2) assessing and mitigating environmental pollution caused by mercury-using plants, with a focus on public health and environmental conservation, and (3) dismantling mercury-contaminated facilities and properly treating and disposing of contaminated sludge and waste. The closure and conversion of chlor-alkali plants using mercury are mandated to be completed by 2025 under this Convention.

This chapter examines the case of inorganic mercury contamination from a chlor-alkali plant in Uruguay, detailing the microbial methylation processes occurring in mercury-contaminated wetlands and the resultant environmental risks. The chapter further discusses the necessary countermeasures and cooperation to mitigate these risks. The microbial methylation of mercury is of significant environmental concern due to the production of methylmercury, a highly toxic compound that bioaccumulates and biomagnifies through food webs, posing serious threats to both human health and wildlife. Given the global distribution and persistence of methylmercury, it is critical to engage in comprehensive monitoring, regulation, and remediation efforts.

2. Mercury pollution in river wetlands - A case history

Between 2015 and 2017, the Dirección Nacional de Medio Ambiente (DINAMA), now known as the Ministry of Environment of Uruguay, in collaboration with the Japan International Cooperation Agency (JICA), conducted a technical cooperation project focused on monitoring mercury contamination and developing countermeasures in the lower Santa Lucia basin in southern Uruguay [15]. This region includes both industrial and agricultural zones, as well as wetlands along the river. Since the 1950s, a chlor-alkali plant in the area, utilizing mercury electrode cells, had been discharging untreated industrial wastewater into the river wetlands, resulting in significant environmental risks associated with mercury contamination [16, 17].

Environmental monitoring by DINAMA indicated that mercury levels in the wetland sediments near the industrial wastewater discharge outlet significantly exceeded the regulatory standards, with total mercury concentrations reaching or exceeding 68 mg/kg. The primary objectives of this project were to build capacity for organic mercury analysis, conduct a comprehensive investigation of mercury contamination in the Santa Lucia River wetlands, assess the associated environmental risks, and develop appropriate countermeasures. This project was a collaborative effort between DINAMA and JICA experts, including the author, during the period from 2015 to 2017. The findings were subsequently published in the DINAMA-JICA report [15].

During the field survey, the author collected sediment samples from mercury-contaminated areas within the wetlands for trace element analysis and to study the microbial communities present in the sediments and water.

2.1 Methodology for mercury contamination analysis

2.1.1 Sample collections

As part of the DINAMA-JICA project, sediment samples from the river wetlands were systematically collected using a five-point sampling method along a predefined 50-meter grid. In addition to sediment samples, fish and shellfish residing in the river wetlands were also collected during the 2015 and 2016 field campaigns. A total of 57 sediment samples (sites) and 235 specimens were obtained. Furthermore, 12 additional biota samples, including crabs, mollusks, earthworms, and plants, were collected within the study area.

2.1.2 Chemical analysis

The original analytical data reported in this paper are ICP-MS analysis and microbial analysis, but other chemical analysis data including total mercury analysis and methylmercury analysis were reported by the DINAMA-JICA Project in 2017 [15].

1. *Total mercury analysis*

   The collected samples were analyzed for total mercury (T-Hg) using atomic absorption spectrometry (AAS) by cold vapor generation with a flow injection system in the Environmental Analysis Laboratory of DINAMA (now the Ministry of Environment), Montevideo, Uruguay. The analysis was carried out as part of the DINAMA-JICA project. Analytical methodologies are used as follows:

   i. Pretreatment: The samples collected were stored in a low-temperature storage (<6°C) until chemical analysis. Prior to drying, remove from the sample any foreign component. The sediment samples were individually homogenized and dried at room temperature in the lab. Digestion of solids was performed in a closed system applying the method of National Reference 3262UY [18] based on the USEPA Method 3051A [19] with a microwave Anton Paar Multiwave 3000.

   ii. Determination of total mercury: The method of National Reference 3141UY [20] based on the international reference ISO 12846 [21] was applied for the total mercury (T-Hg) analysis. Equipment used was a Perkin Elmer FIMS-100 Flow Injection Mercury system. The mercury content in the sample is determined by means of a calibration curve defined by standard sample.

   iii. Analytical quality control was performed based on the procedure of Article 13 of 3141UY [20], where measurement of the reference materials, sample duplication, and blank control were performed.

   Detailed information regarding the sample processing methods and data can be found in DINAMA-JICA [15]. Additionally, the author conducted inductively coupled plasma mass spectrometry (ICP-MS) analysis to supplement the AAS results. The ICP-MS pretreatment involved an aqua regia extraction method, which utilized a mixture of nitric acid and hydrochloric acid.

2. *Methylmercury analysis*

   Methylmercury analysis was conducted using gas chromatography with an electron capture detector (GC-ECD) at the DINAMA laboratory in Montevideo, Uruguay. Pretreatment and accuracy control were executed based on the method of National Reference 3262UY [18]. Analytical method of methylmercury is developed based on the Mercury Analysis Manual issued by the National Institute of Minamata Disease (NIMD), Japan [22]. Extraction and cleanup of organic mercury compounds were applied based on the formation of complexes with dithizone/toluene. The subsequent determination was done by GC-ECD. An important original feature developed by the DINAMA-JICA Project was the incorporation of a 2-cm packed NaCl head column within the GC. The detailed conditions and process for methylmercury analysis applied were documented in the Annex 3 of DINAMA-JICA Technical Cooperation Project report [15].

2.2 State of mercury pollution

Figure 2 presents the contour map illustrating the total mercury concentrations in the wetland sediments as determined by the survey. The analysis revealed that areas with elevated total mercury concentrations radiate in a fan-like pattern from the factory drain outlet toward the marsh, forming multiple hotspot zones with concentrations reaching up to 64 mg/kg. The contamination spans over an area of 15,000 square meters, exceeding regulated limits. Additionally, mercury was predominantly found in the fine sediments (bottom sediment) on the surface of the wetland, suggesting that the contamination likely originated from inorganic mercury adsorbed onto the surface of these sediment particles.

Analysis of organic mercury (methylmercury) content in the sediments from the high-concentration hotspot area revealed that methylmercury constituted up to approximately 0.17% of the total mercury present. In contrast, methylmercury levels in biota collected from the marsh showed significantly higher concentrations, with up to 0.5 mg/kg detected in shellfish and small fish, and up to 0.8 mg/kg in plants such as reeds. The ratio of methylmercury to total mercury in these biological samples ranged from 68 to 100%, indicating a high degree of selective absorption and bioaccumulation of methylmercury within the organisms.

Although plankton were not analyzed in this study, the small fish, carb and shellfish examined are preyed upon by larger predators higher up the food chain, such

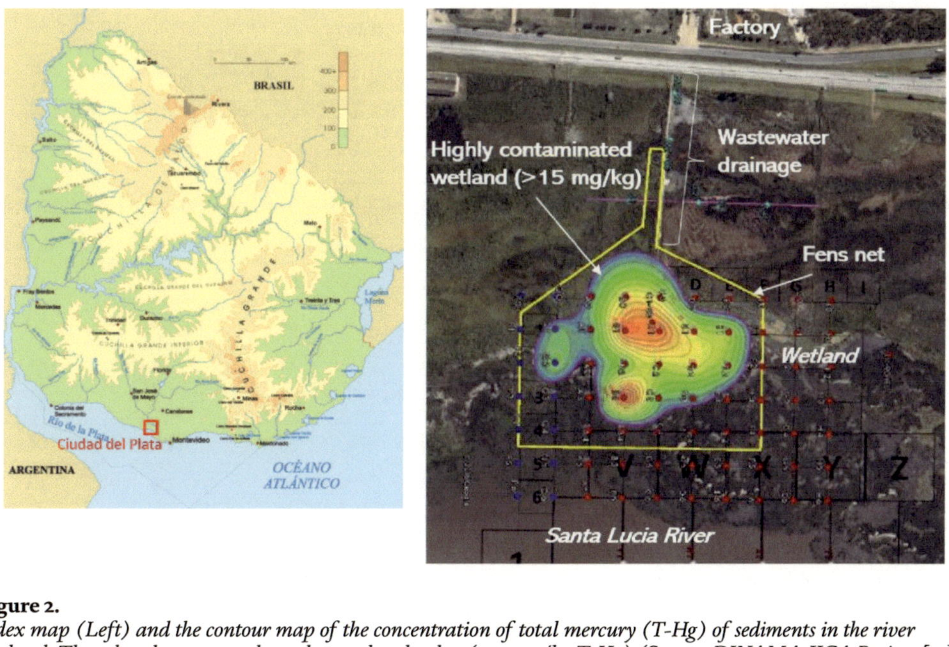

Figure 2.
Index map (Left) and the contour map of the concentration of total mercury (T-Hg) of sediments in the river wetland. The colored parts are above the regulated value (> 15 mg/kg T-Hg) (Source: DINAMA-JICA Project [15]).

as larger fish species, potentially leading to further biomagnification of methylmercury. For example, the Corvina, a large fish known to migrate around the estuary of the Santa Lucia River and the Atlantic coast, is commonly caught, consumed by local residents, and sold in markets. If this situation remains unaddressed, the continued consumption of such large fish poses a significant risk of mercury poisoning due to the bioaccumulation of methylmercury.

According to the results of the trace element analysis conducted using ICP-MS on the wetland sediment, as presented in **Table 1**, the total mercury concentration in the sediment is notably high, measuring 29.1 mg/kg. This finding is consistent with the previous data of total mercury concentrations of the wetland sediments obtained through AAS analysis given by the DINAMA-JICA Project (**Table 2**).

Element	**Mo**	**Cu**	**Pb**	**Zn**	**Ag**	**Ni**	**Co**	**Mn**	**Fe**	**As**
Unit	mg/kg	mg/kg	mg/kg	mg/kg	mg/kg	mg/kg	mg/kg	mg/kg	%	mg/kg
DTL	0.1	0.1	0.1	1	0.1	0.1	0.1	1	0.01	0.5
C3	0.5	20.7	9.9	46	0.2	10.7	6.2	133	1.47	3.3
Element	**Au**	**Th**	**Sr**	**Cd**	**Sb**	**Bi**	**V**	**Ca**	**P**	**La**
Unit	mg/kg	mg/kg	mg/kg	mg/kg	mg/kg	mg/kg	mg/kg	%	%	mg/kg
DTL	0.5	0.1	1	0.1	0.1	0.1	2	0.01	0.001	1
C3	6.9	4.3	39	0.2	0.1	0.1	35	0.23	0.049	18
Element	**Cr**	**Mg**	**Ba**	**Ti**	**B**	**Al**	**Na**	**K**	**W**	**Hg**
Unit	mg/kg	%	mg/kg	%	mg/kg	%	%	%	mg/kg	mg/kg
DTL	1	0.01	1	0.001	20	0.01	0.001	0.01	0.1	0.01
C3	20	0.35	56	0.024	<20	1.45	0.156	0.27	<0.1	29.1
Element	**Sc**	**Tl**	**S**	**Ga**	**Se**	**Te**				
Unit	mg/kg	mg/kg	%	mg/kg	mg/kg	mg/kg				
DTL	0.1	0.1	0.05	1	0.5	0.2				
C3	4.6	0.2	0.28	4	<0.5	<0.2				

Table 1.
Trace element composition of the wetland sediment collected from the location C3, analyzed by ICP-MS. DTL: Detection Limit.

Sample ID	**T-Hg (mg/kg)**	**Sample ID**	**T-Hg (mg/kg)**	**Sample ID**	**T-Hg (mg/kg)**	**Sample ID**	**T-Hg (mg/kg)**
A0	18	**A1**	17	**A2**	12	**A3**	0.74
A4	3.8	**A5**	1.8	**A6**	0.38	**B0**	40
B1	57	**B2**	23	**B3**	66	**B4**	2.3
B5	1.9	B6A	0.42	**C0**	34	**C1**	56
C2	32	**C3**	20	**C4**	6.3	**C5**	2.5
C6	1.1	**D0**	3.8	**D1**	35	**D2**	36
D3	18	**D4**	3.7	**D5**	2.2	**D6**	1.4
E0	2.8	**E1**	33	**E2**	28	**E3**	7.8
E4	3.1	**E5**	1.8	**E6**	1.4	**F0**	2.1
F1	5.3	**G1**	1.8	**H1**	7.6	**Q0**	2.0
Q1	21	**Q2**	24	**Q3**	11	**P0**	3.5

Sample ID	T-Hg (mg/kg)	Sample ID	T-Hg (mg/kg)	Sample ID	T-Hg (mg/kg)	Sample ID	T-Hg (mg/kg)
P1	5.9	**P2**	9.7	**P3**	8.0	**P4**	1.9
P5	0.5	**V1**	0.071	**W1**	0.10	**W2**	0.095
X1	0.22	**X2**	0.12	**Y1**	0.076	**Y2**	0.17
S1–00	3.0						

Table 2.
Result of total mercury (T-Hg, dry basis) analysis of wetland/river sediment samples (collected in 2015 and 2016) analyzed by the DINAMA Laboratory, Uruguay (Source: DINAMA-JICA Project, 2017, Project Final Report [15]).

Sample ID	T-Hg (mg/kg)	Methylmercury (mg/kg)	Ratio of methylmercury (%)
B1	57	LD < x < LC	0.05 < x < 0.1
B3	66	110	0.17
C1	56	LD < x < LC	0.05 < x < 0.1
D2	36	<LD	<0.08

Table 3.
Comparison between T-Hg and methyl mercury concentrations (dry basis) in wetland sediments, 2016, analyzed by the DINAMA Laboratory, Uruguay (Source: DINAMA-JICA Project, 2017, Project Final Report [15]).

Sample ID	Type	T-Hg (mg/kg)	Sampling date
A4-Wb	mollusca	0.38	2015/11/20
C1-Sp	earthworm	0.47	2016/09/20
C1-3mNb	crab	0.45	2015/11/20
D1-Nb	crab	0.25	2016/11/20

Table 4.
Result of T-Hg analysis of biota (mollusca, earthworm and crab) samples in the wetland, analyzed by the DINAMA Laboratory, Uruguay (Source: DINAMA-JICA Project, 2017, Project Final Report [15]).

Additionally, the sediment exhibits relatively high levels of iron, manganese and sulfur. These elevated concentrations suggest conditions that are conducive to the presence and activity of iron-reducing and sulfur-reducing bacteria. A comparison of the total mercury (T-Hg) and methylmercury contents in the wetland sediments is compiled in **Table 3**, the T-Hg concentrations in individual organisms (mollusca, earthworm and crab) living in the wetland are summarized in **Table 4**, and T-Hg concentrations in plants (leaves and roots of reeds) growing in the wetland are given in **Table 5**. These analytical data were also reported by the DINAMA-JICA Project [15].

LD = 0.03 mg/kg, LC = 0.1 mg/kg.

3. Microbial community in the mercury-contaminated wetland

The microbial community analysis of samples from the mercury-contaminated wetlands, as described above, was conducted at the laboratory of Tropical Technology Plus Inc., Okinawa, Japan. The results were compiled and analyzed by the authors.

Sample ID	Type	Sampling date	T-Hg (mg/kg)
C1-3mWL/1	Reed leaves	2015/09/09	<0.02
C1-3mWL/2	Reed leaves	2015/09/09	<0.02
C1-3mWL/3	Reed leaves	2015/09/09	<0.02
C1-3mWL/4	Reed leaves	2015/09/09	0.038
C1-3mNW	Mud of the base of the reed C1-3mWL	2015/09/09	35
H1-90mEL/1	Reed leaves	2015/09/09	<0.02
H1-90mEL/2	Reed leaves	2015/09/09	<0.02
H1-90mEL/3	Reed leaves	2015/09/09	<0.02
H1-90mEL/4	Reed leaves	2015/09/09	<0.02
H1-90mEL/5	Reed leaves	2015/09/09	<0.02
H1-90mE	Mud of the base of the reed H1-90mEL	2015/09/09	8.7
H1-90mEP	Reed roots	2015/09/09	0.81

Table 5.
Result of T-Hg analysis of wetland plants (reed leaves, roots, associated sediments) analyzed by the DINAMA Laboratory, Uruguay (Source: DINAMA-JICA Project, 2017, Project Final Report [15]).

3.1 Methodology for microbial analysis

3.1.1 DNA extraction

Deoxyribonucleic acid (DNA) extraction was carried out using the FastDNA SPIN Kit for Soil (Q-BioGene), following the standard environmental DNA (eDNA) analysis protocol established by the National Institute for Agro-Environmental Sciences [23]. The samples were homogenized in a homogenizer at 1500 rpm for 60 seconds to ensure efficient DNA extraction.

3.1.2 PCR

The extracted DNA was subjected to gene amplification through polymerase chain reaction (PCR). The PCR process was conducted using a PCR Thermal Cycler SP (Takara), with reaction conditions based on protocols outlined by the National Institute for Agro-Environmental Sciences, Japan, as well as those described by Muyzer et al. and Yasumoto-Hirose et al. [23–25].

3.1.3 DGGE

DGGE was performed using the D-code system from Bio-Rad. The preparation of the gel plate followed the methods specified by the National Institute for Agro-Environmental Sciences, Japan, Muyzer et al., and Yasumoto-Hirose et al. [23–25]. For the electrophoresis, 10 μL of the PCR-amplified sample was mixed with an equal volume of 2x loading buffer and loaded into the wells of the gel plate. The electrophoresis was carried out under the following conditions: tank temperature of 58°C, voltage of 50 V and an electrophoresis duration of 18 hours. To facilitate pattern comparison, a marker lane (DGGE Marker III, Nippon Gene) and a negative control (without template DNA) were included. The gel was

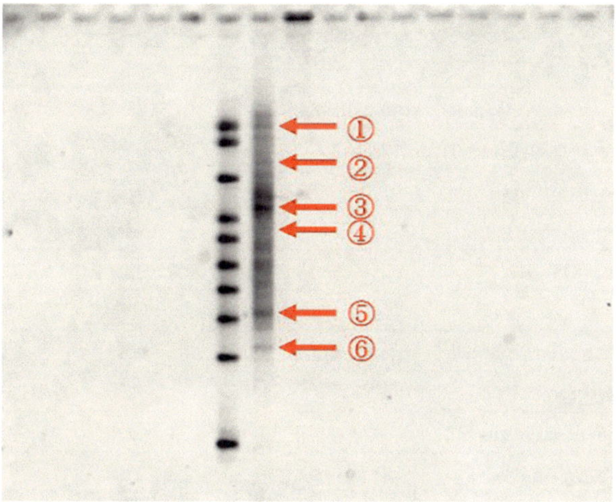

Figure 3.
Six bands obtained by DGGE. M indicates the marker, while A represents the DGGE result of the sample. The numbers (1 to 6) correspond to the identified bands.

stained with SYBR Green and imaged using a luminescent image analyzer LAS-3000 (FUJIFILM). The bands observed in the DGGE were excised with a scalpel, placed into sterilized 1.5-mL tubes, and 100 μL of TE buffer (Tris/EDTA buffer) was added. The tubes were then shaken overnight at 30°C to elute the DNA. The eluted DNA was transferred to new tubes, purified and extracted using ethanol precipitation to prepare the samples (**Figure 3**).

3.1.4 Gene amplification and sequencing

The DNA purified from the excised DGGE bands was subjected to further gene amplification by PCR. This was accomplished using the Takara PCR Thermal Cycler SP with a primer set (F984, R1378) to amplify the target DNA fragments. A negative control, containing no template DNA, was included in the procedure. After amplification and confirmation of the target gene region, the DNA was prepared for sequence analysis. Sequencing was performed using the ABI PRISM 3130xl Genetic Analyzer (Applied Biosystems). The obtained sequences were then compared against available databases using the Basic Local Alignment Search Tool (BLAST) to identify similar sequence groups exceeding a certain threshold score.

3.2 Microbial community

3.2.1 Results of BLAST search

Out of the six bands analyzed, bands 1, 4, 5, and 6 were poorly preserved, making it impossible to perform a BLAST search on their DNA. Consequently, only bands 2 and 3 were successfully subjected to BLAST analysis. **Table 6** presents the results of the BLAST search, which identified sequences with a homology of 92% or higher. The results indicate that the sequences from bands 2 and 3 exhibit very similar genetic profiles.

Genus/species	Homology	Accession number
<Band 2>		
Geobacter sulfurreducens subsp. ethanolicus strain OSK2A	93%	NR_132673
Geobacter sulfurreducens strain PCA	93%	NR_075009
Geobacter metallireducens strain GS-15	92%	NR_075011
Geobacter psychrophilus strain P35	92%	NR_043075
Geobacter luticola strain OSK6	92%	NR_114303
<Band 3>		
Geobacter sulfurreducens subsp. ethanolicus strain OSK2A	92%	NR_132673
Geobacter sulfurreducens strain PCA	92%	NR_075009
Desulfuromonas svalbardensis strain 112	92%	NR_043213
Geobacter psychrophilus strain P35	92%	NR_043075
Nitrospina gracilis strain 3/211	92%	NR_104821

Table 6.
Result of basic local alignment search tool (BLAST) search.

3.2.2 Microbial methylation potential

Analysis of surface sediments from the mercury-contaminated wetland indicated the possible presence of a microbial community primarily composed of six species, as detailed in **Table 6**:

Geobacter sulfurreducens belongs to the Geobacteraceae family, which is known for its iron-reducing capabilities [26, 27]. This bacterium plays a significant role in the methylation of mercury, particularly in iron-rich, anoxic environments such as sediments, wetlands, and groundwater systems. It is capable of reducing various metals and has been identified as a key organism in the conversion of inorganic mercury (Hg(II)) to methylmercury (MeHg) through its possession of the hgcA and hgcB genes, which are essential for this enzymatic process [28]. Additionally, the presence of algae has been shown to enhance Hg methylation by *G. sulfurreducens* PCA, with increased algal organic matter and thiols contributing to elevated MeHg production, even with the presence of dead algae [29]. The proliferation of algae observed in the rivers surveyed in this study may similarly contribute to mercury methylation.

Geobacter metallireducens, another member of the *Geobacter* genus, is known for its ability to reduce metals such as iron and manganese under anoxic conditions [29, 30]. While traditionally not recognized as a primary mercury methylator, recent studies have shown that *G. metallireducens* possesses the genetic capability to methylate mercury under specific environmental conditions. *G. metallireducens* is a motile, nitrogen-fixing, gram-negative anaerobe that oxidizes short-chain fatty acids, alcohols and monoaromatic compounds to carbon dioxide, transferring electrons to various extracellular electron acceptors, including Hg(II) [31].

Geobacter psychrophilus is another metal-reducing bacterium within the *Geobacter* genus, adapted to cold environments (4–30°C) [32]. While its role in mercury methylation is not as well-documented as that of other *Geobacter* species, it may contribute to this process in colder environments, such as the Earth's polar regions where mercury deposition occurs.

Geobacter luticola is known for its role in metal reduction, particularly of iron and manganese, in anaerobic, sediment-rich environments [33]. However, specific information regarding its involvement in mercury methylation is limited and not well-established.

Desulfuromonas svalbardensis, part of the Desulfuromonas genus, is capable of reducing sulfur compounds and metals in anaerobic environments [34]. Closely related to the *Geobacter* genus, *D. svalbardensis* was isolated from Arctic Ocean sediments and is adapted to cold, anaerobic conditions. However, its specific role in mercury methylation remains poorly documented.

Nitrospina gracilis is a marine bacterium involved in the nitrogen cycle, particularly in the oxidation of nitrite to nitrate [35]. This species is primarily found in oceanic environments, and its main metabolic functions do not relate to mercury methylation. Therefore, *N. gracilis* is unlikely to contribute directly or significantly to mercury methylation in the marine or estuary wetlands studied.

Based on the microbial community analysis, it is suggested that mercury methylation in the wetlands occurred predominantly through the metabolic activities of *Geobacter sulfurreducens* and *Geobacter metallireducens*. These species are likely responsible for the conversion of inorganic mercury to methylmercury in this environment. Analysis of surface sediments from a mercury-contaminated wetland suggested possible presence of a microbial community mainly consisting of six species as follows: (see **Table 2**).

3.3 Mechanism of mercury contamination and methylation in the wetland

The microbial methylation of mercury in wetland environments is a critical process, leading to the formation of methylmercury (MeHg), a highly toxic form of mercury that can bioaccumulate within aquatic food webs. Wetlands, with their abundant organic matter and low redox potential, offer ideal conditions for this process. The following outlines the estimated mechanism of mercury methylation in the wetland (refer to **Figure 4**).

3.3.1 Contamination of inorganic mercury

Mercury enters the wetland primarily through industrial wastewater discharged from the chlor-alkali plant. In the environment, mercury mainly exists in inorganic

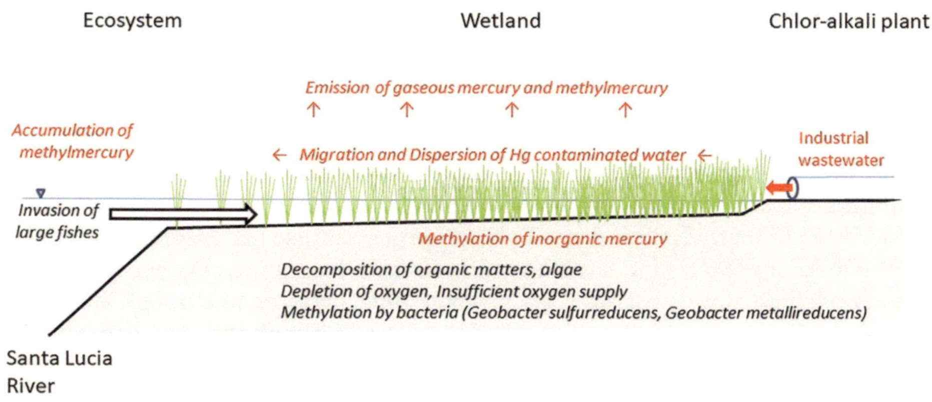

Figure 4.
A schematic model of mercury methylation in wetlands.

forms, such as mercuric ion (Hg(II)) and elemental mercury (Hg(0)), which are the primary precursors for methylmercury (MeHg) production. In the river wetland near the river mouths to the Atlantic Sea, thick growths of reeds and accumulated organic sediments are characteristic, with fluctuating water levels creating alternating anaerobic and aerobic conditions. Seasonal climatic changes between cold and warm temperatures also influence these conditions. The repetition of aerobic and anaerobic states, driven by fluctuating river water levels, has contributed to the formation of mercury-contaminated hotspots in the sediment. These conditions facilitate the release, leaching, migration and fixation of mercury, leading to the gradual expansion of mercury contamination within the wetland.

3.3.2 Microbial community involvement

Anaerobic microorganisms, particularly *Geobacter sulfurreducens* and *Geobacter metallireducens,* are likely the key agents responsible for mercury methylation in the wetland. These microorganisms uptake inorganic mercury (Hg(II)) from their environment and convert it into methylmercury (MeHg) through enzymatic processes. Once formed, MeHg is released from microbial cells into the surrounding environment. Due to its higher solubility and mobility compared to inorganic mercury, MeHg readily binds to organic and inorganic particles in the water and sediments, leading to its accumulation in wetland sediments and biota.

3.3.3 Methylmercury absorption by biota

Methylmercury is readily absorbed by living organisms and tends to bioaccumulate, particularly in higher trophic levels through the food chain. As MeHg accumulates in predators, including fish consumed by humans, it poses a significant risk of organic mercury poisoning, especially in top predators like humans. This bioaccumulation highlights the critical public health risks associated with mercury methylation in contaminated wetlands.

3.3.4 Environmental factors influencing methylation

The availability of organic matter is one of the most critical factors influencing mercury methylation in wetland environments. Wetlands rich in organic matter, such as those with dense reed growth, provide a substantial carbon source for microbial metabolism. This organic matter not only supports microbial activity but also affects the redox conditions, thereby promoting methylation. Wetlands typically exhibit anoxic conditions, which are favorable for the activity of anaerobic microbes responsible for mercury methylation. The river wetland in Uruguay exemplifies these conditions, making it particularly conducive to methylmercury production.

In addition to methylation, demethylation processes also occur, reducing the levels of methylmercury (MeHg) in wetlands and other environments. Demethylation can occur through biotic processes, such as microbial activity, or through abiotic processes, including photo-demethylation (mediated by sunlight) and non-photochemical mechanisms [36]. It has been reported that the ligands of MeHg species and dissolved organic matter are key factors influencing the photo-demethylation of MeHg. These factors play a crucial role in determining the overall balance between methylation and demethylation in wetland ecosystems.

4. Environmental risks

Environmental risk refers to the actual or potential threat of adverse effects on living organisms, ecosystems, and the environment due to pollutants such as effluents, emissions, and solid wastes generated by industrial activities. In the case history of the mercury-contaminated wetland, based on field surveys and chemical analyses, several environmental risks associated with the discharge of inorganic mercury from the chlor-alkali plant were identified. These risks can be categorized into four main types: exposure risk, inhalation risk, intake risk and ecosystem risk (see **Figure 5**).

- Exposure risk: This risk involves direct exposure of living organisms, including local residents, to contaminated sediments and soils. It also includes indirect exposure through the use of contaminated reeds and sediments as construction materials.

- Inhalation risk: This risk arises from the inhalation of fine particulate matter suspended in the air, which originates from dried, contaminated sediments or soils. Additionally, sublimated mercury or gas-phase methylmercury emitted from the contaminated wetland, especially under microbial metabolic activity, poses a significant inhalation risk.

- Intake risk: This risk is related to the consumption of fish and other organisms that have accumulated mercury through the process of biomagnification within the food chain.

- Ecosystem risk: This risk pertains to the direct disruption of ecosystems, impacting all plants, animals and ecosystem in and around the wetland, beyond just the human population.

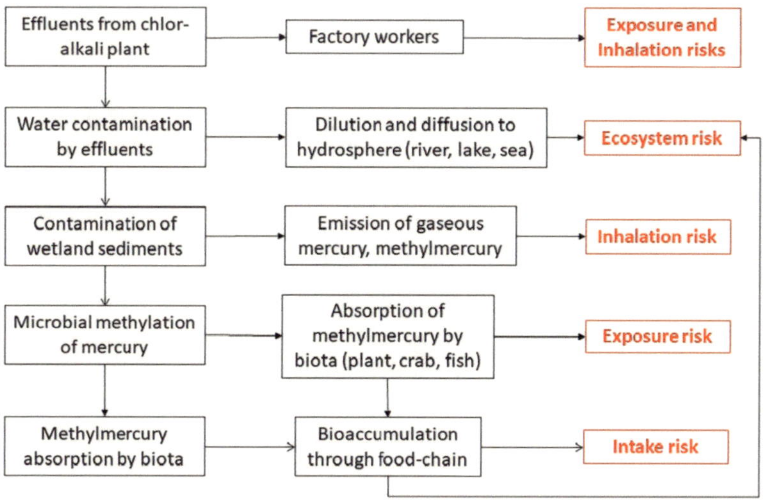

Figure 5.
Environmental risks associated with mercury contamination in the wetland in the case study, categorized into exposure risk, inhalation risk, intake risk and ecosystem risk.

5. Countermeasures taken in the case of uruguay

The environmental risks posed by mercury-contaminated river wetlands extend to local residents who may come into contact with or inhale contaminated particles within these areas. Additionally, the formation of methylmercury by microorganisms in the wetland significantly increases the risk of mercury entering the food chain [37–39]. Methylmercury, once absorbed and accumulated by fish and shellfish, can bio-magnify through the food chain, potentially leading to human consumption, especially in larger fish that serve as top predators.

To mitigate these risks, measures were implemented in Uruguay to restrict access to contaminated areas and prevent the harvesting of fish and shellfish from the wetlands. A fence net was installed around the wetlands with high concentrations of total mercury (see **Figure 2**). Moreover, steps were taken to prevent large fishes from entering the contaminated wetlands, thereby disrupting the food chain and reducing the risk of methylmercury entering higher trophic levels. Specifically, the mercury-contaminated zones were enclosed with both underwater and above-water fences. The costs associated with these measures were borne by the chlor-alkali plant, in accordance with the polluter-pays principle (PPP) [15]. The Ministry of Environment of Uruguay continues to monitor the wetlands and, if deemed necessary, additional actions, such as covering contaminated areas with clean soil, may be considered to further mitigate the risks [15].

6. Conclusions

This chapter presents a case study of mercury contamination in river wetlands in Uruguay, emphasizing the role of wetland-dwelling bacteria, such as *Geobacter sulfurreducens* and *Geobacter metallireducens*, in the methylation of inorganic mercury. Although the initial production of methylmercury through this bacterial process is small, the compound's potential for bioaccumulation greatly amplifies the associated environmental risks. When wetlands are hydrologically connected to open water bodies, this bioaccumulation can propagate further through the food chain, leading to increased levels of methylmercury.

The methylation of mercury in wetland environments is a critical environmental issue due to its toxicity, bio-accumulative properties, and the consequent ecological and human health risks [37–39]. Wetlands can act as both conduits and sinks for mercury, with their overall impact determined by the balance between microbial methylation and demethylation processes [40]. These processes are influenced by various environmental factors and are mediated by specific anaerobic bacteria, making it essential to understand these controlling factors to manage mercury pollution effectively and protect wetland ecosystems.

Given the dual role of wetlands as repositories for mercury contamination and as sites conducive to methylation, comprehensive environmental monitoring is crucial. This includes mapping the distribution of wetlands in riverine and lacustrine systems and focusing on monitoring organic mercury levels within these wetlands and their surrounding ecosystems. Upon detecting anomalies in organic mercury concentrations, it is necessary to investigate the underlying contamination mechanisms and implement appropriate measures to mitigate environmental risks. Such measures may include preventing the inflow of inorganic mercury, inhibiting methylmercury formation and disrupting the food chain within the ecosystem. Additionally, raising awareness among industries and local communities that rely on wetland ecosystem services is vital for promoting the understanding and reduction of environmental risks.

Particularly in South America, Asia and Africa, artisanal and small-scale gold mining (ASGM) is a significant source of inorganic mercury contamination [41–44], with wetlands often located in river basins that receive mercury discharges. In response to the mercury contamination case in Uruguay, the Uruguayan Ministry of Environment is collaborating with South American countries through South-South cooperation to share knowledge and information on mercury contamination in wetlands. However, addressing this persistent issue of mercury contamination in wetlands will require international collaboration and action [40].

Acknowledgements

The author expresses gratitude to Luis Reolón, Magdalena Hill, Gabriel Yorda, and Alejandro Cendón of DINAMA (now the Ministry of Environment, Uruguay); Natalia Barboza, Alejandro Mangarelli, and Vivian Muñoz of the Environmental Analysis Laboratory of DINAMA; Yuji Kashima of the Japan Environment and Sanitation Center (JAES); Makoto Taniguchi and Naomi Hiroi of the Uruguay Office of Japan International Cooperation Agency (JICA); and Mario Sawada, for their various support during the field survey and analysis. The author also thanks Yuji Miyashiro of Tropical Technology Plus Ltd., Okinawa, for his work on microbial analysis. The samples for this study were obtained through the DINAMA-JICA Technical Cooperation Project on the Diagnosis of Actual Situation and Action Plan Design for the Remediation of Coastal Strip of La Plata River burdened by mercury-contaminated sediments. The microbiological study was conducted as part of a research project funded by the International Network for Environmental and Humanitarian Cooperation (iNehc), Nonprofit Inc., Tokyo, Japan. The valuable comments of anonymous reviewers helped to improve this chapter.

Conflict of interest

The author declares no conflict of interest.

Author details

Mitsuo Yoshida[1,2]

1 Global Environment Department, Japan International Cooperation Agency (JICA), Tokyo, Japan

2 International Network for Environmental and Humanitarian Cooperation (iNehc), Nonprofit Inc., Tokyo, Japan

*Address all correspondence to: yoshida.mitsuo@jica.go.jp; mitsuoyoshida.jp@gmail.com

References

[1] Mance G. Pollution Threat of Heavy Metals in Aquatic Environments. London & New York: Elsevier Applied Science; 1987. eISBN-13:978-94-009-3421-4. DOI: 10.1007/978-94-009-3421-4

[2] Zwolak A, Sarzyńska M, Szpyrka E, Stawarczyk K. Sources of soil pollution by heavy metals and their accumulation in vegetables: A review. Water Air Soil Pollution. 2019;**230**:164. DOI: 10.1007/s11270-019-4221-y

[3] Gworek B, Bemowska-Kałabun O, Kijeńska M, Wrzosek-Jakubowska J. Mercury in marine and oceanic waters—A review. Water Air Soil Pollution. 2016;**227**:371. DOI: 10.1007/s11270-016-3060-3

[4] Gworek B, Dmuchowski W, Baczewska-Dąbrowska AH. Mercury in the terrestrial environment: A review. Environmental Sciences Europe. 2020;**32**:128. DOI: 10.1186/s12302-020-00401-x

[5] United Nations Environment Programme (UNEP). Global Mercury Assessment. UNEP; 2018. Available from: https://www.unep.org/topics/chemicals-and-pollution-action/pollution-and-health/heavy-metals/mercury/global-mercury-2 [Accessed: September 12, 2024]

[6] United Nations Environment Programme (UNEP). Global Mercury Assessment Technical Background Report. UNEP; 2018. Available from: https://www.unep.org/topics/chemicals-and-pollution-action/pollution-and-health/heavy-metals/mercury/global-mercury-2 [Accessed: September 12, 2024]

[7] World Health Organization (WHO), editor. Preventing Disease through Healthy Environments Exposure to Mercury: A Major Public Health Concern. 2nd ed. World Health Organization (WHO); 2021. Available from: https://www.who.int/publications/i/item/9789240023567 [Accessed: September 12, 2024]

[8] Guidelines for Drinking-Water Quality: Fourth Edition, Incorporating the First Addendum. Geneva: World Health Organization; 2017. Available from: https://apps.who.int/iris/handle/10665/254637 [Accessed: September 12, 2024]

[9] Air Quality Guidelines for Europe. 2nd ed. Copenhagen: World Health Organization Regional Office for Europe; 2000. Available from: https://apps.who.int/iris/handle/10665/107335 [Accessed: September 12, 2024]

[10] Elemental Mercury and Inorganic Mercury Compounds: Human Health Aspects. CICAD 50. Geneva: World Health Organization; 2003. Available from: https://apps.who.int/iris/handle/10665/42607 [Accessed: September 12, 2024]

[11] Sakamoto M, Tatsuta N, Izumo K, Phan P, Vu LD, Yamamoto M, et al. Health impacts and biomarkers of prenatal exposure to methylmercury: Lessons from Minamata, Japan. Toxics. 2018;**6**:45. DOI: 10.3390/toxics6030045

[12] Eto K. Minamata disease. Neuropathology. 2002;**20**(S1):14-19. DOI: 10.1046/j.1440-1789.2000.00295.x

[13] Harada M. Minamata disease: Methylmercury poisoning in Japan caused by environmental pollution. Critical Reviews in

Toxicology. 1995;**25**(1):1-24. DOI: 10.3109/10408449509089885

[14] UNEP. The Secretariat of the Minamata Convention on Mercury. Available from: https://minamataconvention.org/en [Accessed: September 12, 2024]

[15] DINAMA-JICA. Final Report on the Project for Technical Cooperation on Diagnosis of Actual Situation and Action Plan Design for the Remediation of Coastal Strip of La Plata River Having an Environmental Burden of Mercury Sediments. JICA Open Report Library. 2017. Available from: https://openjicareport.jica.go.jp/619/619/619_711_1000033338.html [Accessed: September 12, 2024]

[16] María Isabel Cárcamo. Informe sobre la situación del mercurio en Uruguay. IPEN Uruguay & RAPAL; 2018. Available from: https://ipen.org/sites/default/files/documents/final_es_uruguay_mercury_country_status_report_ipen_rapal_2018.pdf [Accessed: September 12, 2024]

[17] UNEP-DTIE Chemical Branch. Chloralkali Project - Uruguay Final Report. UN Environment Document Repository. Available from: https://wedocs.unep.org/bitstream/handle/20.500.11822/31227/Chloralkali:Uruguay.pdf; 2011 [Accessed: September 12, 2024]

[18] Ministry of Environment, Uruguay. Mercury 3262UY: Digestión de matrices solidas (suelos, sedi-mentos) para la determinación de todos aquellos metales capaces de ser extraídos bajo las condiciones del ensayo, 2017. Available from: https://www.gub.uy/ministerio-ambiente/sites/ministerio-ambiente/files/documentos/publicaciones/3262UY_0.pdf

[19] U.S. Environmental Protection Agency (EPA). Method 3051A (SW-846): Microwave Assisted Acid Digestion of Sediments, Sludges, and Oils, Revision 1. Washington, DC; 2007. Available from: https://www.epa.gov/esam/us-epa-method-3051a-microwave-assisted-acid-digestion-sediments-sludges-and-oils

[20] Ministry of Environment, Uruguay. Mercury 3141UY: Determinación de Mercurio total (orgánico e inorgánico) en matrices ambientales digeri-das. 2017. Available from: https://www.gub.uy/ministerio-ambiente/comunicacion/publicaciones/mercurio-3141uy

[21] ISO 12846:2012, Water quality—Determination of mercury—Method using atomic absorption spectrometry (AAS) with and without enrichment. Available from: https://www.iso.org/standard/51964.html

[22] National Institute of Minamata Disease (NIMD), Ministry of Environment, Japan. Mercury Analysis Manual. 2004. Available from: http://nimd.env.go.jp/kenkyu/docs/march_mercury_analysis_manual(e).pdf

[23] National Institute for Agro-Environmental Sciences. Analysis of Soil Bacteria and Fungi using PCR-DGGE Ver. 3. Vol. 32010. Available from: https://www.naro.affrc.go.jp/archive/niaes/project/edna/edna_jp/manual_bacterium.pdf [Accessed: September 12, 2024]

[24] Muyzer G, de Waal EC, Uitterlinden AG. Profiling of complex microbial populations by denaturing gradient gel electrophoresis analysis of polymerase chain reaction-amplified genes coding for 16S rRNA. Applied and Environmental Microbiology. 1993;**59**(3):695-700. DOI: 10.1128/aem.59.3.695-700.1993

[25] Yasumoto-Hirose M, Nishijima M, Ngirchechol MK, Kanoh K, Shizuri Y, Miki W. Isolation of marine bacteria by in situ culture on media-supplemented polyurethane foam. Marine Biotechnology (New York, N.Y.). 2006;**8**(3):227-237. DOI: 10.1007/s10126-005-5015-3

[26] Coppi MV, Leang C, Sandler SJ, Lovley DR. Development of a genetic system for *Geobacter sulfurreducens*. Applied and Environmental Microbiology. 2001;**67**(7):3180-3187. DOI: 10.1128/AEM.67.7.3180-3187.2001

[27] Caccavo F Jr, Lonergan DJ, Lovley DR, Davis M, Stolz JF, McInerney MJ. *Geobacter sulfurreducens* sp. nov., a hydrogen- and acetate-oxidizing dissimilatory metal-reducing microorganism. Applied and Environmental Microbiology. 1994;**60**(10):3752-3759

[28] Parks JM et al. The genetic basis for bacterial mercury methylation. Science. 2013;**339**:1332-1335. DOI: 10.1126/science.1230667

[29] Bravo AG, Zopfi J, Buck M, et al. Geobacteraceae are important members of mercury-methylating microbial communities of sediments impacted by waste water releases. The ISME Journal. 2018;**12**:802-812. DOI: 10.1038/s41396-017-0007-7

[30] Zhao Q, Wang J, OuYang S, Chen L, Ming Liu Y, Li FJ. The exacerbation of mercury methylation by *Geobacter sulfurreducens* PCA in a freshwater algae-bacteria symbiotic system throughout the lifetime of algae. Journal of Hazardous Materials. 2021;**415**:125691. DOI: 10.1016/j.jhazmat.2021.125691

[31] Lovley DR. Microbe profile: *Geobacter metallireducens*: A model for novel physiologies of biogeochemical and technological significance. Microbiology (Review). 2022;**168**:001138. DOI 10.1099/mic.0.001138

[32] Nevin KP, Holmes DE, Woodard TL, Hinlein ES, Ostendorf DW, Lovley DR. *Geobacter bemidji*ensis sp. nov. and *Geobacter psychrophilus* sp. nov., two novel Fe(III)-reducing subsurface isolates. International Journal of Systematic and Evolutionary Microbiology. 2005;**55**(4):1667-1674. DOI: 10.1099/ijs.0.63417-0

[33] Viulu S, Nakamura K, Okada Y, Saitou S, Takamizawa K. *Geobacter luticola* sp. nov., an Fe(III)-reducing bacterium isolated from lotus field mud. International Journal of Systematic and Evolutionary Microbiology. 2013;**63**(Pt 2):442-448. DOI: 10.1099/ijs.0.039321-0

[34] Vandieken V, Mußmann M, Niemann H, Jørgensen BB. *Desulfuromonas svalbardensis* sp. nov. and *Desulfuromusa ferrireducens* sp. nov., psychrophilic, Fe(III)-reducing bacteria isolated from Arctic sediments, Svalbard. International Journal of Systematic and Evolutionary Microbiology. 2006;**56**(5):1133-1139. DOI: 10.1099/ijs.0.63639-0

[35] Lücker S, Nowka B, Rattei T, Spieck E, Daims H. The genome of *Nitrospina gracilis* illuminates the metabolism and evolution of the major marine nitrite oxidizer. Frontiers in Microbiology. 2013;**4**:27. DOI: 10.3389/fmicb.2013.00027

[36] Chenghao Y, Peng M, Wang X, Pan X. Photochemical demethylation of methylmercury (MeHg) in aquatic systems: A review of MeHg species, mechanisms, and influencing factors. Environmental Pollution. 2024;**344**:123297. DOI: 10.1016/j.envpol.2024.123297

[37] Clarkson TW, Magos L. The toxicology of mercury and its chemical compounds. Critical Reviews in Toxicology. 2006;**36**(8):609-662

[38] Mergler D et al. Methylmercury exposure and health effects in humans: A worldwide concern. Ambio. 2007;**36**(1):3-11

[39] Driscoll CT et al. Mercury as a global pollutant: Sources, pathways, and effects. Environmental Science & Technology. 2013;**47**(10):4967-4983

[40] Zhang J, Li C, Tang W, Mengjie W, Chen M, He H, et al. Mercury in wetlands over 60 years: Research progress and emerging trends. Science of the Total Environment. 2023;**869**:161862. DOI: 10.1016/j.scitotenv.2023.161862

[41] UN Environment, Global Mercury Partnership. Available from: https://www.unep.org/globalmercurypartnership/what-we-do/artisanal-and-small-scale-gold-mining-asgm [Accessed: September 12, 2024]

[42] Taux K, Kraus T, Kaifie A. Mercury exposure and its health effects in Workers in the Artisanal and Small-Scale Gold Mining (ASGM) sector—A systematic review. International Journal of Environmental Research and Public Health. 2022;**19**:2081. DOI: 10.3390/ijerph19042081

[43] Yoshimura A, Suemasu K, Veiga MM. Estimation of mercury losses and gold production by artisanal and small-scale gold mining (ASGM). Journal of Sustainable Metallurgy. 2021;**7**:1045-1059. DOI: 10.1007/s40831-021-00394-8

[44] Dossou Etui IM, Stylo M, Davis K, et al. Artisanal and small-scale gold mining and biodiversity: A global literature review. Ecotoxicology. 2024;**33**:484-504. DOI: 10.1007/s10646-024-02748-w

Section 2

Risks of Heavy Metals

Chapter 4

Perspective Chapter: Exploring the Toxicity Effect of Heavy Metals on Aquatic Organisms – A Comprehensive Analysis

Mahdi Banaee

Abstract

Heavy metals are naturally occurring elements with high atomic weight and density, and they are toxic to aquatic species even at low concentrations. They enter aquatic ecosystems through natural processes like rock weathering and volcanic eruptions, as well as human activities such as industrial processes and the use of fertilizers. Due to their non-biodegradable nature, heavy metals accumulate in the environment, affecting their reactivity, bioavailability, and toxicity. These metals can cause oxidative stress, enzyme inhibition, and DNA damage in aquatic organisms, leading to bioaccumulation, histopathological changes, reproductive toxicity, and behavioral alterations. This chapter explores heavy metals' sources, pathways, and toxicological effects in aquatic ecosystems, aiming to understand their ecological and health impacts.

Keywords: heavy metals, aquatic organisms, oxidative stress, bioaccumulation, toxicity

1. Introduction

Heavy metals, such as mercury (Hg), cadmium (Cd), lead (Pb), arsenic (As), chromium (Cr), and zinc (Zn) [1], are natural elements with high atomic weight and density [2, 3]. In low concentrations, they can be toxic to various aquatic species [4, 5]. Studies have shown that these pollutants may enter aquatic ecosystems through various natural and human activities. Rock weathering, volcanic eruptions, and natural forest fires are among the most important natural sources of heavy metals. In contrast, mining, smelting, and refining activities, use of pesticides and fertilizers containing heavy metals, leaching from waste dumps and landfills, and emissions from fossil fuel combustion and incineration are known as anthropogenic sources [6, 7]. Heavy metals are non-biodegradable and can accumulate in the environment and organisms, leading to long-term toxic effects. They can be found in various chemical forms, influencing their reactivity, bioavailability, and toxicity. They may

 IntechOpen

interact with sediments, organic matter, and colloidal particles and be translocated by water currents, sedimentation, and resuspension. The physicochemical parameters, including pH, redox potential, water hardness, and the presence of other ions and organic matter, can affect their bio-availabilities and toxicities [8]. Therefore, understanding their sources, concentration, behavior, and effect on water environments and aquatic health is essential in evaluating their toxicity effects on aquatic organisms. In this chapter, we will discuss heavy metal sources, waterborne heavy metals, and the effects of heavy metal pollution on water ecosystems and aquatic health.

2. Sources of heavy metal pollution

Heavy metal pollution in aquatic systems may originate from natural processes and human activities. These sources introduce metals into the environment, where they can persist, accumulate, and exert toxic effects on aquatic organisms and ecosystems [9–11]. Water currents, wave actions, runoff, and floods can break down and erode rocks and release many elements, including heavy metals, into water bodies. A volcanic eruption can emit a lot of ash containing metals such as mercury, arsenic, and lead. Forest fires sometimes lead to the release of metals into the atmosphere, which can be deposited into water systems through rainfall. Moreover, metals from soils and sediments enter water bodies through leaching and erosion [4, 11]. However, human activities significantly release heavy metals and pollute water ecosystems compared to natural processes. Anthropogenic sources of heavy metals include industrial activities, agriculture practices, combustion of fossil fuels, traffic and transport industries, waste dumps and landfills, and war [12]. Finally, heavy metals are carried by waste water, sewage effluents, surface runoff, garbage leachate, and acid rainfall and enter water ecosystems. Therefore, heavy metals may enter water bodies from different sources, including point and non-point sources. We can monitor and manage point sources of heavy metals. At the same time, we may not control and assay the risks of non-point sources of heavy metals that are penetrated into aquatic environments [12, 13].

3. Mechanisms of heavy metal toxicity

Heavy metals have toxic effects on aquatic organisms through various biochemical and physiological mechanisms. These mechanisms can disrupt cellular functions, leading to adverse health effects and ecological imbalances [6, 14]. Heavy metals can bind to sulfhydryl groups in proteins and enzymes, inhibiting their normal function. Some trace elements, such as mercury, bind to thiol groups, disrupting enzymatic activity [15]. Some heavy metals can replace essential metals in biomolecules, impairing their function. For instance, cadmium can replace zinc in metalloproteins, affecting their structural integrity and activity [16]. Heavy metals can trigger a range of biochemical and physiological responses in organisms, including oxidative stress, changes in enzymatic activities, and alterations in metal accumulation and detoxification pathways. Studies have shown that heavy metals can induce the generation of reactive oxygen species [17], such as superoxide radicals, hydrogen peroxide, and hydroxyl radicals. Increasing cell ROS levels leads to oxidative stress, damaging cellular components like lipids, proteins, and DNA. Moreover, ROS can cause the peroxidation of membrane lipids, resulting in loss of membrane integrity and function

[18]. Finally, cell membrane destruction leads to cell necrosis and pathological tissue damage. Heavy metals exposure may deplete cellular antioxidants such as glutathione, reducing the cell's ability to neutralize ROS. Furthermore, Turan et al. [19] showed that some trace elements could inhibit antioxidant enzyme activities such as superoxide dismutase (SOD), catalase (CAT), and glutathione peroxidase (GPx), exacerbating oxidative stress in *Clarias gariepinus*. Abdel-Gawad et al. [20] found that exposure to heavy metals could change oxidative biomarkers in *Oreochromis niloticus*. Oxidative stress in fish exposed to heavy metals could cause an imbalance in different physiological functions [21]. Heavy metals can inhibit key metabolic enzymes involved in processes such as carbohydrate metabolism, protein synthesis, and energy production. Depending on the metal and concentration, heavy metal exposure can lead to changes in enzyme activity, such as increased or decreased activity. The interaction of heavy metals with DNA can lead to strand breaks, DNA adducts, and mutations. ROS generated by metal exposure can also cause oxidative damage to DNA, leading to mutations and chromosomal aberrations. Toxicogenic damage was reported in European eels, *Anguilla anguilla*, exposed to heavy metals [22]. El-Agri et al. [23] found that exposure to heavy metals could induce genotoxicity in *Solea aegyptiaca*. Studies have shown that exposure to heavy metals could disrupt cellular membrane functions. They can affect cell membrane permeability, disrupting ion gradients and homeostasis. Moreover, trace elements may interfere with the function of membrane transport proteins, affecting nutrient uptake and waste excretion [24]. Penetrating heavy metals into cellular organelles can disrupt their biological functions. Shahjahan et al. [3] indicated that metals can disrupt the mitochondrial membrane potential, impairing energy metabolism and inducing apoptosis. They can also inhibit components of the mitochondrial electron transport chain, leading to reduced ATP production and increased ROS generation [14]. Research has revealed that heavy metals can potentially interfere with cellular signaling pathways. Some metals such as lead and cadmium can interfere with calcium signaling pathways, affecting processes such as muscle contraction, neurotransmission, and cell proliferation [25]. Moreover, they can often activate stress response pathways, such as heat shock and unfolded protein responses, leading to altered gene expression and cellular stress [6]. Therefore, histopathological damages in aquatic animals exposed to heavy metals are critical biomarkers in assaying metals' toxicity effects [26]. The histopathological damages were reported in various tissues of *Catla catla* exposed to heavy metals [27]. Exposure to heavy metals can suppress the immune response in aquatic animals, making organisms more susceptible to infections [28, 29]. Metals can induce inflammation by activating immune cells and promoting the release of pro-inflammatory cytokines [5, 30, 31]. Di Paola et al. [17] found that heavy metals can disrupt endocrine systems in aquatic animals. Some heavy metals can mimic or inhibit hormones, disrupting endocrine functions. Sometimes, trace elements can bind to hormone receptors, altering their normal signaling pathways. Moreover, heavy metals can hurt endocrine tissues and organs [1]. Studies have shown that aquatic organisms can accumulate heavy metals in various tissues, such as the liver, kidneys, and bones. Accumulation often depends on the metal's chemical properties and the organism's ability to uptake and store it [2]. Metals can be preferentially stored in specific tissues. For example, cadmium often accumulates in the liver and kidneys, while lead tends to accumulate in bones. Like other organisms, aquatic animals use metal-binding proteins to sequester and detoxify heavy metals. Moreover, specific transport proteins facilitate heavy metal uptake, distribution, and excretion. Finally, they can excrete heavy metals through urine, feces, and sweat. Efficient excretion helps minimize metal accumulation and toxicity [2].

4. Heavy metal toxicity in different aquatic organisms

Heavy metal toxicity affects various aquatic organisms differently, depending on species, ecological position, behavioral and nutritional characteristics, habitat type, developmental stage, exposure duration, and the specific heavy metal involved [32]. Therefore, we have examined each group separately to understand better the toxic effect of heavy metals on different aquatic species.

4.1 Impact of heavy metals on fish species

Heavy metal toxicity affects fish species at multiple levels, including physiological, biochemical, behavioral, genetic, and reproductive aspects [33]. The extent and nature of these effects can vary based on the type of heavy metal, exposure concentration, duration, and fish species [34–37]. Fish may uptake heavy metals through their gills, skin, and digestive systems. These metals can cause physical damage to gill tissues, leading to lamellar edema, necrosis, and hypertrophy. This impairs respiration and reduces the efficiency of gas exchange [38]. Moreover, heavy metals can interfere with ion transport mechanisms, leading to imbalances in essential ions like sodium, potassium, and calcium. Therefore, exposure to heavy metals can cause osmoregulatory dysfunction, affecting overall homeostasis in fish [39, 40]. Saglam et al. [41] and Jasim et al. [42] found that exposure to heavy metals could induce histopathological damage to gills, change Na+/K + -ATPase activity in the gills, increase cortisol and glucose levels, and decrease the triiodothyronine (T3) and thyroxine (T4) levels in fish plasma. In this situation, fish cannot manage the osmoregulation process. Although the detoxification mechanisms try to omit a significant portion of heavy metals from fish's bodies, heavy metals may accumulate in vital organs such as the liver, which can cause oxidative stress by generating ROS [43, 44]. Furthermore, metals such as mercury and lead impair the liver's detoxification enzymes, such as cytochrome P450, reducing the fish's ability to metabolize and excrete toxins. Increasing ROS rate production in the hepatocytes of fish exposed to heavy metals can lead to oxidative damage to lipids, proteins, and DNA. At the same time, the depletion of antioxidants such as glutathione and inhibition of antioxidant enzymes such as superoxide dismutase (SOD) and catalase (CAT) may lead to the collapse of antioxidant defense mechanisms [45]. Histopathological damage in in the liver, including hepatocyte vacuolation, fibrosis, and necrosis following exposure to heavy metal, can disrupt physiological and metabolically dysfunctions in fish. At the cellular level, heavy metals can inhibit enzymes crucial for metabolic processes and disrupt biochemical hemostasis [46]. Eroglu et al. [47] found that disruption of glutathione metabolism in the hepatocyte of *O. niloticus* exposed to heavy metals can lead to the collapse of the detoxification system and antioxidant defense. The bioaccumulation of some metals in the kidneys can induce nephrotoxicity and impair the excretion of metabolic wastes. Studies have shown that exposure to heavy metals can damage renal tubules, impairing reabsorption and secretion functions [3]. Rahman et al. [48] showed that increased creatinine and urea levels could be bioindicators of nephrotoxicity in *O. niloticus* exposed to heavy metals. Exposure to heavy metals can induce neurotoxicity due to inhibiting essential enzymes involved in the metabolism of neurotransmitters such as acetylcholine and destruction of central and peripheral nervous tissue [49]. Thus, alteration in swimming patterns in fish exposed to heavy metals can be related to neurotoxicity [50]. Inhibiting cholinesterase activities and damaging the nervous systems can result in erratic, reduced swimming activity,

Species	Type of heavy metals	Changes observed	References
Rainbow Trout (*Oncorhynchus mykiss*)	Heavy metals	Ovarian deformities, apoptosis, oxidative stress, hormonal changes, DNA damage.	[54]
Various fish species	Heavy metals	Heavy metals cause reduced GSI, fecundity, and hatching rates; abnormalities in reproductive organs.	[21]
Zebrafish (*Danio rerio*)	Heavy metals	Damage to zebrafish reproductive organs, affecting overall reproductive health	[58]
Spotted Snakehead (*Channa punctatus*)	Mercuric chloride (HgCl2)	Oxidative stress, genotoxicity, hematological and histopathological changes in liver and kidney tissues	[59]
Various fish species	Hg, Pb, Cr, Cd	Endocrine disruption, reproductive impairment, altered gene expression related to reproduction	[60]

Table 1.
Example of heavy metal accumulation and biological effects in fish species.

a change in feeding behavior, and a decrease in fish appetite. LeFauve and Connaughton [51], Heffern et al. [52], and Volz et al. [53] showed that damage to the olfactory, visual, and gustatory systems and reduced ability to evade predators can affect fish survival.

Taslima et al. [21] and Bhat et al. [54] found that heavy metal exposure can decrease fish fertility potential. The bioaccumulation of trace elements in gonads can affect endocrine functions, leading to altered levels of sex hormones such as estrogen and testosterone. Moreover, they can induce damage and impair gamete production. Decreased hatching success, increased mortality rates, and developmental abnormalities were observed in the fish embryos and larvae exposed to heavy metals. Moreover, reduced reproductive success and increased mortality rates can result in population declines and affect the overall biodiversity of aquatic ecosystems. Sfakianakis et al. [55] and Green et al. [49] showed that acute and chronic exposure to heavy metals can lead to increased mortality in fish populations. Moreover, Chen et al. [56] and Jiaxin et al. [57] reported immunosuppression and inflammatory response in *Cyprinus carpio* exposed to cadmium. Exposure to heavy metals can increase fish's susceptibility to infections and diseases (**Table 1**) [29].

4.2 Effects of heavy metals on crustaceans

Studies have indicated that heavy metal toxicity significantly impacts crustaceans, affecting their growth, development, behavior, immune system, and overall survival. The degree of these effects can vary based on the type and concentration of the heavy metal, the duration of exposure, and the specific species of crustacean [54, 55]. Some trace elements such as zinc, cadmium, and lead can interfere with molting, which is crucial for crustaceans' growth and development. A disruption in molting can lead to incomplete or abnormal molts, hindering growth and leading to mortality. Exposure to heavy metals can affect the endocrine system that regulates molting by altering hormone levels, such as ecdysteroids [61, 62]. A significant decrease in the hemolymph osmotic pressure of white shrimp, *Litopenaeus vannamei* was reported after exposure to cadmium and zinc by Wu and Chen [63]. Brooks and Mills [64] found that exposure to copper could decrease Na^{+}/K^{+}

ATPase activity in gills and sodium levels in the hemolymph of *Gammarus pulex*. Studies have shown that exposure to heavy metals can lead to physical deformities, such as malformed appendages and carapaces, and decrease growth performance. Developmental delays, abnormalities, and reduced survival rates were reported in the larvae of crustaceans exposed to heavy metals [58]. Oxidative damage and collapse of antioxidant defense systems were reported in crustaceans exposed to heavy metals [59]. Genotoxicity, including DNA strand breaks, DNA adducts, and mutations, was also found in the *Scylla paramamosain* and *Talitrus saltator* exposed to heavy metals such as cadmium and lead [60, 65]. Heavy metals can suppress the immune system, reducing crustaceans' ability to fight infections and diseases. Moreover, metals can impair the function and viability of hemocytes, which are critical for crustaceans' immune responses (**Table 2**) [71].

Exposure to heavy metals can induce chronic inflammation, leading to tissue damage and impaired physiological functions [61]. Feeding behavior, including decreasing appetites, ability to find feed, reduced feeding rates, and changing foraging behavior of crustaceans, was observed after exposure to heavy metals [62]. Heavy metals can affect the neuromuscular system, leading to reduced or erratic movement and impaired ability to escape predators. Furthermore, exposure to trace elements can cause changes in social behaviors, mating behaviors, and predator-prey interactions [54]. The bioaccumulation of heavy metals in the reproductive organs of crustaceans can reduce fecundity, fertility potential, and quality of egg and sperm and decrease fertilization success [63, 64, 66]. Histopathological damage was reported in the gonad tissues of crustaceans exposed to heavy metals [31, 54]. The larvae and embryos of crustaceans exposed to heavy metals showed reduced survival and viability of offspring, and embryotoxic effects, such as abnormality development and decreased chances of survival [67–69]. Wang et al. [72] found that exposure to mercury, cadmium, and lead could affect embryogenesis in *Meretrix meretrix*, and decrease survival chance, growth performance rate, and disrupt in metamorphosis of larvae. The disruption in parthenogenesis was reported in *Daphnia magna* exposed to cadmium by Djekoun et al. [73].

Species	Type of heavy metals	Changes observed	References
Crayfish (*Procambarus clarkii*)	Heavy metals	Increased antioxidant enzyme activity and metallothionein intensity in crayfish from more polluted areas	[66]
Isopod (*Porcellio laevis*)	Cd, Zn	Significant hepatopancreas alterations due to Cd and Zn exposure	[67]
Mud crab (*Scylla paramamosain*)	Cd	Intestinal damage and oxidative stress increased with cadmium exposure; altered microbial diversity and gene expression	[68]
Crabs (*Ucides cordatus*)	Heavy metals	Elevated enzymes and histological differences in crabs from more polluted areas	[69]
Crayfish (*Procambarus clarkii*)	Pb	Pb exposure caused dose-dependent damage to the hepatopancreas and intestine, disrupted gut microbiota, and increased pathogenic bacteria	[70]

Table 2.
Example of heavy metal accumulation and its biological impact on crustaceans.

4.3 Influence of heavy metals on mollusks

Mollusks, including bivalves (e.g., clams, oysters, and mussels) and gastropods (e.g., snails), are particularly sensitive to heavy metal pollution. Heavy metals can affect mollusks' physiological, biochemical, reproductive, and behavioral processes, leading to adverse health effects and ecological consequences [70, 74, 75]. Although Mollusks can biotransform and execrate a significant portion of heavy metals, studies have shown that mollusks can accumulate heavy metals in their soft tissues and shells [76]. Therefore, they are important in transiting heavy metals throughout the food chain. They usually use metallothioneins to bind heavy metals and remove their bodies [77, 78]. Moreover, they can store metals in intracellular granules, which are then excreted or remain in the organism's tissues [79]. Documents have indicated that a significant portion of heavy metals are also reserved in their shells [78]. Exposure to various metals can induce histopathological changes in the gills (e.g., epithelial lifting, necrosis, and lamellar fusion) and in the hepatopancreas (e.g., vacuolation, necrosis, and fibrosis) of mollusks [80]. Lipid peroxidation, protein oxidation, DNA damage, and cellular antioxidant defense reactions were reported in mollusks exposed to heavy metals [81]. Studies have shown that trace elements can induce genotoxicity and cause direct DNA damage, leading to mutations, strand breaks, and chromosomal aberrations in mollusks [82, 83]. The embryotoxicity and genotoxicity were observed in the Pacific oyster, *Crassostrea gigas*, exposed to cadmium and copper by Mai et al. [84]. Accumulation of heavy metals in the gonads can induce histological damage and reproductive toxicity [85, 86]. Studies have shown that heavy metals can interfere with endocrine functions, leading to altered levels of reproductive hormones [87]. Decreasing fecundity, gamete quality, and fertility rates are common signs reported in the mollusks inhabiting polluted sites [72]. Exposure to heavy metals can change spawning cycles and hatching success [73]. Developmental abnormalities and increased mortality in embryos and larvae are reported in mollusks exposed to heavy metals [88]. Sawasdee and Köhler [89] studied the survival rate, formation of tentacles and eyes, heart rate, hatching rate, and weight of *Marisa cornuarietis* larvae after exposure to heavy metals. They found that heavy metals could induce embryotoxicity in the ramshorn snail.

Exposure to metals can reduce feeding rates due to damage to sensory organs and alterations in foraging behavior [90]. Metals can affect the neuromuscular system, reducing mobility and altering behaviors such as burrowing and shell closure [91]. Furthermore, metals can impair digestive enzyme activity, reducing nutrient absorption and assimilation. Studies have shown that exposure to heavy metals can induce chronic inflammatory responses, suppress the immune system, and change the hemocytes and their functions, making mollusks more susceptible to infections and diseases (**Table 3**) [96, 98].

4.4 Impact of heavy metal on amphibians

Heavy metal contamination significantly impacts amphibians, affecting their growth, development, behavior, immune system, and reproductive health. Amphibians are particularly vulnerable due to their permeable skin, complex life cycles, and dependence on both aquatic and terrestrial habitats [99]. Amphibians directly absorb water and dissolved substances, including trace elements, through their skin, which can lead to metal accumulation and toxicity. Therefore, heavy metals can damage skin and gill tissues, impair respiration, and lead to hypoxia. The accumulation

Species	Type of heavy metals	Changes observed	References
Ruditapes philippinarum, Paphia undulata, Meretrix meretrix, Sinonovacula constricta, Meretrix lyrata	Cu, Pb, Zn, Cd, Cr, Hg, As	Concentrations varied among species; carcinogenic risks from Cd and Cr; non-carcinogenic risks were low.	[92]
Various bivalve mollusks	As, Cd, Cr, Hg, Ni, Pb	Accumulation of heavy metals; Cd posed a potential cancer risk	[93]
Various bivalve mollusks	Heavy metals	Hemocytes used to assess toxicity	[94]
Magallana bilineata, Perna viridis, Villorita cyprinoides	Trace metals	Histological alterations in digestive glands and gills; higher histopathological indices in post-monsoon season and regions with higher metal pollution	[95]
Unio tigridis	Cu, Zn, Cd, Pb	Histopathological lesions (necrosis, atrophy, etc.) in digestive glands	[96]
Crassostrea sp. *(oysters)*	Various metals	Changes in enzymatic activity (GST and CAT); histological alterations in gills	[97]

Table 3.
Example of heavy metal accumulation and biological effects in mollusks.

of heavy metals in amphibians' vital organs can lead to oxidative stress, histopathological damage, and physiological dysfunctions [100]. Some heavy metals, such as mercury and cadmium, can cause developmental abnormalities and increased mortality rates in amphibian embryos. Studies showed that exposure to heavy metals can delay hatching and metamorphosis, leading to prolonged larval stages and reduced survival rates [100, 101]. Heavy metals can impair growth by disrupting nutrient uptake and metabolism, leading to inhibited growth and smaller body size. Studies showed that exposure to metals can result in physical deformities, such as malformed limbs and skeletal abnormalities in the Egyptian toad *Sclerophrys regularis* [102]. Increasing the production of ROS was reported in the amphibians after exposure to heavy metals. Therefore, any change in the oxidative biomarkers, including depletion of non-enzyme antioxidants such as glutathione and disruption in SOD, GPx, GR, and CAT activities, exacerbates oxidative stress and causes oxidative damage to lipids, proteins, and DNA [103]. The genotoxic effects, including DNA strand breaks and mutations, were observed in the amphibians exposed to heavy metals (**Table 4**) [103, 109–111].

Taiwo et al. [112] reported oxidative damage in the liver of wild African tiger frogs, *Hoplobatrachus occipitalis*, after exposure to various heavy metals. The immunotoxicity was observed in the Indian green frog, *Euphlyctis hexadactylus,* exposed to heavy metals [113]. Jayawardena et al. [114] found that exposure to heavy metals could change micronucleus and comet rates in *E. hexadactylus*. Exposure to heavy metals can reduce feeding rates and alter foraging behavior, impacting growth and energy reserves. Heavy metals can affect taste and smell, leading to changes in food preference and reduced dietary intake [109]. Studies have shown that heavy metals can affect the neuromuscular system, leading to reduced mobility and altered behaviors such as reduced escape responses and impaired predator avoidance [84, 115, 116].

Species	Type of heavy metals	Changes observed	References
Various frog species	As, Cd, Cr, Pb	Accumulation of heavy metals in muscles, bioaccumulation factors (BAFs), potential human health risks *via* frog consumption	[104]
Pelophylax ridibundus, Xenopus laevis, Xenopus tropicalis	Pb^{2+}, Hg^{2+}, Cd^{2+}	Changes in sulfur metabolism-related enzymes, antioxidant response, gene expression changes, oxidative stress indicators	[105]
Sclerophrys regularis	Fe, Pb, Cd, Cr	Genotoxicity (micronucleated erythrocytes), fluctuating asymmetry in limbs, and regional differences in pollution impact	[106]
Pelophylax bedriagae	Zn^{2+}, Hg^{2+}	Accumulation of microplastics and heavy metals in the upper gastrointestinal tract, correlation between MPs and heavy metals	[107]
Pelophylax ridibundus	Heavy metals	Nuclear abnormalities in erythrocytes (micronuclei, notched nuclei, nuclear buds, and blebbed nuclei), cyto- and genotoxicity	[108]

Table 4.
Example of heavy metal toxicity and biological effects in different amphibian species.

Metals can cause changes in social behaviors, mating behaviors, and predator-prey interactions [117]. Metals accumulate in reproductive organs, causing histological damage and affecting gamete production [118]. Furthermore, heavy metals can interfere with endocrine functions, leading to altered levels of reproductive hormones [101]. Yee-Duarte et al. [116] found that heavy metals could cause developmental abnormalities and increased mortality in embryos, larvae, and *Bufo raddei* tadpoles. After exposure to heavy metals, reduced fecundity, altered spawning cycles, and lower hatching success were reported in various amphibians [116]. Studies have shown that heavy metals can suppress the immune system, making amphibians more susceptible to infections and diseases. Persistent exposure to heavy metals can induce chronic inflammatory responses, leading to tissue damage and impaired physiological functions [115].

4.5 Effect of heavy metals on reptiles

Heavy metal contamination in their habitats affects reptiles, including turtles, lizards, snakes, and crocodilians [119, 120]. Heavy metals can accumulate in their tissues, leading to various physiological, biochemical, reproductive, and behavioral effects [121–123]. Ma and Wang [124] showed that aquatic reptiles are good bioindicators for studying environmental pollution. The bioaccumulation of heavy metals was reported in different species of reptiles that live in water ecosystems. Studies have shown that metals can be stored in their tissues, particularly in the liver, kidneys, and bones, for long periods, leading to chronic exposure and potential biomagnification up the food chain. However, reptiles can detoxify heavy metals using metallothioneins [89]. These proteins bind heavy metals, reducing their toxicity by sequestering them away from vital cellular processes. Furthermore, they can excrete some metals through feces, urine, and shedding of skin. Liu et al. [125] and Yee-Duarte et al. [120] showed that heavy metals can accumulate in turtles and *Caretta caretta* livers, causing oxidative stress and impairing detoxification functions. Moreover, exposure to heavy metals can induce structural damage in the kidneys and impair the excretion of metabolic wastes [126]. The bioaccumulation of heavy metals in bones causes structural weakness and affects locomotion. Weng et al. [127] and Huiping [128] reported the bioaccumulation of heavy metals and shell deformities in the skeletal muscles of softshell and hard-shell freshwater and marine turtles. Oxidative stress was reported in various species of aquatic reptiles inhabited at polluted sites with heavy metals [92]. Studies have shown that heavy metals could induce the production of ROS, causing oxidative damage to lipids, proteins, and DNA [93, 124]. Moreover, trace elements can cause direct DNA damage, leading to mutations, strand breaks, and chromosomal aberrations (**Table 5**) [94, 143].

The accumulation of heavy metals in reproductive organs can cause histological damage and affect gamete production [95]. They also can interfere with endocrine functions, leading to altered levels of reproductive hormones in aquatic reptiles [97, 144]. Exposure to some trace elements can cause developmental abnormalities and increased mortality in embryos and hatchlings. Reduced fecundity, altered nesting behaviors, and lower hatching success were observed in various species exposed to heavy metals [145]. Exposure to heavy metals can reduce feeding rates and alter foraging behavior, impacting growth and energy reserves in aquatic reptiles. Furthermore, metals can affect taste and smell, leading to changes in food preference and reduced dietary intake [144]. Metals can cause changes in social behaviors, mating behaviors,

Species	Type of heavy metals	Changes observed	References
Loggerhead Sea Turtle (*Caretta caretta*)	As, Cd, Fe, Zn, Barium, Boron, Cu	Higher bioaccumulation in the liver, worrying levels of As and Cd in kidneys and liver, and different metal dominance in eggs and tissues.	[129, 130]
Freshwater Turtles, Crocodilians	Heavy metals	Bioaccumulation of heavy metals in the blood of turtles and crocodilians	[131–136]
Broad-Snouted Caiman (*Caiman latirostris*), Testudines	Fe, Cu, Cr, Cd, Pb, Ni, Al	Bioaccumulation of heavy metals in the blood	[137]
Neotropical Aquatic Snake (*Helicops pastazae*)	Cd, Cr, Pb	Bioaccumulation of heavy metals in the liver and muscle	[138]
Freshwater Turtle (*Phrynops geoffroanus*)	Al, Cd, Co, Cr, Cu, Fe, Mn, Mo, Ni, Pb, Zn	Bioaccumulation of heavy metals in the serum, muscle, liver, and kidney	[139]
Sea Snakes (*Hydrophis* species)	Al, Cd, Pb, Cr, Mn, Fe, Ni, Cu, Zn	Bioaccumulation of heavy metals in the serum, muscle, liver, and kidney	[140]
Khorat Snail-Eating Turtle (*Malayemys khoratensis*)	Fe, Mn, Zn, Cu, Cr, Ni, As, Pb, Cd, Hg	Bioaccumulation of heavy metals in the serum, muscle, liver, and kidney	[141]
Softshell and Hard-shell Freshwater Turtles	Cu, Zn, Cd, Ni, Co, Pb, Cr	Bioaccumulation of heavy metals in the skeletal muscles	[142]
Mugger Crocodile (*Crocodylus palustris*)	Cd, Pb, As, Hg, Cu, Cr, Fe	Bioaccumulation of heavy metals in the blood serum and feces	[119]

Table 5.
Example of heavy metal toxicity and biological effects in different reptile species.

and predator-prey interactions between aquatic reptiles. They can also affect the neuromuscular system, leading to reduced mobility and altered behaviors such as reduced escape responses and impaired predator avoidance. Studies have shown that exposure to heavy metals can induce chronic inflammatory reactions, leading to tissue damage and impaired physiological functions in aquatic reptiles [144, 146, 147]. Heavy metals can suppress the immune system, making reptiles more susceptible to infections and diseases [106, 148].

4.6 Impact of heavy metals on aquatic plants and algae

Aquatic plants, including algae, macrophytes, and submerged vegetation, are primary producers and play a crucial role in aquatic ecosystems. Furthermore, they provide habitat and shelter for various aquatic organisms. Aquatic plants contribute to nutrient cycling and sediment stabilization. Also, healthy aquatic plants contribute to water oxygenation through photosynthesis. Heavy metals can significantly affect plant growth, physiological processes, reproduction, and overall ecosystem health [149–152]. Heavy metal-induced growth and photosynthesis reduction can impact primary production and food web dynamics. Studies have shown that heavy metals can inhibit the growth of aquatic plants by disrupting key physiological processes [114]. Moreover, they can interfere with the uptake and assimilation of essential nutrients, leading to nutrient deficiencies and stunted growth. Exposure to heavy metals can cause morphological abnormalities such as reduced leaf size, deformed stems, and abnormal root growth. Furthermore, they can impair root development and branching, affecting the plant's ability to anchor and absorb water and nutrients [153]. Heavy metals can damage chlorophyll, reducing photosynthetic efficiency and overall plant health. Moreover, metals may inhibit key enzymes involved in metabolic processes, such as those involved in carbon fixation and respiration. Thus, changes in these enzyme activities can lead to altered metabolic pathways and reduced efficiency in energy production [4, 112, 153].

Oxidative stress, DNA damage, and harm to roots and leaves can interfere with the uptake of essential nutrients, leading to nutrient imbalances and deficiencies in aquatic plants exposed to heavy metals. Therefore, heavy metals can alter the availability of nutrients in the surrounding water or sediment, affecting plant growth and health. Studies have shown that the bioaccumulation of heavy metals in specific tissues depends on the plant species and metal type [112, 113]. Exposure to heavy metals can disrupt reproductive functions in aquatic plants. Heavy metals can affect the development of flowers and seeds, leading to reduced reproductive success and lower seed viability. Moreover, they can impact the health of pollinators and the process of pollination, affecting plant reproduction. Also, heavy metal contamination can reduce seed germination rates and seedling survival [4, 16, 28, 112, 153].

4.7 Impact of heavy metals on microorganisms

Heavy metals can profoundly affect microorganisms, including bacteria, fungi, and protozoa, which play crucial roles in ecosystems, such as nutrient cycling, decomposition, and disease regulation. The impact of heavy metals on microorganisms can affect their growth, metabolism, community structure, and ecological functions [104]. Some trace elements such as mercury, cadmium, and lead can inhibit the growth and reproduction of microorganisms by interfering with essential cellular processes [154–160]. Moreover, they can inhibit key enzymes involved in metabolic pathways, affecting energy production and other cellular functions [104, 105, 107]. Studies have shown that exposure to sub-lethal concentrations can reduce microorganisms' viability and functional activity. Metals induce the production of ROS, leading to oxidative damage to lipids, proteins, and DNA [108]. Heavy metal contamination can exert selective pressure, leading to shifts in microbial community structure and reduced diversity [107]. Heavy metals can affect microorganisms involved in decomposition and nutrient cycling, impacting water fertility and ecosystem health. Moreover, heavy metals can affect the growth and pathogenicity of microorganisms, including potential pathogens in sediment and water environments. Thus, some trace elements can enhance the virulence of pathogenic microorganisms by promoting their survival and proliferation [108, 161]. Heavy metal exposure can promote horizontal gene transfer, spreading resistance genes among microbial populations [162]. Microorganisms exposed to heavy metals may develop resistance mechanisms that can also confer resistance to antibiotics, complicating treatment efforts and public health concerns [161, 163].

5. Conclusion

The study showed that heavy metal pollution poses significant risks to aquatic ecosystems, affecting a wide range of organisms, from fish and crustaceans to mollusks, amphibians, reptiles, and aquatic plants. These metals, introduced through natural processes and human activities, can persist in the environment and lead to bioaccumulation and long-term ecological consequences. Mechanisms of toxicity are complex, including oxidative stress, enzyme dysfunction, DNA damage, and interference with reproductive and immune systems. Effects vary among species, depending on factors such as duration of exposure, type of metal, and physiology of the organism. For example, fish experience osmotic dysregulation, neurotoxicity, and impaired reproductive success, while crustaceans face challenges in molting, immunosuppression, and behavioral changes. Mollusks accumulate metals in their tissues, leading to physiological and reproductive damage, while amphibians and reptiles suffer from developmental abnormalities, oxidative stress, and compromised immune responses. Aquatic plants and algae, which are critical to ecosystem health, are not immune, and exposure to metals leads to reduced growth, photosynthetic efficiency, and overall productivity. As a result, heavy metal pollution is a serious threat to aquatic ecosystems, affecting a wide range of organisms and leading to ecological imbalance. This study emphasizes the importance of monitoring, managing, and reducing heavy metal pollution to protect aquatic health and biodiversity. Therefore, the persistence and accumulation of heavy metals in aquatic environments emphasizes the need for careful monitoring and management strategies to reduce their impact. Protection of aquatic health is necessary not only for the survival of these species, but also for maintaining biodiversity and ecological balance of water bodies. As heavy metals continue to be a global environmental challenge, more research and more effective pollution control measures are necessary to protect aquatic ecosystems and their dependent organisms.

Acknowledgements

The author of this chapter acknowledges that he used Grammarly and ChatGPT 3.5 to polish the language and check the grammar of the manuscript.

Conflict of interest

The authors declare no conflict of interest.

Author details

Mahdi Banaee
Department of Aquaculture, Faculty of Natural Resources, Behbahan Khatam Alanbia University of Technology, Iran

*Address all correspondence to: mahdibanaee2@gmail.com

References

[1] Yu G, Wu L, Su Q, Ji X, Zhou J, Wu S, et al. Neurotoxic effects of heavy metal pollutants in the environment: Focusing on epigenetic mechanisms. Environmental Pollution. 2024;**134**:123563

[2] Ali H, Khan E. What are heavy metals? Long-standing controversy over the scientific use of the term 'heavy metals'–proposal of a comprehensive definition. Toxicological and Environmental Chemistry. 2018;**100**(1):6-19

[3] Shahjahan M, Taslima K, Rahman MS, Al-Emran M, Alam SI, Faggio C. Effects of heavy metals on fish physiology–a review. Chemosphere. 2022;**300**:134519

[4] Asiminicesei D-M, Fertu DI, Gavrilescu M. Impact of heavy metal pollution in the environment on the metabolic profile of medicinal plants and their therapeutic potential. Plants. 2024;**13**(6):913

[5] Bagheri S, Gholamhosseini A, Hoseinifar SH, Banaee MJAoAS. Investigation of the effects of heavy metals (copper, cobalt, manganese, selenium, and zinc) on fish immune systems: An overview. Annals of Animal Science. 2024 [Article in press]

[6] Paithankar JG, Saini S, Dwivedi S, Sharma A, Chowdhuri DK. Heavy metal associated health hazards: An interplay of oxidative stress and signal transduction. Chemosphere. 2021;**262**:128350

[7] Banaee M, Sagvand S, Sureda A, Amini M, Haghi BN, Sopjani M, et al. Evaluation of single and combined effects of mancozeb and metalaxyl on the transcriptional and biochemical response of zebrafish (*Danio rerio*). Comparative Biochemistry and Physiology. Toxicology and Pharmacology: CBP. 2023;**268**:109597

[8] Liu Y, Wang P, Gojenko B, Yu J, Wei L, Luo D, et al. A review of water pollution arising from agriculture and mining activities in Central Asia: Facts, causes and effects. Environmental Pollution. 2021;**291**:118209

[9] Huang Z, Liu C, Zhao X, Dong J, Zheng B. Risk assessment of heavy metals in the surface sediment at the drinking water source of the Xiangjiang River in South China. Environmental Sciences Europe. 2020;**32**:1-9

[10] Singh A, Sharma A, Verma R, Chopade R, Pandit P, Nagar V, et al. Heavy metal contamination of water and their toxic effect on living organisms. In: The Toxicity of Environmental Pollutants. Vol. 1. London, UK: IntechOpen; 2020. pp. 1-19

[11] Mahmoud N, Al-Shahwani D, Al-Thani H, Isaifan RJ. Risk assessment of the impact of heavy metals in urban traffic dust on human health. Atmosphere. 2023;**14**(6):1049

[12] An J, Zha X, Wang H, Deng L, Yang Y, Wang Y, et al. Analysis of influencing factors of heavy metals and non-point source pollution in typical areas of Tethys Himalayan tectonic domain. Water. 2024;**16**(2):291

[13] Jiang T, Wang M, Zhang W, Zhu C, Wang F. A comprehensive analysis of agricultural non-point source pollution in China: Current status, risk assessment and management strategies. Sustainability. 2024;**16**(6):2515

[14] Sun Q, Li Y, Shi L, Hussain R, Mehmood K, Tang Z, et al. Heavy metals induced mitochondrial dysfunction in animals: Molecular mechanism of toxicity. Toxicology. 2022;**469**:153136

[15] Ajsuvakova OP, Tinkov AA, Aschner M, Rocha JB, Michalke B, Skalnaya MG, et al. Sulfhydryl groups as targets of mercury toxicity. Coordination Chemistry Reviews. 2020;**417**:213343

[16] Li D, He T, Saleem M, He G. Metalloprotein-specific or critical amino acid residues: Perspectives on plant-precise detoxification and recognition mechanisms under cadmium stress. International Journal of Molecular Sciences. 2022;**23**(3):1734

[17] Di Paola D, Capparucci F, Lanteri G, Cordaro M, Crupi R, Siracusa R, et al. Combined toxicity of xenobiotics bisphenol a and heavy metals on zebrafish embryos (*Danio rerio*). Toxics. 2021;**9**(12):344

[18] Ullah S, Li Z, Hassan S, Ahmad S, Guo X, Wanghe K, et al. Heavy metals bioaccumulation and subsequent multiple biomarkers based appraisal of toxicity in the critically endangered *Tor putitora*. Ecotoxicology and Environmental Safety. 2021;**228**:113032

[19] Turan F, Eken M, Ozyilmaz G, Karan S, Uluca H. Heavy metal bioaccumulation, oxidative stress and genotoxicity in African catfish *Clarias gariepinus* from Orontes river. Ecotoxicology. 2020;**29**:1522-1537

[20] Abdel-Gawad FK, Khalil WK, Bassem SM, Kumar V, Parisi C, Inglese S, et al. The duckweed, lemna minor modulates heavy metal-induced oxidative stress in the Nile Tilapia, *Oreochromis niloticus*. Water. 2020;**12**(11):2983

[21] Taslima K, Al-Emran M, Rahman MS, Hasan J, Ferdous Z, Rohani MF, et al. Impacts of heavy metals on early development, growth and reproduction of fish–a review. Toxicology Reports. 2022;**9**:858-868

[22] Turan F, Karan S, Ergenler A. Effect of heavy metals on toxicogenetic damage of European eels *Anguilla anguilla*. Environmental Science and Pollution Research International. 2020;**27**:38047-38055

[23] El-Agri AM, Emam MA, Gaber HS, Hassan EA, Hamdy SM. Integrated use of biomarkers to assess the impact of heavy metal pollution on *Solea aegyptiaca* fish in Lake Qarun. Environmental Sciences Europe. 2022;**34**(1):74

[24] De Caroli M, Furini A, DalCorso G, Rojas M, Di Sansebastiano G-P. Endomembrane reorganization induced by heavy metals. Plants. 2020;**9**(4):482

[25] Kocadal K, Alkas F, Battal D, Saygi S. Cellular pathologies and genotoxic effects arising secondary to heavy metal exposure: A review. Human and Experimental Toxicology. 2020;**39**(1):3-13

[26] Onita B, Albu P, Herman H, Balta C, Lazar V, Fulop A, et al. Correlation between heavy metal-induced histopathological changes and trophic interactions between different fish species. Applied Sciences. 2021;**11**(9):3760

[27] Naz S, Hussain R, Ullah Q, Chatha AMM, Shaheen A, Khan RU. Toxic effect of some heavy metals on hematology and histopathology of major carp (*Catla catla*). Environmental Science and Pollution Research. 2021;**28**:6533-6539

[28] Okereafor U, Makhatha M, Mekuto L, Uche-Okereafor N,

Sebola T, Mavumengwana V. Toxic metal implications on agricultural soils, plants, animals, aquatic life and human health. International Journal of Environmental Research and Public Health. 2020;**17**(7):2204

[29] Falco F, Banaee M, Mauro M, Faggio C, Kollath A, Elumalai P. Immunomodulators: An Introduction Immunomodulators in Aquaculture and Fish Health. Boca Raton: CRC Press; 2024. pp. 3-17

[30] Ebrahimi M, Khalili N, Razi S, Keshavarz-Fathi M, Khalili N, Rezaei N. Effects of lead and cadmium on the immune system and cancer progression. Journal of Environmental Health Science and Engineering. 2020;**18**:335-343

[31] Banaee M, Di Paola D, Cuzzocrea S, Cordaro M, Faggio C. Biomarkers in aquatic ecotoxicology: Understanding the effects of xenobiotics on the health of aquatic organisms. In: Biochemical and 4 Physiological Response during Oxidative Stress - From Invertebrates to Vertebrates [Working Title]. London, UK: IntechOpen; 16 Jul 2024. DOI: 10.5772/intechopen.1006063

[32] Vajargah MFJFS. A review on the effects of heavy metals on aquatic animals. Journal of Fisheries Research. 2021;**2766**(2276):854

[33] Kar S, Sangem P, Anusha N, Senthilkumaran B. Endocrine disruptors in teleosts: Evaluating environmental risks and biomarkers. Aquaculture and Fisheries. 2021;**6**(1):1-26

[34] Banaee M, Beitsayah A, Jorabdoz I. Assessment of mercury bioaccumulation in zebra cichlid (*Cichlasoma Nigrofasciatum*) exposed to sublethal concentrations of permethrin. Iranian Journal of Toxicology. 2015;**8**(27):1168-1173

[35] Banaee M, Soltanian S, Sureda A, Gholamhosseini A, Haghi BN, Akhlaghi M, et al. Evaluation of single and combined effects of cadmium and micro-plastic particles on biochemical and immunological parameters of common carp (*Cyprinus carpio*). Chemosphere. 2019;**236**:124335

[36] Banaee M, Beitsayah A, Prokić MD, Petrović TG, Zeidi A, Faggio C. Effects of cadmium chloride and biofertilizer (Bacilar) on biochemical parameters of freshwater fish, *Alburnus mossulensis*. Comparative Biochemistry and Physiology. Toxicology and Pharmacology: CBP. 2023;**268**:109614

[37] Garai P, Banerjee P, Mondal P, Saha NJJCTS. Effect of heavy metals on fishes: Toxicity and bioaccumulation. Journal of Clinical Toxicology. 2021;**S18**(001):1-18

[38] Evaz-Zadeh Samani H, Banaee M, Shoukat P, Noori A, Mousavi Dehmoredi L. Protective effects of dietary Spirulina platensis against cadmium-induced oxidative stress in gills of rainbow trout. Iranian Journal of Toxicology. 2017;**11**(4):5-12

[39] Shah N, Khan A, Ali R, Marimuthu K, Uddin MN, Rizwan M, et al. Monitoring bioaccumulation (in gills and muscle tissues), hematology, and genotoxic alteration in *Ctenopharyngodon idella* exposed to selected heavy metals. BioMed Research International. 2020;**2020**(1):6185231

[40] Solgi E, Mirmohammadvali S. Comparison of the heavy metals, copper, iron, magnesium, nickel, and zinc between muscle and gills of four benthic fish species from Shif Island (Iran). Bulletin of Environmental Contamination and Toxicology. 2021;**106**(4):658-664

[41] Saglam D, Atli G, Canli M. Investigations on the osmoregulation of freshwater fish (*Oreochromis niloticus*) following exposures to metals (Cd, Cu) in differing hardness. Ecotoxicology and Environmental Safety. 2013;**92**:79-86

[42] Jasim SA, Golgouneh S, Jaber MM, Indiaminov SI, Alsaikhan F, Hammid AT, et al. Effects of short-term exposure to the heavy metal, nickel chloride (Nicl2) on gill histology and osmoregulation components of the gray mullet, *Mugil cephalus*. Comparative Biochemistry and Physiology. Toxicology and Pharmacology: CBP. 2022;**258**:109361

[43] Banaee M, Mohammadipour S, Madhani S. Effects of sublethal concentrations of permethrin on bioaccumulation of cadmium in zebra cichlid (*Cichlasoma nigrofasciatum*). Toxicological and Environmental Chemistry. 2015;**97**(2):200-207

[44] Gholamhosseini A, Shiry N, Soltanian S, Banaee M. Bioaccumulation of metals in marine fish species captured from the northern shores of the Gulf of Oman, Iran. Regional Studies in Marine Science. 2021;**41**:101599

[45] Elhetawy AI, Abdel-Rahim MM, Sallam AE, Shahin SA, Lotfy AM, El Basuini MF. Dietary wood and activated charcoal improved ammonium removal, heavy metals detoxification, growth performance, blood biochemistry, carcass traits, and histopathology of European seabass. Aquaculture Nutrition. 2023;**2023**(1):8860652

[46] Savassi LA, Paschoalini AL, Arantes FP, Rizzo E, Bazzoli N. Heavy metal contamination in a highly consumed Brazilian fish: Immunohistochemical and histopathological assessments. Environmental Monitoring and Assessment. 2020;**192**:1-14

[47] Eroglu A, Dogan Z, Kanak E, Atli G, Canli M. Effects of heavy metals (Cd, Cu, Cr, Pb, Zn) on fish glutathione metabolism. Environmental Science and Pollution Research International. 2015;**22**:3229-3237

[48] Rahman ANA, ElHady M, Hassanin ME, Mohamed AA-R. Alleviative effects of dietary Indian lotus leaves on heavy metals-induced hepato-renal toxicity, oxidative stress, and histopathological alterations in Nile tilapia, *Oreochromis niloticus* (L.). Aquaculture. 2019;**509**:198-208

[49] Green AJ, Planchart A. The neurological toxicity of heavy metals: A fish perspective. Comparative Biochemistry and Physiology. Part A, Molecular and Integrative Physiology. 2018;**208**:12-19

[50] Calvo-Ochoa E, Byrd-Jacobs CA. The olfactory system of zebrafish as a model for the study of neurotoxicity and injury: Implications for neuroplasticity and disease. International Journal of Molecular Sciences. 2019;**20**(7):1639

[51] LeFauve MK, Connaughton VP. Developmental exposure to heavy metals alters visually-guided behaviors in zebrafish. Current Zoology. 2017;**63**(2):221-227

[52] Heffern K, Tierney K, Gallagher EP. Comparative effects of cadmium, zinc, arsenic and chromium on olfactory-mediated neurobehavior and gene expression in larval zebrafish (*Danio rerio*). Aquatic Toxicology. 2018;**201**:83-90

[53] Volz SN, Hausen J, Nachev M, Ottermanns R, Schiwy S, Hollert H. Short exposure to cadmium disrupts the olfactory system of zebrafish (*Danio rerio*)–relating altered gene expression in

the olfactory organ to behavioral deficits. Aquatic Toxicology. 2020;**226**:105555

[54] Bhat RA, Bakhshalizadeh S, Guerrera MC, Kesbiç OS, Fazio FJ. Toxic effect of heavy metals on ovarian deformities, apoptotic changes, oxidative stress, and steroid hormones in rainbow trout. Journal of Trace Elements in Medicine and Biology. 2023;**75**:127106

[55] Sfakianakis D, Renieri E, Kentouri M, Tsatsakis A. Effect of heavy metals on fish larvae deformities: A review. Environmental Research. 2015;**137**:246-255

[56] Chen J, Chen D, Li J, Liu Y, Gu X, Teng X. Cadmium-induced oxidative stress and immunosuppression mediated mitochondrial apoptosis via JNK-FoxO3a-PUMA pathway in common carp (*Cyprinus carpio* L.) gills. Aquatic Toxicology. 2021;**233**:105775

[57] Jiaxin S, Shengchen W, Yirong C, Shuting W, Shu L. Cadmium exposure induces apoptosis, inflammation and immunosuppression through CYPs activation and antioxidant dysfunction in common carp neutrophils. Fish and Shellfish Immunology. 2020;**99**:284-290

[58] Koçak Ş, Bağdatlı S, İkican K, Ertuğ NDY. Histopathological effects of environmental pollutants on the reproductive system of zebrafish. Düzce Üniversitesi Bilim ve Teknoloji Dergisi. 2023;**11**(4):2103-2136

[59] Trivedi SP, Singh S, Trivedi A, Kumar MJ. Mercuric chloride-induced oxidative stress, genotoxicity, haematological changes and histopathological alterations in fish *Channa punctatus* (B loch, 1793). Journal of Fish Biology. 2022;**100**(4):868-883

[60] Chakraborty SB. Non-essential heavy metals as endocrine disruptors: Evaluating impact on reproduction in teleosts. Proceedings of the Zoological Society. 2021;**74**:417-431. DOI: 10.1007/s12595-021-00399-x

[61] Ramos RJ, Tadokoro CE, de Carvalho Gomes L, Leite GR. Efficiency in heavy metal purge in crustaceans during the ecdysis. Environment Development and Sustainability. 2021;**23**(10):14878-14907

[62] de Almeida Rodrigues P, Ferrari RG, Kato LS, Hauser-Davis RA, Conte-Junior CAJBTER. A systematic review on metal dynamics and marine toxicity risk assessment using crustaceans as bioindicators. Biological Trace Element Research. 2022;**200**(2):881-903

[63] Wu JP, Chen H-C. Effects of cadmium and zinc on oxygen consumption, ammonium excretion, and osmoregulation of white shrimp (*Litopenaeus vannamei*). Chemosphere. 2004;**57**(11):1591-1598

[64] Brooks SJ, Mills CL. The effect of copper on osmoregulation in the freshwater amphipod *Gammarus pulex*. Comparative Biochemistry and Physiology. Part A, Molecular and Integrative Physiology. 2003;**135**(4):527-537

[65] Banaee M, Zeidi A, Mikušková N, Faggio CJBTER. Assessing metal toxicity on crustaceans in aquatic ecosystems: A comprehensive review. Biological Trace Element Research. 2024:1-19. DOI: 10.1007/s12011-024-04122-7 [Article in press]

[66] Mona MM, Younis ML, Atlam AI. Evaluation of freshwater heavy metals accumulation effect on oxidative stress, metallothionein biosynthesis and histopathology of *Procambarus clarkii* (Girard, 1985) collected from three locations in the Delta region, Egypt. BMC Zoology. 2023;**8**(1):21

[67] Ghemari C, Jelassi R, Khemaissia H, Waterlot C, Raimond M, Souty-Grosset C, et al. Histopathological changes in the hepatopancreas of *Porcellio laevis* (Crustacea, Isopoda) after exposure to Cd and Zn mixture. In: Ksibi M, et al. Recent Advances in Environmental Science from the Euro-Mediterranean and Surrounding Regions. 2nd ed. EMCEI 2019. Environmental Science and Engineering. Cham: Springer; 2021. pp. 587-592

[68] Cheng C-H, Ma H-L, Liu G-X, Fan S-G, Deng Y-Q, Jiang J-J, et al. Toxic effects of cadmium exposure on intestinal histology, oxidative stress, microbial community, and transcriptome change in the mud crab (*Scylla paramamosain*). Chemosphere. 2023;**326**:138464

[69] de Jesus WB, de Oliveira Mota TdS, Soares SH, Pinheiro-Sousa DB, de Oliveira SRS, Torres HS, et al. Biomarkers and occurrences of heavy metals in sediment and the bioaccumulation of metals in crabs (*Ucides cordatus*) in impacted mangroves on the Amazon coast, Brazil. Chemosphere. 2021;**271**:129444

[70] Li Y, Zhou X, Guo W, Fu Y, Ruan G, Fang L, et al. Effects of lead contamination on histology, antioxidant and intestinal microbiota responses in freshwater crayfish, *Procambarus clarkii*. Aquatic Toxicology. 2023;**265**:106768

[71] Gholamhosseini A, Banaee M, Zeidi A, Multisanti CR, Faggio CJ. Individual and combined impact of microplastics and lead acetate on the freshwater shrimp (*Caridina fossarum*): Biochemical effects and physiological responses. Journal of Contaminant Hydrology. 2024;**262**:104325

[72] Wang Q, Liu B, Yang H, Wang X, Lin Z. Toxicity of lead, cadmium and mercury on embryogenesis, survival, growth and metamorphosis of *Meretrix meretrix larvae*. Ecotoxicology. 2009;**18**:829-837

[73] Djekoun M, Bensoltane S, Bourechrouche A, Bourechrouche M, Berrebah HJ. In-vitro toxicity of cadmium on the development of parthenogenetic eggs of a freshwater cladoceran: Daphnia magna. Journal of Materials and Environmental Sciences. 2015;**6**(4):957-962

[74] Waqas W, Yuan Y, Ali S, Zhang M, Shafiq M, Ali W, et al. Toxic effects of heavy metals on crustaceans and associated health risks in humans: A review. Environmental Chemistry Letters. 2024;**22**(3):1391-1411

[75] Frías-Espericueta MG, Bautista-Covarrubias JC, Osuna-Martínez CC, Delgado-Alvarez C, Bojórquez C, Aguilar-Juárez M, et al. Metals and oxidative stress in aquatic decapod crustaceans: A review with special reference to shrimp and crabs. Aquatic Toxicology. 2022;**242**:106024

[76] Di Donato G, Davolos D, Chimenti C, De Matthaeis E. A genotoxicity study on *Talitrus saltator* (Crustacea, Amphipoda) exposed to lead and benzo (a) pyrene. The European Zoological Journal. 2020;**87**(1):235-241

[77] Cheng C-H, Ma H-L, Deng Y-Q, Feng J, Jie Y-K, Guo Z-X. Oxidative stress, cell cycle arrest, DNA damage and apoptosis in the mud crab (*Scylla paramamosain*) induced by cadmium exposure. Chemosphere. 2021;**263**:128277

[78] Tong R, Pan L, Zhang X, Li Y. Neuroendocrine-immune regulation mechanism in crustaceans: A review. Reviews in Aquaculture. 2022;**14**(1):378-398

[79] Wang L, Feng J, Wang G, Guan T, Zhu C, Li J, et al. Effects of cadmium on antioxidant and non-specific immunity of *Macrobrachium nipponense*. Ecotoxicology and Environmental Safety. 2021;**224**:112651

[80] Tang D, Liu R, Shi X, Shen C, Bai Y, Tang B, et al. Toxic effects of metal copper stress on immunity, metabolism and pathologic changes in Chinese mitten crab (Eriocheir japonica sinensis). Ecotoxicology. 2021;**30**:632-642

[81] Yogeshwaran A, Gayathiri K, Muralisankar T, Gayathri V, Monica JI, Rajaram R, et al. Bioaccumulation of heavy metals, antioxidants, and metabolic enzymes in the crab *Scylla serrata* from different regions of Tuticorin, Southeast Coast of India. Marine Pollution Bulletin. 2020;**158**:111443

[82] Liu J, Wang E, Jing W, Dahms H-U, Murugan K, Wang L. Mitigative effects of zinc on cadmium-induced reproductive toxicity in the male freshwater crab *Sinopotamon henanense*. Environmental Science and Pollution Research International. 2020;**27**:16282-16292

[83] Rodríguez EM, Medesani DA, Fingerman M. Endocrine disruption in crustaceans due to pollutants: A review. Comparative Biochemistry and Physiology Part A: Molecular and Integrative Physiology. 2007;**146**(4):661-671

[84] Mai H, Cachot J, Brune J, Geffard O, Belles A, Budzinski H, et al. Embryotoxic and genotoxic effects of heavy metals and pesticides on early life stages of Pacific oyster (*Crassostrea gigas*). Marine Pollution Bulletin. 2012;**64**(12):2663-2670

[85] Lavolpe M, Greco LL, Kesselman D, Rodríguez E. Differential toxicity of copper, zinc, and lead during the embryonic development of *Chasmagnathus granulatus* (Brachyura, Varunidae). Environmental Toxicology and Chemistry. 2004;**23**(4):960-967

[86] Güner U. Heavy metal effects on P, Ca, Mg, and total protein contents in embryonic pleopodal eggs and stage-1 juveniles of freshwater crayfish *Astacus leptodactylus* (Eschscholtz, 1823). Turkish Journal of Biology. 2010;**34**(4):405-412

[87] López Greco L, Rodriguez E, Bolaños J, Hernandez G, Fingerman M. Toxicity of copper sulphate during early and late embryonic development of a palaemonid shrimp (Crustacea). Invertebrate Reproduction and Development. 2002;**41**(1-3):165-170

[88] Banaee M, Sureda A, Taheri S, Hedayatzadeh F. Sub-lethal effects of dimethoate alone and in combination with cadmium on biochemical parameters in freshwater snail, *Galba truncatula*. Comparative Biochemistry and Physiology. Toxicology and Pharmacology: CBP. 2019;**220**:62-70

[89] Sawasdee B, Köhler H-R. Embryo toxicity of pesticides and heavy metals to the ramshorn snail, *Marisa cornuarietis* (Prosobranchia). Chemosphere. 2009;**75**(11):1539-1547

[90] Banaee M, Taheri SJ. Metal bioaccumulation, oxidative stress, and biochemical alterations in the freshwater snail (*Galba truncatula*) exposed to municipal sewage. Journal of Advances in Environmental Health Research. 2019;**7**(1):8-17

[91] Vidyalakshmi D, Yesudas A, Sivan G, Prakash EA, Priyaja P. Heavy metal accumulation analysis using bivalve, sponge, sea urchin, and gastropod species as bioindicators. Marine Pollution Bulletin. 2024;**202**:116374

[92] Qin L-y, Zhang R-c, Liang Y-d, Wu L-c, Zhang Y-j, Mu Z-l, et al. Concentrations and health risks of heavy metals in five major marketed marine bivalves from three coastal cities in Guangxi, China. Ecotoxicology and Environmental Safety. 2021;**223**:112562

[93] Pan X-D, Han J-L. Heavy metals accumulation in bivalve mollusks collected from coastal areas of Southeast China. Marine Pollution Bulletin. 2023;**189**:114808

[94] Impellitteri F, Curpăn A-S, Plăvan G, Ciobica A, Faggio CJI. Hemocytes: A useful tool for assessing the toxicity of microplastics, heavy metals, and pesticides on aquatic invertebrates. Journal of Environmental Research and Public Health. 2022;**19**(24):16830

[95] Joshy A, Sharma SK, Mini K, Gangadharan S, Pranav P. Histopathological evaluation of bivalves from the southwest coast of India as an indicator of environmental quality. Aquatic Toxicology. 2022;**243**:106076

[96] Abdoul HH, Khalloufi N, Bejaoui M, Al-Hoshani N, Boufahja F, Mahmoudi E. Bioaccumulation of heavy metals in water, sediments, and tissues and their histopathological effects on *Unio tigridis* (Bourguignat 1852). Acta Geophysica. 2024;**72**(4):2653-2662

[97] Noleto KS, De Oliveira SRS, Lima IMA, de Jesus WB, da Silva Castro J, de Santana TC, et al. Biochemical and histological biomarkers in *Crassostrea sp*. (Bivalvia, Ostreidae) for environmental monitoring of a neotropical Estuarine Area (Sao Jose Bay, Northeastern Brazil). Bulletin of Environmental Contamination and Toxicology. 2021;**106**:614-621

[98] Dallinger RJB. Metals and metallothionein evolution in snails: A contribution to the concept of metal-specific functionality from an animal model group. Biometals. 2024;**37**(3):671-696

[99] Dhiman V, Pant D. Environmental biomonitoring by snails. Biomarkers. 2021;**26**(3):221-239

[100] Gnatyshyna L, Khoma V, Martinyuk V, Matskiv T, Pedrini-Martha V, Niederwanger M, et al. Sublethal cadmium exposure in the freshwater snail *Lymnaea stagnalis* meets a deficient, poorly responsive metallothionein system while evoking oxidative and cellular stress. Comparative Biochemistry and Physiology. Toxicology and Pharmacology: CBP. 2023;**263**:109490

[101] El Mageed YSA, Ghobashy AEFA, Al-Thomali AW, Soliman MF, Mohammadein A, El-Shenawy NS. Impact of heavy metals contamination on biology, biochemical, and histology of *Eobania vermiculata* and *Monacha obstructa*. Toxicology and Environmental Health Sciences. 2023;**15**(1):19-30

[102] Karim RMJR. Evaluation of DNA damage of cadmium in gastropod *Lymnaea radix cor* (Annandale and Prashad, 1919) from Euphrates river. Journal of Pharmacy and Technology. 2022;**15**(6):2761-2764

[103] de Siqueira WN, de França EJ, Pereira DR, de Vasconcelos Lima M, Silva HAMF, de Araújo HDA, et al. Study of genotoxic and cytotoxic effects after acute and chronic exposures to industrial sewage sludge on *Biomphalaria glabrata* hemocytes. Chemosphere. 2020;**249**:126218

[104] Thanomsangad P, Tengjaroenkul B, Sriuttha M, Neeratanaphan LHERAAIJ. Heavy metal accumulation in frogs surrounding an e-waste dump site and human health risk assessment.

Human and Ecological Risk Assessment. 2020;**26**(5):1313-1328

[105] Kaczor-Kamińska M, Sura P, Wróbel M. Multidirectional changes in parameters related to sulfur metabolism in frog tissues exposed to heavy metal-related stress. Biomolecules. 2020;**10**(4):574

[106] Khattab NMA, Saber SA, El-Salkh BA, Said REM. Genotoxicity and limbs asymmetry in the Egyptian toad (*Sclerophrys regularis*) as biomarkers for heavy metals toxicity. Egyptian Journal of Aquatic Biology and Fisheries. 2021;**25**(4):705-717

[107] Najibzadeh M, Kazemi A, Hassan HU, Esmaeilbeigi MJER. Hazard assessment of microplastics and heavy metals contamination in Levant frogs (*Pelophylax bedriagae*): A bioindicator in Western Iran. Environmental Research. 2024;**126**(1):119774

[108] Mitkovska VI, Dimitrov HA, Chassovnikarova TGJEB. Chronic exposure to heavy metals induces nuclear abnormalities and micronuclei in erythrocytes of the marsh frog (*Pelophylax ridibundus Pallas*, 1771). Journal of Experimental Biology. 2021;**4**:97-108

[109] Slobodskova VV, Chelomin VP, Kukla SP, Mazur AAJ. Copper induced DNA damage in the gills of the mussel *Mytilus trossulus* and reversibility after depuration. Journal of Marine Science and Engineering. 2022;**10**(11):1570

[110] dos Santos Carvalho C, Utsunomiya HSM, Pasquoto-Stigliani T, Costa MJ, Fernandes MN. Biomarkers of the oxidative stress and neurotoxicity in tissues of the bullfrog, *Lithobates catesbeianus* to assess exposure to metals. Ecotoxicology and Environmental Safety. 2020;**196**:110560

[111] Zhelev ZM, Arnaudova DN, Popgeorgiev GS, Tsonev SV. In situ assessment of health status and heavy metal bioaccumulation of adult *Pelophylax ridibundus* (Anura: Ranidae) individuals inhabiting polluted area in southern Bulgaria. Ecological Indicators. 2020;**115**:106413

[112] Taiwo IE, Henry AN, Imbufe AP, Adetoro OO. Heavy metal bioaccumulation and biomarkers of oxidative stress in the wild African tiger frog, *Hoplobatrachus occipitalis*. African Journal of Environmental Science and Technology. 2014;**8**(1):6-15

[113] Priyadarshani S, Madhushani W, Jayawardena U, Wickramasinghe D, Udagama P. Heavy metal mediated immunomodulation of the Indian green frog, *Euphlyctis hexadactylus* (Anura: Ranidae) in urban wetlands. Ecotoxicology and Environmental Safety. 2015;**116**:40-49

[114] Jayawardena UA, Wickramasinghe DD, Udagama PV. Cytogenotoxicity evaluation of a heavy metal mixture, detected in a polluted urban wetland: Micronucleus and comet induction in the Indian green frog (*Euphlyctis hexadactylus*) erythrocytes and the *Allium cepa* bioassay. Chemosphere. 2021;**277**:130278

[115] Chelyadina N, Popov M, Pospelova N, Smyrnova L. Effects of heavy metals on sex inversion of the mussel *Mytilus galloprovincialis lam.*, 1819 in coastal zone of the Black Sea. Marine Pollution Bulletin. 2022;**185**:114323

[116] Yee-Duarte JA, Ceballos-Vázquez BP, Arellano-Martínez M, Camacho-Mondragón MA, Uría-Galicia EJ. Histopathological alterations in the gonad of *Megapitaria squalida* (Mollusca: Bivalvia) inhabiting a heavy metals polluted environment.

Journal of Aquatic Animal Health. 2018;**30**(2):144-154

[117] Guo X, Huang M, Shi B, You W, Ke C. Effects of ocean acidification on toxicity of two trace metals in two marine molluscs in their early life stages. Aquaculture Environment Interactions. 2020;**12**:281-296

[118] Yee-Duarte JA, Racotta IS, Camacho-Mondragón MA, Roldán-Wong NT, Carreño-León DP, Shumilin E, et al. Contrasting reproductive health of female clams *Megapitaria squalida* from two nearby metal-polluted sites in the Gulf of California: Potential effects of copper, lead, and cobalt. Marine Pollution Bulletin. 2020;**160**:111583

[119] Gholamhosseini A, Banaee M, Soltanian S, Sakhaie F. Heavy metals in the blood serum and feces of mugger crocodile (*Crocodylus palustris*) in Sistan and Baluchistan Province, Iran. Biological Trace Element Research. 2022;**200**(7):3336-3345

[120] Yee-Duarte JA, Arellano-Martínez M, Roldán-Wong NT, Kidd KA, Ceballos-Vázquez BP. Reduction in reproductive activity from degeneration of testicular follicles in *Megapitaria squalida* (Mollusca: Bivalvia) exposed to metal pollution in the Gulf of California. Marine Pollution Bulletin. 2024;**205**:116648

[121] de Oliveira CPA, Carneiro AA, Ervilha LOG, Machado-Neves M, Souza ACF, Carvalho RPR. Does environmental pollution affect male reproductive system in naturally exposed vertebrates? A systematic review. Theriogenology. 2023;**198**:305-316

[122] Milnes MR. Environmental Influences on Hormones and Reproduction in Reptiles: Hormones and Reproduction of Vertebrates. Vol. 3. Elsevier, Academic Press; 2024. pp. 341-365

[123] Orós J, Camacho M, Luzardo OPNDPR. Environmental and miscellaneous toxicoses in reptiles. In: Garner MM, Jacobson ER, editors. Noninfectious Diseases and Pathology of Reptiles: Color Atlas and Text, Diseases and Pathology of Reptiles. 1st ed. Vol. 2. Boca Raton: CRC Press; 2020. pp. 273-329

[124] Ma L, Wang W-X. Subcellular metal distribution in two deep-sea mollusks: Insight of metal adaptation and detoxification near hydrothermal vents. Environmental Pollution. 2020;**266**:115303

[125] Liu X, Yu S, Chen P, Hong B, Zhang Y, Lin X, et al. Metal loadings in estuarine bivalve and gastropod shellfish in response to socioeconomic development in watershed. Marine Environmental Research. 2022;**176**:105593

[126] Nour HES. Distribution and accumulation ability of heavy metals in bivalve shells and associated sediment from Red Sea coast, Egypt. Environmental Monitoring and Assessment. 2020;**192**(6):353

[127] Weng N, Meng J, Huo S, Wu F, Wang W-X. Hemocytes of bivalve mollusks as cellular models in toxicological studies of metals and metal-based nanomaterials. Environmental Pollution. 2022;**312**:120082

[128] Huiping YJPottUBoJFR, Agency EBJFREA. Immunological assays of hemocytes in molluscan bivalves as biomarkers to evaluate stresses for aquaculture. Bulletin of Japan Fisheries Research and Education Agency. 2021;**50**:31-45

[129] Savoca D, Arculeo M, Arizza V, Pace A, Melfi R, Caracappa S, et al.

Impact of heavy metals in eggs and tissues of C. Caretta along the sicilian coast (Mediterranean Sea). Environments. 2022;**9**(7):88

[130] Esposito M, De Roma A, Sansone D, Capozzo D, Iaccarino D, di Nocera F, et al. Non-essential toxic element (Cd, As, Hg and Pb) levels in muscle, liver and kidney of loggerhead sea turtles (*Caretta caretta*) stranded along the southwestern coasts of Tyrrhenian Sea. Comparative Biochemistry and Physiology. Toxicology and Pharmacology. 2020;**231**:108725

[131] Dos Santos RL, de Sousa Correia JM, Dos Santos EM. Freshwater aquatic reptiles (Testudines and Crocodylia) as biomonitor models in assessing environmental contamination by inorganic elements and the main analytical techniques used: A review. Environmental Monitoring and Assessment. 2021;**193**:1-23

[132] Buenfil-Rojas A, Alvarez-Legorreta T, Cedeño-Vazquez J, Rendón-von Osten J, González-Jáuregui M. Distribution of metals in tissues of captive and wild Morelet's crocodiles and the potential of metallothioneins in blood fractions as a biomarker of metal exposure. Chemosphere. 2020;**244**:125551

[133] Dong A, He H, Zhang T, Jing X, Ma Y, Wang X, et al. Effects of cadmium on liver function in turtle *Mauremys reevesii*. Environmental Science and Pollution Research. 2023;**30**(59):123827-123831

[134] Dong A, Dong H, He H, Dong A, Yan J, Huo J. Effects of cadmium on kidney function of the freshwater turtles *Mauremys reevesii*. Biological Trace Element Research. 2023;**201**(6):3000-3005

[135] Rivera BI. Difference in Heavy Metal Concentrations in the American Crocodile (*Crocodylus acutus*) in National Park Coiba Island and the Montijo Gulf, Panama. 2020;**136**:387-401

[136] Romero-Calderón A, Alvarez-Legorreta T, von Osten JR, González-Jáuregui M, Cedeño-Vázquez J. Stress responses in captive Crocodylus moreletii associated with metal exposure. Environmental Pollution. 2022;**308**:119685

[137] Dos Santos RL, de Sousa Correia JM, Paim APS, da Silva Selva de Oliveira LL, Diniz GTN, dos Santos EM. Metallic elements in aquatic herpetofauna (Crocodylia; Testudines) from a lentic Atlantic rainforest environment in northeastern Brazil. Indiana Journal of Global Legal Studies. 2021;**193**(5):314

[138] Hurtado-Morales MJ, Susa MR, Amézquita A. Heavy metal concentration in neotropical aquatic snakes (*Helicops pastazae*) and its potential as a bioindicator of water pollution. Archives of Environmental Contamination and Toxicology. 2022;**82**(3):367-378

[139] da Fonseca CF, da Silva IJS, da Silva Rodrigues M, de Souza Silva BH, Soares PC, de Moura GJB, et al. Evaluation of metal exposure through the composition of essential and toxic micro-minerals in freshwater turtles (*Phrynops geoffroanus*) from a Brazilian river. Environmental Science and Pollution Research. 2023;**30**(19):54871-54884

[140] Yaghmour F, Samara F, El Sayed Y, Mohammed A, Maio E, Philip S, et al. Investigating heavy metal concentrations in sea snakes (Elapidae: Hydrophiinae) as an outcome of oil spill exposure. Heliyon. 2024;**10**(16):e35954

[141] Mokarat M, Ruangsakunporn A, Chaianunporn T, Lomthaisong K,

Robson MG, Keithmaleesatti S. Heavy metal contamination and bioaccumulation in Khorat snail-eating turtles (*Malayemys khoratensis*) from an e-waste recycling activity site in Thailand. Human and Ecological Risk Assessment. 2022;**28**(1):133-153

[142] Sultana S, Al-Ghanim K, Khan QF, Al-Misned F, Atique U, Ahmed Z, et al. Comparative assessment of heavy metal bioaccumulation in skeletal muscles of softshell and hard-shell freshwater turtles. Journal of King Saud University—Science. 2021;**33**(5):101463

[143] Odetti L, Paravani E, Simoniello MF, Poletta GJMRGTM. Micronucleus test in reptiles: Current and future perspectives. Mutation Research—Genetic Toxicology and Environmental Mutagenesis. 2024;**897**:503772

[144] Kloas W, Stöck M, Lutz I, Ziková-Kloas A. Endocrine disruption in teleosts and amphibians is mediated by anthropogenic and natural environmental factors: Implications for risk assessment. Philosophical Transactions of the Royal Society of London. 2024;**379**(1898):20220505

[145] Arslan E, Güngördü A. Subacute toxicity and endocrine-disrupting effects of Fe2O3, ZnO, and CeO2 nanoparticles on amphibian metamorphosis. Journal of Environmental Science and Research Pollution. 2024;**31**(3):4174-4195

[146] Araújo G, Silva I, Fernandes T, Valença Y, Lopes A, Costa FJ. Osteoporosis and bone deformities in free-living Geoffroy's Toadhead turtles (*Phrynops geoffroanus*) rescued from urban areas in the metropolitan region of Recife (Pernambuco, Brazil). Journal of Herpetological Medicine and Surgery. 2024;**34**(2):104-109

[147] Dong A, Huo J, Yan J, Dong A. Oxidative stress in liver of turtle *Mauremys reevesii* caused by cadmium. Environmental Science and Pollution Research. 2021;**28**:6405-6410

[148] Toledo FADO, Santos DOD, Vasconcelos IMA, Oliveira AR, Cabral JAG, Toledo RARD, et al. Heavy metals bioaccumulation in free-ranging South American rattlesnakes (*Crotalus durissus*) in Southeastern Brazil. Environmental Science and Pollution Research. 2024;**31**:32339-32349

[149] Azeez NMJBJoBD. Bioaccumulation and phytoremediation of some heavy metals (Mn, Cu, Zn and Pb) by bladderwort and duckweed. Biodiversitas Journal of Biological Diversity. 2021;**22**(5):2993-2998

[150] Ceschin S, Bellini A, Scalici M. Aquatic plants and ecotoxicological assessment in freshwater ecosystems: A review. Environmental Science and Pollution Research. 2021;**28**:4975-4988

[151] Kosakivska IV, Babenko LM, Romanenko KO, Korotka IY, Potters G. Molecular mechanisms of plant adaptive responses to heavy metals stress. Cell Biology International. 2021;**45**(2):258-272

[152] Nowicka B. Heavy metal–induced stress in eukaryotic algae—Mechanisms of heavy metal toxicity and tolerance with particular emphasis on oxidative stress in exposed cells and the role of antioxidant response. Environmental Science and Pollution Research. 2022;**29**(12):16860-16911

[153] Bejaoui S, Rabeh I, Telahigue K, Tir M, Chetoui I, Fouzai C, et al. Assessment of oxidative stress, genotoxicity and histopathological responses in the digestive gland of Ruditapes decussatus collected from

northern Tunisian lagoons. Scientia Marina. 2020;**84**(4):403-420

[154] Priya A, Gnanasekaran L, Dutta K, Rajendran S, Balakrishnan D, Soto-Moscoso M. Biosorption of heavy metals by microorganisms: Evaluation of different underlying mechanisms. Chemosphere. 2022;**307**:135957

[155] Kakade A, Sharma M, Salama E-S, Zhang P, Zhang L, Xing X, et al. Heavy metals (HMs) pollution in the aquatic environment: Role of probiotics and gut microbiota in HMs remediation. Environmental Research. 2023;**223**:115186

[156] Kadim MK, Risjani Y. Biomarker for monitoring heavy metal pollution in aquatic environment: An overview toward molecular perspectives. Emerging Contaminants. 2022;**8**:195-205

[157] Diaconu M, Pavel LV, Hlihor R-M, Rosca M, Fertu DI, Lenz M, et al. Characterization of heavy metal toxicity in some plants and microorganisms—A preliminary approach for environmental bioremediation. New Biotechnology. 2020;**56**:130-139

[158] Wang S, Wang J, Liu Z, Zhang B. Unraveling diverse survival strategies of microorganisms to vanadium stress in aquatic environments. Water Research. 2022;**221**:118813

[159] Li Z, Junaid M, Chen G, Wang J. Interactions and associated resistance development mechanisms between microplastics, antibiotics and heavy metals in the aquaculture environment. Reviews in Agriculture. 2022;**14**(2):1028-1045

[160] Komijani M, Shamabadi NS, Shahin K, Eghbalpour F, Tahsili MR, Bahram M. Heavy metal pollution promotes antibiotic resistance potential in the aquatic environment. Journal of Environmental Pollution. 2021;**274**:116569

[161] Norris DO. Environmental Influences on Hormones and Reproduction in Amphibians: Hormones and Reproduction of Vertebrates. Vol. 2. Elsevier, Academic Press; 2024. pp. 257-289

[162] Ruthsatz K, Glosv JJToAT. Effects of pollutants on the endocrine system of tadpoles. In: Alves de Almeida E, Freitas JS, editors. Toxicology of Amphibian Tadpoles. 1st ed. Boca Raton: CRC Press; 2023. pp. 78-110

[163] Zhang W, Zhi H, Sun H. Effects of heavy metal pollution on fitness and swimming performance of *Bufo raddei tadpole*. Bulletin of Environmental Contamination and Toxicology. 2020;**105**:387-392

Chapter 5

Relationship between Heavy Metal Accumulation in Freshwater Fish Hosts and Parasites in Aquatic Environments

Tímea Brázová, Tomáš Faragó and Jozef Hančuľák

Abstract

Certain parasitic species may enhance fish fitness in contaminated environments by bioaccumulating pollutants from their hosts. The fate of trace elements and the combined effects of parasitism and pollution were studied in European perch infected with acanthocephalans *Acanthocephalus lucii* and cestodes *Proteocephalus percae* from the water reservoir Ružín in Slovakia. Most trace elements were found at higher levels in parasites than in fish matrices, except for mercury, with the highest levels in the fish muscle. The concentration of toxic mercury in fish muscle exceeded the maximum level allowed in European foodstuffs by more than twice. Notably, higher levels of certain elements in both parasite species corresponded to lower element concentrations in the fish. Similarly, concentrations of certain essential elements were lower at high infection intensities in both parasite species, indicating competitive interactions within the host-parasite system. The high percentage of morphological malformations in the tapeworms suggests a negative impact of toxic elements on parasites. These findings support the idea that intestinal fish parasites may provide benefits to their hosts in dealing with trace elements. Moreover, using fish intestinal parasites as bioindicators of trace element contamination can help identify environmental issues and highlight the relationship between parasitism and ecological conditions.

Keywords: pollution, fish, parasites, aquatic environment, Slovakia

1. Introduction

The global increase in heavy-metal pollution in nature represents a growing concern. Both natural processes and human activities, such as urbanization, mining, ore processing, and industrialization, contribute to heightened concentrations of metals in the environment, posing significant threats to living organisms [1]. This includes disrupting metabolic functions, displacing essential minerals, causing cell damage, and increasing mortality rates [2, 3].

Aquatic environments can be evaluated for pollutants using water and sediment monitoring or bioindicators, such as phytoplankton, macrophytes, invertebrates, and fish [4]. However, recent studies have suggested that certain parasitic species can also be valuable for environmental biomonitoring [5]. Certain parasites can indicate the chemical condition of the environment based on their presence or absence, while others can accumulate pollutants in their tissues, absorbing significantly higher levels of contaminants than their host.

Host-parasite relationships are complex and not well understood, especially in contaminated environments, where pollutants may affect various biological factors, that is, parasitism [6]. While parasites are generally considered burdensome to their hosts, several studies suggest that in polluted environments, parasites can detoxify their hosts by removing contaminants from their tissues [7, 8]. This phenomenon arises from the sequestering of toxic trace elements like cadmium and lead instead of essential elements like calcium, copper, magnesium, and zinc [9, 10] as a consequence of chemical similarities among these elements, leading to competition between the fish hosts and their parasites [9]. Such heightened absorption of toxic metals from host tissues is sometimes perceived as a beneficial aspect of parasitism [11–14].

However, the potential impacts of this capability on the survival and morphological abnormalities in parasites due to pollution have not been extensively investigated [15, 16] and data on morphological deformities in fish parasites resulting from field contamination are limited [17, 18].

To elucidate the host-parasite relationships in a contaminated environment, we used perch (*Perca fluviatilis*) and their predominant intestinal parasites (acanthocephalan *Acanthocephalus lucii* and cestode *Proteocephalus percae*) as a model. We examined the distribution of heavy metals in specific matrices of perch and parasites based on the type of infection (single versus mixed infections). We also studied the competitive interactions in the accumulation of trace elements between the fish and parasites and the size of parasitic populations. Additionally, we investigated the potential occurrence and frequency of morphological abnormalities in the tapeworm *P. percae* in a polluted environment.

2. Material and methods

2.1 Study area

The Ružín water reservoir in eastern Slovakia (48°40'N, 20°53'E) covers 3.9 km^2 with a volume of 59 million m^3 (**Figure 1**). Created by dams on the Hornád River in 1967 and 1972, it is fed by the Hornád and Hnilec Rivers from the Spiš-Gemer Rudohorie Mountains. The region's intensive mining and ore processing history has led to sediment accumulation with high heavy metal concentrations, particularly near river inflows [19]. These pollutants originate from the geological environment and the processing of ores and mining materials sourced from nearby areas such as Krompachy, Rudňany, Spišská Nová Ves, and Smolník.

The map (**Figure 1**) was created using the program Quantum GIS 3.16 [20] and visualized on georeferenced raster base topographic maps provided by the Geodetic and Cartographic Institute in Bratislava, freely available on the ZBGIS map portal [21]. The background map layer was imported into the QGIS program in the S-JTSK/Krovak East-North coordinate system (EPSG:5514).

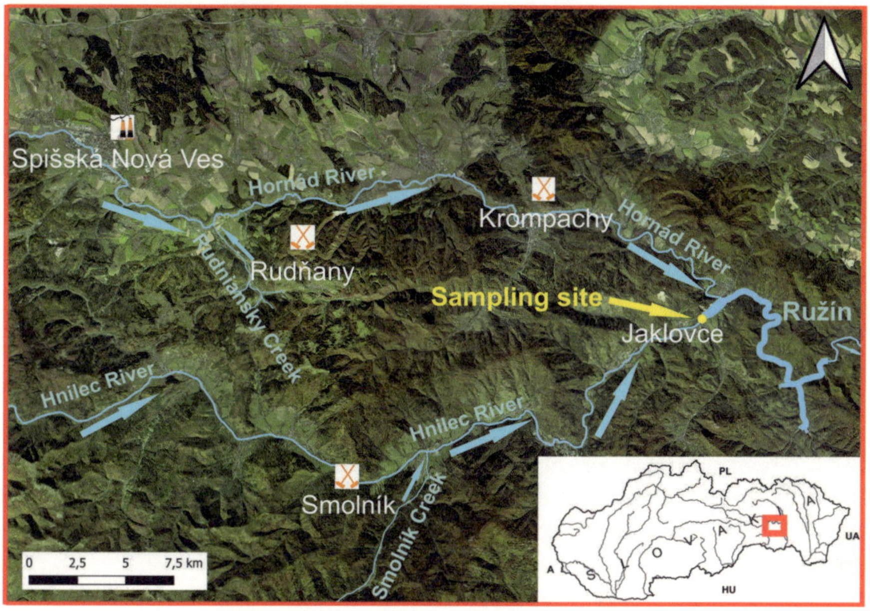

Figure 1.
Location of the sampling site (yellow dot) of European perch (Perca fluviatilis) from the Ružín reservoir. The origin of pollution and the pathway of heavy metals entering the reservoir are indicated by pictograms and blue arrows.

2.2 Collection of samples and elemental analysis

A total of 43 specimens of perch (*P. fluviatilis*) were collected from the vicinity of Jaklovce, where the Hnilec River flows into the reservoir (**Figure 1**). The fish were live-transferred to the laboratory, euthanized, and dissected. Matrices (muscle, kidney, liver, brain, hard or soft roe, and adipose tissue) were frozen and stored individually for trace element analysis. Parasites found in the gastrointestinal tracts of fish included acanthocephalans (*Acanthocephalus lucii*) and cestodes (*Proteocephalus percae*). The prevalence (82.3%) and mean intensity of infection (17.5 ± 13.3) were determined for the adult intestinal acanthocephalan *A. lucii*. Similarly, for the cestode *P. percae*, the prevalence (64.7%) and mean intensity of infection (6.03 ± 1.63) were calculated. All parasites were carefully isolated from the fish intestines, washed in ultrapure water, and individually frozen in preparation for further trace element analysis. Of the 43 examined fish, 25 infected perch were selected for further study and categorized into three groups based on their parasitic infections: those infected exclusively with acanthocephalans (9), exclusively with tapeworms (6), and those harboring both parasite species simultaneously (10). The size of the parasitic infrapopulation (i.e., the number of parasites of a single species within an individual fish host) was assessed according to the methodology outlined by Bush et al. [22]. Fish and parasite samples were analyzed for nine trace elements (As, Cd, Cr, Cu, Hg, Mn, Ni, Pb, and Zn) using an inductively coupled plasma mass spectrometer according to Brázová et al. [23]. The analysis included standard reference materials and blanks for quality control. The concentrations of all measured elements are reported in μg g^{-1} wet weight.

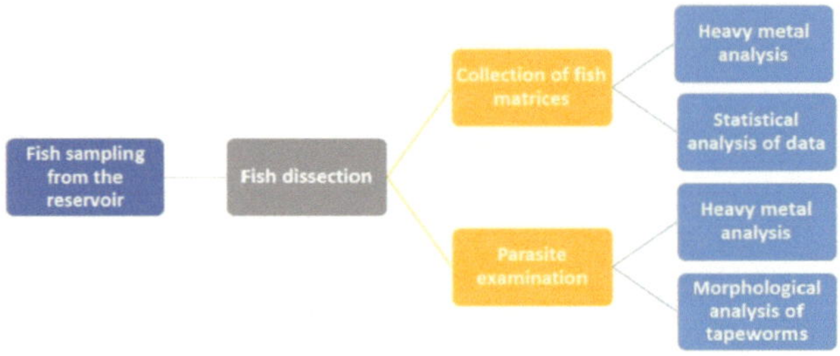

Figure 2.
Schematic diagram of preparation of biological samples from sampling to analysis.

2.3 Morphological analysis of Proteocephalus percae

Adult tapeworms (*P. percae*, N – 144) were examined for morphological abnormalities, with specimens prepared and stained for detailed examination, according to Oros et al. [24]. Morphological deviations were assessed in mature proglottids using light microscopy.

Mature proglottids are characterized by the presence of both male and female genital systems, with the uterus, if present, not predominating over the gonads. Terminologies "proglottization" (development of proglottids) and "strobilation" (somatic compartmentalization of the proglottids) were as defined by Olson et al. [25].

Typical mature proglottids have numerous testes in a single layer, an irregularly opening cirrus sac, a bilobed ovary near the posterior margin, lateral vitelline follicles, and a uterus with lateral diverticula. The strobila consists of proglottids that vary in shape, predominantly wider than long, with marked segmentation [26].

2.4 Data analysis

Bioconcentration factors (BCFs) were calculated as the ratio of each trace element's concentration in the parasite to that in the fish host matrix (BCF = c[parasite]/c[host matrix]) for each infection group, following Sures et al. [9]. Differences in element concentrations between parasites and fish matrices were assessed using the Kruskal-Wallis ANOVA with post-hoc Neményi multiple comparisons [27, 28]. Spearman's rank correlation was used to examine associations between element concentrations in fish vs. parasite tissues/parasite infrapopulations and element concentrations in parasites vs. parasite infrapopulation size, considering only negative significant correlations. The significance level was set at $\alpha = 0.05$, and analyzes were conducted using Statistica 12.0 [29].

Figure 2 illustrates material acquisition and processing methods.

3. Results and discussion

3.1 Distribution of trace elements in fish and parasites

Figure 3 presents the concentrations of various elements in perch and their parasites.

Trace element concentrations exhibited variability across different matrices, with the kidney showing the highest levels, followed by the liver, adipose tissue, hard

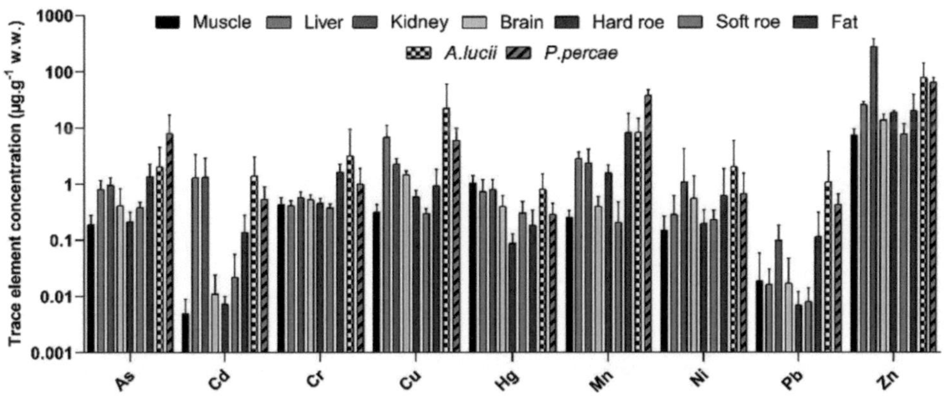

Figure 3.
Heavy metal concentrations (mean ± standard deviation) measured in the internal organs (μg.g^{-1} wet weight) of perch and parasites.

roe, brain, muscle, and soft roe. Statistical analysis revealed significant differences among tissue samples ($P < 0.001$). Nickel and zinc concentrations reached the highest amounts in the kidney (Ni = 1.103, Zn = 281 μg.g^{-1} wet weight, **Figure 3**). Cadmium was detected in the highest levels in the liver and kidney. Copper primarily accumulated in the liver, while arsenic, chromium, manganese, and lead were predominantly found in adipose tissue. Mercury levels in muscle were higher (1.05 μg.g^{-1} wet weight) than in other matrices (e.g., hard roe = 0.087 μg.g^{-1} wet weight, **Figure 3**).

Several studies have confirmed the preferential accumulation of heavy metals in different fish organs [30, 31]. The liver and kidney, being metabolically active organs, play a significant role in storing, redistributing, detoxifying, and eliminating heavy metals and other harmful substances in the body [32]. In contrast, fish muscle typically contains the lowest metal levels compared to other matrices, yet it is important to monitor due to its role as an essential human protein source [33, 34].

Mercury is a significant human and animal toxin, with fish consumption being a major source of Hg [35]. This study found that perch muscle from the Ružín reservoir had mercury levels (1.05 μg g^{-1} wet wt) exceeding European food safety limits [36], highlighting potential health risks for local fish consumers. Similar findings in other studies, like those by Farkas et al. [34], suggest that muscle's high protein content, which binds mercury, may explain these elevated levels [37].

Recent studies have demonstrated that intestinal fish parasites accumulate trace elements at significantly higher concentrations than their fish hosts [38, 39]. Our findings support previous statements as most trace elements were found to be primarily accumulated by parasites. The most pronounced disparity was observed in cadmium accumulation between fish muscle and the parasites acanthocephalan *A. lucii* and cestode *P. percae*, with the parasites exhibiting 270-fold and 120-fold higher cadmium concentrations, respectively, compared to fish muscle.

Parasites similarly to fish matrices, expressed distinct elemental preferences. Cestodes contained higher concentrations of arsenic and manganese, whereas acanthocephalans accumulated cadmium, copper, chromium, lead, nickel, and mercury.

Consistent with our findings, Turčeková et al. [13] reported higher concentrations of arsenic in the cestode *P. percae* and cadmium in *A. lucii*. However, contrary to our results, they observed that copper and lead concentrations were higher in cestodes than in acanthocephalans. They attributed these differences in element absorption capacities to the specific properties of the cestode and acanthocephalan tegument or the unique aspects of their life cycles.

3.2 Bioconcentration factors (BCFs)

Trace element concentrations in perch matrices were significantly lower than in parasites across different infection types, except mercury, which was predominantly concentrated in fish muscle (**Table 1**).

Notably, the liver and kidney accumulated elements in amounts often comparable to those found in parasites, with bioconcentration factors (**Table 1**) showing no significant differences between the heavy metal concentrations in parasites and these organs in most cases. Conversely, muscle exhibited the lowest concentrations of heavy metals than parasites (see **Table 1**), serving as the primary host matrix for our comparative analyzes.

Statistical analyzes confirmed significant differences in trace element concentrations between fish (muscle) and parasites for each infection type, except for mercury and nickel in parasite monoinfections and chromium in cestode monoinfections (Kruskal-Wallis ANOVA; $P < 0.01$; **Table 1**).

Muscle	As	Cd	Cr	Cu	Hg	Mn	Ni	Pb	Zn
AL	12.6***	502**	10.4**	118***	0.909	29.1***	6.4	93.7**	10.4***
PP	35.4***	234**	2.54	20.4**	0.364*	188***	8.39	67.3**	11.4**
ALPP	75.5***	571***	9.42***	93.6***	1.13	201***	90.4***	180***	18.2***
Liver									
AL	3.06	1.93	10.4**	7.83	1.39	3.15	11	120**	3.71
PP	7.16	0.775	2.41	0.75	0.487	12.4	5.66	65.9*	2.76
ALPP	117	1.97	5.97***	4.67	15.1	13	15.1	86.6***	4.27
Kidney									
AL	2.3	1.23	6.87	11.7	1.11	3.84	0.907	14.4	0.3
PP	5.44	0.653	1.65	1.98	0.437	16.6	1.06	7.19	0.271
ALPP	12.8	2.03	6.11	9.8	1.43	26.9**	8.61	21.1	0.424
Hard roe									
AL	10.9***	255	11.1*	79.7***	9.7*	6.32	3.31	370***	5.43
PP	25**	79.1	2.44	9.63	4.03	21.9	32	24.4*	3.91
ALPP	50.4***	225**	8.83	33.5***	13.4	35.8**	21.7	157***	7.77*
Brain									
AL	6.43*	227	7.01	23.9**	2.69	23.5***	6.1	259***	6.27**
PP	24.8*	79.4	1.93	3.61	0.918	111***	1.76	36**	5.25
ALPP	29.4**	228**	6.82	15.3**	2.96**	116***	7.36	116***	9.13**

The asterisk (*) indicates significant differences between the calculated ratio and the threshold value of 1. $^{*}p = 0.05$, $^{**}p = 0.01$, $^{***}p = 0.001$ *from Kruskal-Wallis analysis of variation with multiple comparisons of mean ranks for all groups. AL – Acanthocephalus lucii monoinfection, PP – Proteocephalus percae monoinfection, ALPP – A. lucii, P. percae mixed infections.*

Table 1.
Bioconcentration factors of heavy metals in parasites relative to perch matrices (BF = c[parasite]/c[host matrix]).

Among all types of infections, the highest bioconcentration factors (BCFs) were observed for cadmium. Notably, BCF values for arsenic, nickel, lead, and zinc were higher in mixed infections than in monoinfections (see **Table 1**). The most pronounced difference was observed for nickel, showing a more than tenfold increase between mixed- and monoinfections (BCF_Ni values of 90.4, 6.40, and 8.39, respectively).

Conversely, BCFs for cadmium, chromium, and copper in *A. lucii*, and manganese in *P. percae*, did not exhibit significant differences between monoinfections and mixed infections (**Table 1**).

In the present study, nickel concentrations in *A. lucii* and *P. percae* were notably higher in mixed infections than in monoinfections. Ni is essential for blood formation and various enzyme activities but can cause chromosomal aberrations and morphological transformations in cells. It can also alter hemoglobin properties, reducing oxygen affinity and making erythrocytes more fragile and permeable [40, 41]. Consequently, the competition for Ni between fish and parasites, resulting in decreased Ni concentration in host tissues, might positively affect fish health, though further investigation is needed.

Bioconcentration factors (BCFs) for Cd, Cr, and Cu in *A. lucii* and Mn in *P. percae* were similar between mono- and mixed infections. This indicates that species-specific element preference is not influenced by infection type. However, As was more intensively stored in cestodes when *A. lucii* was present, and *A. lucii* concentrated Ni more when *P. percae* was present. Studies involving mixed infections are limited and include parasitic organisms at different developmental stages (adult vs. larvae) or occupying different sites (intestine, body cavity, etc.) within the host, making it difficult to compare with our results [8, 42].

3.3 Competitive interactions

Certain significant negative associations were observed between concentrations in fish matrices and parasites. Specifically, for *A. lucii*, negative correlations were found in cases of monoinfection (e.g., with nickel in the hard roe; $r = -0.723$, $p = 0.05$) and mixed infections (e.g., with nickel in the liver, $r = -0.77$, $p = 0.01$, and kidney, $r = -0.673$, $p = 0.05$; and zinc in the hard roe, $r = -0.663$, $p = 0.05$). Additional significant negative associations were noted between chromium concentrations in fish muscle ($r = -0.729$, $p = 0.05$) and hard roe ($r = -0.72$, $p = 0.05$) and the infrapopulation size of *A. lucii* in mixed infections, as well as between lead in muscle and *Proteocephalus percae* in monoinfections ($r = -0.926$, $p = 0.01$).

Furthermore, we observed that a higher intensity of infection with Acanthocephalans *A. lucii* in monoinfections was associated with lower concentrations of Cr ($r = -0.95$, $p = 0.001$) and Mn ($r = -0.783$, $p = 0.01$) in the parasites. In mixed infections, an increased infrapopulation size of the tapeworm *P. percae* was negatively correlated with Mn ($r = -0.762$, $p = 0.01$) and Cr ($r = -0.667$, $p = 0.05$) levels in the parasites. Additionally, an increasing number of *A. lucii* negatively correlated with Mn concentrations in the cestode *P. percae* ($r = -0.644$, $p = 0.05$). In contrast, a higher number of *P. percae* was associated with lower Cr concentrations in *A. lucii* ($r = -0.793$, $p = 0.01$).

In our analysis, we identified several significant positive and negative correlations. We focused on the negative correlations because they suggest a reduction in heavy metal burden in infected fish hosts, potentially indicating a beneficial aspect of the host-parasite relationship in a polluted environment.

Lower Pb concentrations were observed in fish muscles harboring tapeworms, possibly due to parasites absorbing bile-bound metals, preventing their reabsorption and distribution in the host [43].

Our results also revealed significant negative associations between parasite element concentrations and infection intensity. Cr and Mn concentrations in parasites decreased with higher *A. lucii* or *P. percae* incidence in the host intestine. Similarly to our findings, Sures [11] demonstrated strong competitive interactions among essential elements (Ba, Ca, Fe, Mn, Sr., Zn) in parasites and their infrapopulation size. The levels of these elements in parasites decreased as the number of acanthocephalans increased. The authors suggest that since these elements are of physiological importance for living organisms, it is possible that competition among parasites for these elements could result in increased absorption of other, non-essential, or even toxic elements such as lead and cadmium as long as they can be detoxified.

We found that the intensity of acanthocephalan infection was higher than that of cestodes. Negative associations may help us understand the parasitological effects of crowding in definitive hosts, as the availability of minerals can influence the size of parasite populations. This suggests that one parasite species can reduce the abundance of others by competing for host resources [11, 44]. Therefore, we suppose that the lower population size of cestodes compared to acanthocephalans may be due to competition for minerals between parasites.

3.4 Morphological abnormalities of Proteocephalus percae

Most tapeworms exhibited normal development (**Figure 4a**), however, approximately 30% of specimens displayed abnormal proglottid structures.

The observed anomalies of the proglottids included:

1. Structural abnormalities:

    - Duplicated ovary, with one typically underdeveloped – 2 cases (**Figure 4b**).
    - Partially developed single ovary – 2 cases (**Figure 4c**).
    - Two cirrus sacs within the same proglottid – 5 cases (**Figure 4d**).
    - Ovary located on the opposite side (anterior margin) of the proglottid – 3 cases (**Figure 4e**).
    - Presence of an additional ovarian lobe alongside a fully developed ovary – 7 cases (**Figure 4f**).
    - Absence of an ovary – 1 case (**Figure 4g**).

2. Incomplete strobilation:

    - The process of strobilation did not always correspond to proglottization. During the segmentation and growth of the strobila, a proglottid with typical morphology was connected to another proglottid possessing an incompletely developed ovary, along with the cirrus sac (**Figure 4h**).

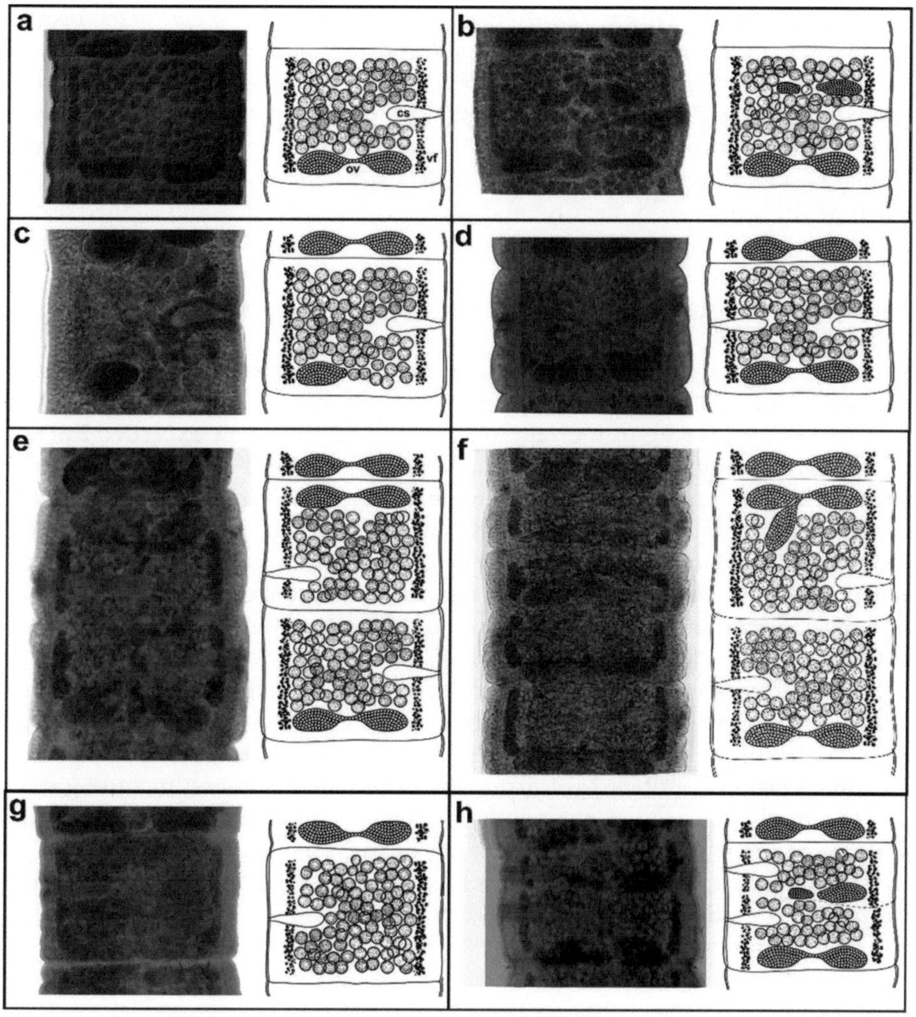

Figure 4.
Microphotography and schematic illustrations of the morphology of the normal mature proglottid (a) and the malformed proglottids (b-h) of tapeworm Proteocephalus percae b—doubled ovary, c—single ovarian lobe, d—two cirrus sacs within the same proglottid, e—misplaced ovary, f—three-lobed ovary (extra ovarian lobe), g—missed ovary, h—incomplete strobilation; cs – cirrus sac, t – testes, ov – ovary, vf – vitelline follicles.

Among all abnormalities observed in the entire sample of *Proteocephalus percae* individuals, incomplete strobilation was the most frequent, occurring 23 times (**Figure 4h**).

In aquatic environments, high concentrations of pollutants persist for long periods, leading to frequent changes in the structure and function of living organisms [45, 46]. Several studies have noted pollution-induced anatomical and physiological defects in fish and amphibians, attributed to toxic trace elements exposure [47–49].

Regarding parasites, certain parasitologists [50–52] provided numerous abnormalities in the cestodes, such as the duplication or partial absence of female and male reproductive organs, the lack of male genital canals, abnormal segmentation, and several variants of unusual positions of the ovary and testes, but not linked to environmental stress.

In the present study, we have confirmed the great ability of *P. percae* to absorb and store toxic trace elements, mainly arsenic and cadmium. Cadmium and arsenic are known for their teratogenic and cytotoxic effects [53, 54]. We assume that the elevated levels of these two elements in parasites may play a significant role in the development of the morphological deformities observed in tapeworms. However, other factors such as developmental instability during the cestode's ontogeny, genetic disorders of the parasites, or detrimental ecological factors, cannot be ruled out.

4. Conclusions

The results of the present study indicate that the Ružín water reservoir remains highly contaminated due to historical pollution from intensive mining and ore processing. The elevated levels of mercury detected in perch muscle exceeded European food safety limits, highlighting potential health risks for local consumers. Therefore, we strongly recommend the establishment of routine monitoring of heavy metal concentrations in this area. Trace element concentrations varied across different fish matrices, with the highest levels found in the kidney and liver. From a bioindication perspective, comprehensive heavy metal risk assessments should include matrices beyond muscle. Parasites, particularly in mixed infections, exhibited higher bioconcentration factors for certain elements than fish matrices. Significant negative correlations between heavy metal concentrations in fish matrices and the presence of parasites suggest that parasites may help reduce the metal burden in their fish hosts, potentially offering a protective effect in polluted environments. A high percentage of tapeworm specimens exhibited abnormal proglottid development, including structural deformities and incomplete strobilation, likely influenced by elevated levels of toxic elements in the parasites. The frequency of these malformations in tapeworms could serve as an alternative indicator for monitoring aquatic contamination. Fish intestinal parasites, due to their efficient accumulation of contaminants, provide a more accurate reflection of the current state of environmental contamination. Additionally, they can influence the levels of pollutants in their hosts. Therefore, we propose that fish intestinal parasites should considered in assessments of aquatic environmental conditions.

Acknowledgements

We gratefully acknowledge the funding of this study by the Grant Agency of the Ministry of Education of the Slovak Republic and Slovak Academy of Sciences (VEGA), project No 2/0052/24, and the Grant Agency for Research and Development, project No APVV-21-0212.

Conflict of interest

The authors declare no conflict of interest.

Author details

Tímea Brázová[1]*, Tomáš Faragó[2] and Jozef Hančuľák[3]

1 Institute of Parasitology, Slovak Academy of Sciences, Košice, Slovakia

2 Faculty of Natural Sciences, Department of Geochemistry, Comenius University, Bratislava, Slovakia

3 Institute of Geotechnics, Slovak Academy of Sciences, Košice, Slovakia

*Address all correspondence to: brazova@saske.sk

References

[1] Järup L. Hazards of heavy metal contamination. British Medical Bulletin. 2003;**68**:167-182. DOI: 10.1093/bmb/ldg032

[2] Singh J, Kalamdhad AS. Effects of heavy metals on soil, plants, human health and aquatic life. International Journal of Research in Chemistry and Environment. 2011;**1**:15-21

[3] Rice KM, Walker EM, Wu M, Gillette C, Blough ER. Environmental mercury and its toxic effects. Journal of Preventive Medicine and Public Health. 2014;**47**:74-83. DOI: 10.3961/jpmph.2014.47.2.74

[4] Rosenberg DM, Resh VH. Freshwater Biomonitoring and Benthic Microinvertebrates. New York NY, USA: Chapman and Hall; 1993. 488 p

[5] Sures B, Nachev M, Selbach C, Marcogliese DJ. Parasite responses to pollution: What we know and where we go in 'environmental parasitology'. Parasites & Vectors. 2017;**10**:65. DOI: 10.1186/s13071-017-2001-3

[6] Squadrone S, Favaro L, Prearo M, Vivaldi B, Brizio P, Abete MC. NDL-PCBs in muscle of the European catfish (*Silurus glanis*): An alert from Italian rivers. Chemosphere. 2013;**93**:521-525. DOI: 10.1016/j.chemosphere.2013.06.037

[7] Filipović Marijić V, Vardić Smrzlić I, Raspor B. Does fish reproduction and metabolic activity influence metal levels in fish intestinal parasites, acanthocephalans, during fish spawning and post-spawning period? Chemosphere. 2014;**112**:449-455. DOI: 10.1016/j.chemosphere.2014.04.086

[8] Khalil M, Furness DN, Zholobenko V, Hoole D. Effectoftapeworm parasitation on cadmium toxicity in the bioindicator copepod *Cyclops strennus*. Ecological Indicators. 2014;**37**:21-26. DOI: 10.1016/j.ecolind.2013.09.033

[9] Sures B, Siddall R, Taraschewski H. Parasites as accumulation indicators of heavy metal pollution. Parasitology Today. 1999;**15**(1):16-21. DOI: 10.1016/s0169-4758(98)01358-1

[10] Jankovská I, Lukešová D, Száková I, Langrová I, Vadlejch J, Čadková Z, et al. Competitionn for minerals (Zn, Mn, Fe, Cu) between sheep tapeworm (*Moniezia expansa*) and its definitive host sheep (*Ovis aries*). Helminthologia. 2011;**48**:237-243. DOI: 10.2478/s11687-011-0033-3

[11] Sures B. Competition for minerals between *Acanthocephalus lucii* and its definitive host perch (*Perca fluviatilis*). International Journal for Parasitology. 2002;**32**:1117-1122. DOI: 10.1016/s0020-7519(02)00083-8

[12] Sures B, Siddall R. *Pomphorhynchus laevis:* The intestinal acanthocephalan as a lead sink for its fish host, chub (*Leuciscus cephalus*). Experimental Parasitology. 1999;**93**:66-72. DOI: 10.1006/expr.1999.4437

[13] Turčeková Ľ, Hanzelová V, Špakulová M. Concentrations of heavy metals in perch and its endoparasites in the polluted water reservoir in eastern Slovakia. Helminthologia. 2002;**39**:23-28

[14] Eira C, Torres J, Miquel J, Vaqueiro J, Soares AMVM, Vingada J. Trace element concentrations in *Proteocephalus macrocephalus* (Cestoda) and *Anguillicola crassus* (Nematoda) in comparison to their fish host, *Anguilla anguilla* in Ria de Aveiro, Portugal. Science of the

Total Environment. 2009;**407**:991-998. DOI: 10.1016/j.scitotenv.2008.10.040

[15] Taraschewski H. Host-parasite interactions in acanthocephala: A morphological approach. Advances in Parasitology. 2000;**46**:1-179. DOI: 10.1016/s0065-308x(00)46008-2

[16] Falis M, Krupicer I, Legáth J, Pitoňák V. The effects of heavy metals and pesticides on survival of miracidia stage of *Fasciola hepatica*. Helminthologia. 2013;**50**:287-290. DOI: 10.2478/s11687-013-0142-2

[17] Šebelová Š, Kuperman B, Gelnar M. Abnormalities of the attachment clamps of representatives of the family Diplozoidae. Journal of Helminthology. 2002;**76**:249-259. DOI: 10.1079/JOH2002133

[18] Pečínková M, Vøllestad LA, Koubková B, Huml J, Jurajda P, Gelnar M. The relationship between developmental instability of gudgeon *Gobio gobio* and abundance or morphology of its ectoparasite *Paradiplozoon homoion* (Monogenea). Journal of Fish Biology. 2007;**71**:1358-1370. DOI: 10.1111/j.1095-8649.2007.01599.x

[19] Findoráková L, Šestinová O, Danková Z, Findorák R, Hančuľák J. Thermal and spectral characterization of bottom sediment from the water reservoir Ružín no.1 in eastern Slovakia. Journal of Soils and Sediments. 2015;**15**:1781-1788. DOI: 10.1007/s11368-014-0941-1

[20] QGIS Development Team. QGIS Geographic Information System. Open Source Geospatial Foundation Project; 2018. Available from: https://qgis.org/en/site/ [Accessed: June 22, 2024]

[21] GKÚ Bratislava, NLC. Geodetic and Cartographic Institute Bratislava, National Forest Centre (NLC). 2024. Available from: https://geoportal.sk/sk/sk/sluzby/mapove-sluzby/wms/wms-zbgis.html [Accessed: June 22, 2024]

[22] Bush AO, Lafferty KD, Lotz JM, Shostak AW. Parasitology meets ecology on its terms: Margolis et al. revised. Journal of Parasitology. 1997;**83**:575-583. DOI: 10.2307/3284227

[23] Brázová T, Torres J, Eira C, Hanzelová V, Miklisová D, Šalamún P. Perch and its parasites as heavy metal biomonitors in a freshwater environment: The case study of the Ružín water reservoir, Slovakia. Sensors. 2012;**12**:3068-3081. DOI: 10.1016/j.ecoenv.2015.09.032

[24] Oros M, Scholz T, Hanzelová V, Mackiewicz JS. Scolex morphology of monozoic cestodes (Caryophyllidea) from the Palearctic region: An useful tool for species identification. Folia Parasitologica. 2010;**57**:37-46. DOI: 10.14411/fp.2010.006

[25] Olson PD, Littlewood DTJ, Bray RA, Mariaux J. Interrelationships and evolution of the tapeworms (Platyhelminthes: Cestoda). Molecular Phylogenetics and Evolution. 2001;**18**:443-467. DOI: 10.1006/mpev.2001.0930

[26] Hanzelová V, Šnábel V, Králová I, Scholz T, D'Amelio S. Genetic and morphological variability in cestodes of the genus Proteocephalus: Geographic variation in *Proteocephalus percae* populations. Canadian Journal of Zoology. 1999;**77**:1450-1458. DOI: 10.1139/cjz-77-9-1450

[27] Siegel S, Castellan NJ. Nonparametric Statistics for the Behavioral Sciences. 2nd ed. New York, NY, USA: McGraw-Hill; 1988. 399 p

[28] Hollander M, Douglas A, Wolfe DW. Nonparametric Statistical Methods. 2nd ed. NewYork: John Willey & Sons; 1999. 816 p

[29] StatSoft, Inc. STATISTICA (Data Analysis Software System), Version 12.0. 2013. Available from: https://www.statsoft.com [Accessed: May 16, 2024]

[30] Pourang N. Heavy-metal bioaccumulation in different tissues of 2 fish species with regards to their feeding-habits and trophic levels. Environmental Monitoring and Assessment. 1995;**35**:207-219. DOI: 10.1007/BF00547632

[31] Jia Y, Wang L, Qu Z, Wang C, Yang Z. Effects on heavy metal accumulation in freshwater fishes: Species, tissues, and sizes. Environmental Science and Pollution Research. 2017;**24**:9379-9386. DOI: 10.1007/s11356-017-8606-4

[32] Al-Yousuf MH, El-Shahawi MS, Al-Ghais SM. Trace metals in liver, skin and muscle of Lethrinus lentjan fish species in relation to body length and sex. Science of the Total Environment. 2000;**256**:87-94. DOI: 10.1016/S0048-9697(99)00363-0

[33] Karakade H, Ünlü E. Concentrations of some heavy metals in water, sediment and fish species from the Atarürk Dam Lake (Euphrates), Turkey. Chemosphere. 2000;**41**:1371-1376. DOI: 10.1016/s0045-6535(99)00563-9

[34] Farkas A, Salánki J, Specziár A. Relation between growth and the heavy metal concentration in organs of bream *Abramis brama L*. populating Lake Balaton. Archives of Environmental Contamination and Toxicology. 2002;**43**:236-243. DOI: 10.1007/s00244-002-1123-5

[35] Zhang Z, He L, Li J, Wu Z. Analysis of heavy metals of muscle and intestine tissue in fish—In Banan section of Chongqing from Three Gorges Reservoir, China. Polish Journal of Environmental Studies. 2007;**16**:949-958

[36] Commission Regulation No 78/2005 of 19 January 2005 Amending Regulation No 466/2001 as Regards Heavy Metals. Available from: http://eurlex.europa.eu/LexUriServ/LexUriServ.do?uri=OJ:L:2005:016:0043:0045:EN:PDF [Accessed: June 19, 2024]

[37] Geeraerts C, Belpaire C. The effects of contaminants in European eel: A review. Ecotoxicology. 2010;**19**:239-266. DOI: 10.1007/s10646-009-0424-0

[38] Leite LAR, Januário FF, Padilha PM, do Livramento ETC, de Azevedo RK, Abdallah VD. Heavy metal accumulation in the intestinal tapeworm *Proteocephalus macrophallus* infecting the butterfly peacock bass (*Cichla ocellaris*), from southeastern Brazil. Bulletin of Environmental Contamination and Toxicology. 2019;**103**:670-675. DOI: 10.1007/s00128-019-02704-z

[39] Mijošek T, Šariri S, Kljaković-Gašpić Z, Fiket Ž, Filipović Marijić V. Interrelation between environmental conditions, acanthocephalan infection and metal(loid) accumulation in fish intestine: An in-depth study. Environmental Pollution. 2024;**356**:124358. DOI: 10.1016/j.envpol.2024.124358

[40] Coen N, Mothersill C, Kadhim M, Wright EG. Heavy metals of relevance to human health induce genomic instability. The Journal of Pathology. 2001;**195**:293-299. DOI: 10.1002/path.950

[41] Vinodhini R, Narayanan M. The impact of toxic heavy metals on the hematological parameters in common carp *(Cyprinus caprio L.)*. Iranian Journal of Health and Environment. 2009;**6**(1):23-28

[42] Nachev M, Schertzinger G, Sures B. Comparison of the metal accumulation capacity between the acanthocephalan *Popmphorhynchus laevis* and larval nematodes of the genus Eustrongylides sp. infecting barbell *(Barbus barbus)*. Parasites & Vectors. 2013;**6**:21. DOI: 10.1186/1756-3305-6-21

[43] Sures B, Siddall R. *Pomphorhynchus laevis* (Palaeacanthocephala) in the intestine of chub *(Leuciscus cephalus)* as an indicator of metal pollution. International Journal for Parasitology. 2003;**33**:65-70. DOI: 10.1016/s0020-7519(02)00249-7

[44] Sures B, Steiner W, Rydlo M, Taraschewski H. Concentrations of 17 elements in the zebra mussel (*Dreissena polymorpha*), in different tissues of perch (*Perca fluviatilis*), and in perch intestinal parasites (*Acanthocephalus lucii*) from the subalpine Lake Mondsee, Austria. Environmental Toxicology and Chemistry. 1999;**18**(11):2574-2579. DOI: 10.1002/etc.5620181126

[45] Kertész V, Fáncsi T. Adverse effects of (surface water pollutants) Cd, Cr and Pb on the embryogenesis of the mallard. Aquatic Toxicology. 2003;**65**:425-433. DOI: 10.1016/S0166-445X(03)00155-3

[46] Amoatey P, Baawain MS. Effects of pollution on freshwater aquatic organisms. Water Environment Research. 2019;**91**:1272-1287. DOI: 10.1002/wer.1221

[47] Linzey DW, Burroughs J, Hudson L, Marini M, Robertson J, Nagarcatti M, et al. Role of environmental pollutants on immune functions, parasitic infections and limb malformations in marine toads and whistling frogs from Bermuda. International Journal of Environmental Health Research. 2003;**13**:125-148. DOI: 10.1080/0960312031000098053

[48] Jezierska B, Lugowska K, Witeska M. The effects of heavy metals on embryonic development of fish (a review). Fish Physiology and Biochemistry. 2009;**35**:625-640. DOI: 10.1007/s10695-008-9284-4

[49] Monroy-Vilchis O, Lizzouli Parra-Lopez L, Beltran-Leon T, Lugo JA, Balderas A, Zarco-Gonzalez MM. Morphological abnormalities in anurans from central Mexico: A case study. Herpetozoa. 2015;**27**:115-121

[50] Honigberg B. A morphological abnormality in the cestode *Dipylidium caninum*. Transactions of the American Microscopical Society. 1944;**63**:342-344. DOI: 10.2307/3223307

[51] Freeman RS. Another hymenolepid with great morphological variation, *Hymenolepis bennetti* n. sp. (Cestoda) from *Napaeozalpus insignis algonquinensis* price. Canadian Journal of Zoology. 1960;**38**:737-743

[52] Macko JK. About some morphological peculiarities and abnormalities in the species of the genus *Diorchis* Clerc, 1903 (Cestoda) from *Fulica atra*. Biologia. 1968;**23**:148-153

[53] Pedlar RM, Ptashynski MD, Evans R, Klaverkamp JF. Toxicological effects of dietary arsenic exposure in lake whitefish *(Coregonus clupeaformis)*. Aquatic Toxicology. 2002;**57**:167-189. DOI: 10.1016/s0166-445x(01)00198-9

[54] Patrick I, Tualla B, Bitacura J. Effects of cadmium and zinc on the gamete viability, fertilization, and embryonic development of *Tripneustes gratilla* (Linnaeus). Scientifica. 2016;**1**:1-10. DOI: 10.1155/2016/8175213

Chapter 6

Impacts of Heavy Metals on Aquatic Dwellers: A Literature Review

Femi Peter Adesina

Abstract

Heavy metals are one of the key contaminants and pollutants of the aquatic ecosystem, that are not only posing a life threat to aquatic dwellers, but also they are of a serious concern for human health. Humans are at risk because man depends on aquatic resources for food and medicines. Although heavy metals can be naturally released into the aquatic environment, their presence is majorly attributed to anthropogenic activities such as the indiscriminate disposal of household and industrial wastes, mining, fertilizer, and pesticide application. Heavy metals are metals that are greater than 5 g/cm^3 in density and additionally, highly soluble and persist in the environment for a long period of time, and therefore easily accumulate in a food chain. Due to the close contact between aquatic organisms and water bodies, aquatic organisms easily take up heavy metals in dissolved form. Therefore, this paper reveals different kinds of aquatic dwellers, the ecological risk assessment and the toxicity effects of heavy metals in aquatic ecosystems. Furthermore, the bioaccumulation of heavy metals on man as a final consumer was discussed and the affected body organs were revealed. Finally, there are discussions on some the bioindicators of heavy metals and their remediation from the aquatic environment.

Keywords: bioindicator, heavy metals, risk assessment, remediation, parasites, insects

1. Introduction

Urbanization, rapid industrialization, and advancement in technology comes with environmental pollution, and this has been the present-day challenges in society [1]. Heavy metals are one of the key contaminants and pollutants that pose a life threat and are of serious concern to human health [2, 3]. Heavy metals are made available in the environment through natural occurrences such as weathering of rocks and volcanic eruptions, and largely by anthropogenic activities such as mining, smelting, fertilizer and pesticide application, and industrial emissions (**Figure 1**). They are defined as metals greater than 5 g/cm^3 in density and additionally, highly soluble and persist in the environment for a long period of time, and therefore easily accumulate in a food chain. Though the presence of some of the metals (such as copper, cobalt, manganese,

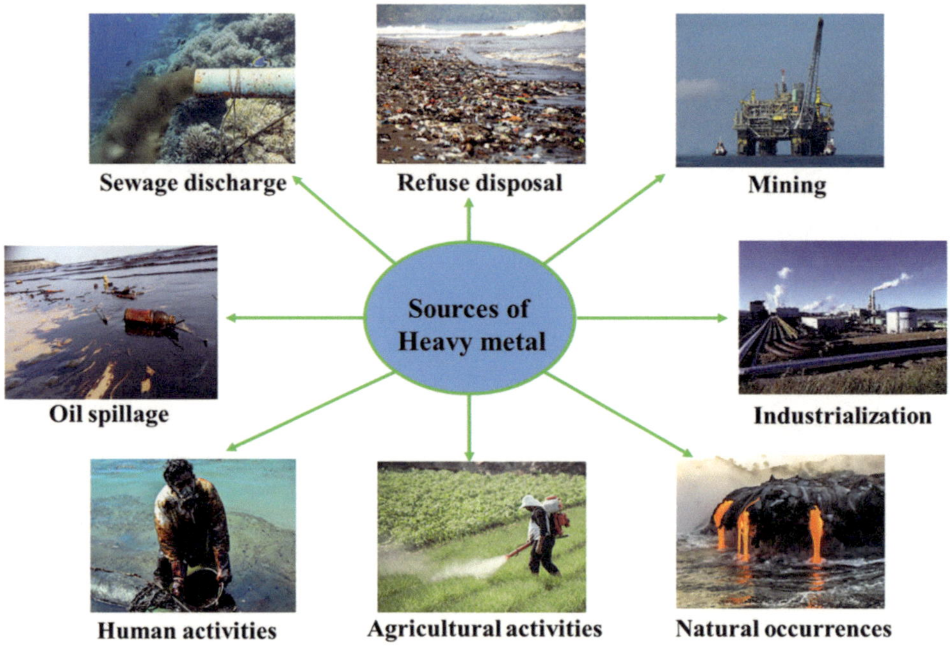

Figure 1.
Sources of heavy metals in aquatic environment (Source: By the author).

iron, and zinc) at a low concentration aids in biochemical and physiological processes in a biological system, their aggravation above acceptable threshold concentrations is lethal and a threat to life [4].

Above 75% of the Earth's surface is composed of aquatic regions, while approximately 2.5% is distributed among the soil, glaciers, polar ice caps, subsurface, and atmosphere. The aquatic environment receives a heavy load of contaminants and pollutants that disgorge heavy metals such as plastics, chemicals, spill oils, sewages, and refuges from surroundings. Seepage, runoff water, and gullies take toxic metals from hillsides, streams, lakes, and rivers, gather them together, and then channel them into the ocean and sea. While this does not exclude the most developed countries in the world, it is also prominent in developing countries. Tao et al. [5] reported heavy metals in seawater and sediment from Daya Bay, South China. According to European Environmental Agency [6], 96% of the assessed area of the Baltic Sea was contaminated with harmful chemicals or metals, 91% of the Black Sea, 87% of the Mediterranean, and 75% of the North-East Atlantic Ocean. Likewise, Akinnifesi et al. [7] revealed the occurrence of heavy metals from water bodies across the Southwestern of Nigeria. Addo-Bediako et al. [8] investigated heavy metals and metalloids in the sediments of the Spekboom River, South Africa, and found high concentrations of Cr and Ni, among other metals at all the sites examined. Cd, Ni, Zn, and Co were also detected by Al-Ghanim et al. [9] at the Indus River, in the Mianwali district of Pakistan. The most commonly found heavy metals in water include cadmium, chromium, cobalt, copper, lead, mecury, nickel, and zinc [10]. It is a point to note that aquatic ecosystems hold a diverse number of organisms, including microorganisms, invertebrates, insects, plants, and fishes [11]. The presence of heavy metals is one of the main types of pollution that may stress the biotic community in aquatic ecosystems. Even, at low concentrations heavy metals can be absorbed by aquatic organisms, plants and into the sediment [12], and then accumulate over time. This can lead to a decrease in productivity, loss of biological diversity, alteration of habitats, and risks for human health. Humans are at risk because of man dependent on aquatic resources for food, medicines, and materials as well as for recreational and commercial purposes such as fishing and tourism [13]. Therefore, maintaining a good aquatic environment, free from contaminants (such as heavy metals and other stressors) is crucial for human health, socioeconomics, and sustainability reasons.

2. Methods

The following search engines were searched: ScienceDirect, Medline/PubMed, and Google Scholar to retrieve literature with the keywords heavy metals, aquatic environment, bioindicator, and remediation. Specifically, papers that were published from 2000 to 2024 were examined. In order to ensure that the literature dataset is specifically centered on the topic of the chapter, a thorough manual screening process was conducted. Initially, a total of 167 articles were collected, and ultimately, 116 were selected for usage. Therefore, papers that did not mainly focus on heavy metals in the aquatic ecosystem were not included.

3. What are aquatic dwellers?

Aquatic dwellers are organisms that inhabit the aquatic environment such as phytoplankton, zooplankton, aquatic plants, insects, fish, birds, mammals, and others. Biodiversity in an aquatic environment depends on the types of aquatic ecosystems. There is a freshwater ecosystem that comprises rivers, lakes, ponds, streams, reserviors, wetlands, and groundwater, and a marine ecosystem that includes estuaries, coral reefs, mangroves, and oceans. In addition, many factors regulate the presence and distribution of dwellers in an aquatic environment such as temperature, pH, current, dissolved oxygen, available food, altitude, season, vegetation, and competition [14, 15]. Different kinds of dwellers in aquatic ecosystems can be microorganisms, plants, insects, and animals.

Microorganisms encompass organisms from the plant kingdom, protozoa, bacteria, and fungi. These organisms are very small in size, so they cannot be seen with the naked eye, except through the aid of a microscope. Though some colonies are visible to the naked eye. Microorganisms are present in vast quantities everywhere and can survive extreme physical and chemical conditions. Microorganisms are an important part of aquatic ecosystems because of the many roles they play, such as their role in decomposition, releasing nutrients stored in organic tissue, capturing the sun's energy through photosynthesis, and being a source of food and nutrients for other organisms, present in the aquatic ecosystem such as protozoa [16]. These microorganisms also play a critical job in remineralizing and restoring the nutrients which influence the materials circulation in aquatic ecosystems [17].

Aquatic plants or macrophytes primarily grow in water and can be grouped into classes based on the positioning of their root and leaves; algae, floating plants, submerged plants, and emerged plants. They are primary producer just like land plants but different in terms of developing adaptable features such as air sacs, flat leaves, and hollow roots that make them survive the aquatic environment. They require light, water, and carbon dioxide for photosynthesis, oxygen to respire, and micro- and macronutrients [18]. Aquatic macrophytes act as a source of food, provide shelter and refuge for small animals, influence water flow, help in biogeochemical cycles, provide habitat structural complexity, and affect the conservation ecology and the diversity and composition of other biotic assemblages [19–21].

Insects represent more than half of animal biodiversity on the earth and represent 90% of the aquatic invertebrate fauna. Some species of insects inhabit or live at a particular stage of their life in aquatic environments [22]. The majority of aquatic insects are found in freshwater habitats such as springs, streams, rivers, lakes, ponds, wetlands, and bogs, while few of them dwell in marine habitats [23]. Aquatic insects may primarily dwell in aquatic habitats or move from terrestrial to aquatic habitats and there are up to 100,000 species of them, from 13 orders [24, 25], with over 200 identified species that are edible [26]. Some of the orders are Coleoptera (5000 species of beetle are considered to be aquatic), Collembola (Springtails), Diptera (Craneflies, mosquitoes, non-biting midges, blackflies), Ephemeroptera (Mayfly), Hemiptera (there are around 3800 known species of aquatic and semiaquatic, such as giant water bugs, water strider, backswimmer), Hymenoptera (diving wasps), Lepidoptera (butterflies and moths), Neuroptera (alderflies, dobsonflies, and spongillaflies), Odonata (damselflies and dragonfly), Plecoptera (stone flies), and Trichoptera (caddisflies). They are at the bottom of the food, serving as a food source for other aquatic animals and their presence in a water body is a good indicator of a healthy environment.

Aquatic animals are diverse and it is estimated that about 1.4–1.6 million marine species exist on earth but 24,000 species are known. Account of animals in freshwater is yet to be well documented, but there are more than 10,000 fish species and 5778 identified species of amphibian's phylum, in freshwater habitat [27, 28]. Based on internal structure there are vertebrate and invertebrate types that may reside in fresh or marine water. Vertebrate types are fishes (bony and cartilaginous such as groupers, tuna, sharks, and rays), reptiles (crocodiles, sea turtles, sea snakes, and marine iguanas), birds (penguins, pelicans, seagulls, and herons) and mammals (whales, dolphin, beluga, hippo, rhinoceros, manatee, dugong, seal, sea elephants, sea lions, walrus, and rodents such as otter and capybara). Examples of invertebrates are marine worms, jellyfish, hydra, coral, polyps, sponges, leechs, mussel, octopuses, cuttlefish, sea anemones, sea cucumbers, sea snails, limpets, clams, starfish, sea urchins, ofiura, sand dollars, crayfish, barnacles, lobsters, crabs, and shrimp. They serve as sources of nutrients and food for man and other predatory animals. Therefore, aquatic animals are an essential component of the aquatic ecosystem.

4. Ecological risk assessment of heavy metals

Ecological risk assessment is the process of evaluating how likely and the degree of effects that exposure to environmental stressors (such as heavy metals, other chemicals, or diseases) which result from human activities, have on organisms or ecosystems. Many methodologies and models (**Table 1**) have been developed to assess the ecological risks of heavy metals.

Procedures/ Models	**Functions**	**Formula**	
Contamination factors (CF)	For monitoring the heavy metal enrichment in sediment over a period of time.	$CF = \frac{Mc}{Bc}$ **Mc** - sample metal concentration. **Bc** - background metal concentration. **CF** ≥ 1 is moderate to very high contamination.	[29]
Geoaccumulation index method	To determine the heavy metal pollution load, i.e. the degree of contamination.	$\text{Igeo} = \log 2\left(\frac{Csample}{1.5\text{x}\,Cbackground}\right)$ C*sample* represents the heavy metal sediment concentration. C*background* represents the heavy metal geochemical background concentration ($\text{mg kg} - 1$). If **Igeo** ≥ 1 is moderately to extremely polluted	[30]
Enrichment factor	To determine the metal abundance in the sediment.	$EF = \frac{\left(\frac{\text{M}}{Fe}\right)\text{sample}}{\left(\frac{\text{M}}{Fe}\right)\text{bacground}}$ **M** - metal investigated **Fe** - iron, a reference metal (Mn or Al can also be used). **EF** < 2 is deficiency to minimal enrichment	[30]
Pollution load Index (PLI)	A sediment quality assessment index. It is a geometric mean of the n number of a metal's CF values at a particular site	$PLI = \sqrt[n]{CF1xCF2xCF3x\ldots\ldots\ldots CFn}$ **n** - represents metal number **CF** - indicates the contamination factor. **PLI** > 1 is progressive deterioration.	[29, 30]
Ecological Risk Index (RI)	E_r^i - To determine the single metal-mediated ecological risk in the sediment. RI - To evaluate the combined pollution risk of an ecological system.	$RI = \sum_{i=1}^{n} E_r^i$ $E_r^i = T_r^i . C_f^i$ $C_f^i = \frac{C\,sample}{C\,background}$ $\boldsymbol{E_r^i}$ is the potential ecological risk associated with the heavy metal i. $\boldsymbol{T_r^i}$ is the toxic response factor for a given metal i. **RI** ≥ 150 is moderate to very high risk	[29, 30]
Bioconcentration ratio (BCR)	To assess the effects of toxic heavy metals on biotic organisms.	$BCR = \frac{C\,organism}{C\,water\,/\,sediment}$ $\boldsymbol{C_{organism}}$ is the metal concentration (mg/kg) in the organism. $\boldsymbol{C_{water/sediment}}$ is the elemental concentration (mg/L) in the sample H_2O/sediment. *BCR* value >1 indicates bioaccumulation.	[30]

Procedures/ Models	Functions	Formula	
Bioaccumulation coefficient (*BAC*)	To assess the effects of toxic heavy metals on plant.	$BAC = \frac{C\ plant}{C water\ /\ sediment}$ C_{plant} is the concentration of metal in the plant $C_{water/sediment}$ is the metal concentration in sediment/water. *BAC* value >1 indicates bioaccumulation.	[30]
Probabilistic risk assessment model, AQUARISK.	The software estimates of the hazardous concentrations of metals in sediments or water likely to affect a certain percent of species at a chosen confidence level can be made.	Utilizing ecotoxicity or dose–response data.	[31]
mean Sediment Quality Guidelines (SQG) quotient (mSQGQs)	For assessing the potential effects of contaminant mixtures in sediments and biota		

Table 1.
Different procedures for assessing the ecological risk of heavy metals in an aquatic ecosystem.

5. Effects of heavy metals on aquatic animals

Due to the close contact between aquatic organisms and water bodies, aquatic organisms easily take up heavy metals in dissolved form. Uptake of heavy metals from the water body by aquatic organisms can take place through gills, their digestive tracts, or subcutaneous layers [32, 33]. Then metal ions bind to specific proteins when entered into the bloodstream, transported to various organs or parts in the organism body (digestive gland, muscle, intestine, kidney, and liver) [34] through the facilitated diffusion across their cell membrane or by active transport or carrier-mediated pathways for utilization, storage, and release. Ali *et al.* [3] stated that gills, liver, and kidneys (metabolically active tissues) accumulate heavy metals higher than other tissues such as skin and muscles. Heavy metals dissolution, availability, uptake, and toxicity depend on water-dissolved organics, temperature, pH, salinity, ionic strength and hardness, sedimentary load, metal speciation and distribution in water, the organism's size, age, assimilation efficiency, feeding strategies, digestion efficiency, intestinal transit time, and reproductive stage [35–38]. For instance, Vajargah [39] revealed that larger and older fish accumulate high concentrations of heavy metals in marine areas with a high density of phytoplankton compared to areas where the amount of phytoplankton is lower. Many researchers have investigated the uptake and bioaccumulation of heavy metals in various aquatic animals and this is shown in **Table 2**.

Likewise, many researchers have documented the effects of heavy metals in aquatic animals at different levels of biological organization, both in a review and original research studies (**Table 3**). Furthermore, it is also a point to note that

Aquatic animal	Part	Heavy metals	Location	Authors
Crocodile	Muscle	As, Pb, Hg, Tl, Mo, Cu, Ba, Zn, U, Au, Ag, Ni, Co, Fe, Pt.	Olifants River of the Kruger National Park (KNP), South Africa.	[40]
Alligators *(Alligator mississippiensis)*	Liver	As, Cd, Cu, Fe, Pb, Se, and Zn.	Louisiana and Florida (legally harvested).	[41]
Sharks	Muscle tissues	Cr, Fe, Cu, Zn, As, Se, Sn, Sb, Pb, Hg, and MeHg. ("As" detected was above the regulatory maximum limits of many organizations, including the Codex standard).	Sea area between Gapa Island and Mara Island.	[42]
	Blood	Hg and As (at levels considered toxic in other vertebrates).	South Africa (Algoa Bay, False Bay, Gansbaai, Mossel Bay, and Struisbaai).	[43]
	Muscle tissue	Cd, Pb, Cr, Mn, Co, Cu, Zn, As, Ag, and THg. The concentrations detected are considered toxic for human consumption.	Bahamas region	[44]
Whales	Humpback skin biopsies.	Elevated levels of Al, Cr, Fe, Mg, Ni, and Zn.	Gulf of Maine	[45]
Striped dolphins (*Stenella coeruleoalba*)	Internal tissues	Hg showed the highest concentration, followed by Se, Cd, As, and Pb.	Murcia Region	[46]
Freshwater fish *Wallago attu* (lanchi)	Muscles, skin, gills, and liver	Cd, Ni, Zn, and Co	Indus River in the Mianwali district of Pakistan.	[9]
Fish *Clarias gariepinus*, Tilapia zillii, and Raimas nigeriensis	Muscle	Fe, Zn, Mn, Cu, Cr, and Pb.	Chanchaga River in Minna, Niger State, Nigeria.	[47]
Benthic cyprinid fish (*Labeo umbratus*)	Muscle, liver, kidney, gill, and spinal cord	As, Cu, Fe, Mn, Se, Sr., Zn Cd, Cr, Hg, and Ni.	The Vaal Dam. South Africa.	[48]
Cultured freshwater fishes *(Labeo rohita, Cirrhinus mrigala, Hypophthalmicthys molitrix, Labeo bata, Puntius sarana)*	Muscle	Zn, Cd, Cr, Pb, Ni, and As.	A commercial fish farm, Natore district, Bangladesh.	[49]

Aquatic animal	Part	Heavy metals	Location	Authors
The Pontic shad *fish (Alosa immaculate)*	Muscle tissue	Cd, As, Cu, Zn, Pb, Fe, Mg, Ni, Hg, Fe, and Cr.	Danube River at Romania, and Serbia. The Black Sea in Bulgaria, Romania, and Turkey.	[50]
Painted turtles (*Chrysemys picta*)	Liver and muscle tissue	Hg (tHg)	Lake Michigan (USA) coastal wetlands.	[51]
Nase (*Chondrostoma nasus*), Barbel (*Barbus barbus*), Chub (*Squalius cephalus*)	Muscle, gill, and liver	Pb, Cd, Cr, Cu, Mn, Sr., Fe, and Zn.	Szamos River, Hungary.	[52]

Table 2.
Instances of heavy metal accumulation in aquatic animals.

Reported effects	Authors
It causes cellular and molecular stress (on proteins, redox homeostasis, cytoskeleton rearrangement, metabolism reshuffle, cellular energy and mitochondrial metabolism, and immunity).	[53]
At the embryonic, alevin, and larvae stages of organisms, they can potentially react with nucleic acids, enzymes, vitamins, hormones, and other substances in the organisms and change chemical structures and biological activities. Leading to the susceptibility to predators (impaired locomotor performance), vertebral deformities, disease, and death, and therefore reduces the survival and growth of fish larvae.	[33, 54–56]
Exposure to copper inhibited skeletal ossification, while lead caused scoliosis.	[57]
Cd alters the reproductive and physiological behaviors of freshwater organisms. It causes cell DNA synthesis inhibition, interferes with many protein and carbohydrate metabolism by inhibiting the enzymes involved in the processes, perturb the ion balance, interacts with calcium metabolism, and causes abnormally low calcium levels (hypocalcemia) in fish. Induces both hepatic and renal injuries in mammals and fish.	[58–63]
Sub-lethal and teratogenic deformities such as growth retardation, lack of tail development, lack of eye lens (placode), yolk sac edema, pericardial edema, hemorrhages, shrinkage of the chorion, and scoliosis were observed in metal-treated embryos (zebrafish embryos).	[64]
Ni exposure promoted the formation and phagocytosis of superoxide anions in the hemolymph of crab but significantly inhibited phenoloxidase activities.	[65]
Hg competes with other substances in the body of an organism.	[33]
Chromium produces malformation effects on organisms at higher concentrations.	[66]

Table 3.
Effects of heavy metals on aquatic animals.

reactions resulting from a combination of metals may pose more effects on an organism compared to the effect of an individual metal. In support of this, Zhou et al. [67] investigated the combine toxicity of eight heavy metals and found that their mixture arrested JB6 cells at the S phase, and also induced the generation of ROS and cell apoptosis. Likewise, this was also highlighted by Li et al. [68] when revealing the cocktail effects in risk assessment of metal toxicity.

6. Effects of heavy metals on aquatic plants

It is commonly recognized that aquatic plants may concentrate and accumulate large amounts of a variety of chemicals, including metals. Their uptake of metal from the environment, concentrates it on the trophic chains with an accumulative effect. Multiple studies have demonstrated that aquatic plants serve as storage for heavy metals in aquatic ecosystems. In 2018, Bai et al. [69] investigated heavy metals accumulation in aquatic plants in rivers and lakes in the Taihu Basin, China, and Ti, V, Cr, Mn, Fe, Co, Cu, Zn, and Ni were detected in *P. malaianus*, *E. crassipes*, and *H. verticillata* and *N. peltata.* Heavy metals were also detected in samples of the aquatic plant species P. crispus and S. natans collected from four rivers in Linyi City, China by Xiuling et al. [70]. Likewise, in the plant, the specificity of metal, concentration, chemical form, pH, and plant species determine the heavy metal toxicity in the plant. Though the plant root is the primary contact site, the whole plant body is exposed to metal ions in the aquatic system [71]. The toxicity of heavy metals may be felt in the physical and biochemical properties, and cellular level of aquatic plants, causing reduction, inhibition, and impairment of growth, enzymes, and mineral nutrients or initiating the formation of free radical and reactive oxygen species or even plant death [72].

Table 4 shows the toxic effects of excessive amounts of metals on aquatic plant.

Metals	Effects	
Cadmium (Cd)	• It disorganizes chloroplast causing the damage of photosynthetic pigments. • It has an adverse influence on the enzymatic systems of cells, oxidative stress, and inducing nutritional deficiency in plants. • It reduces protein content. • Thereby causes growth inhibition or complete death.	[73]
Iron (Fe)	• It inhibits the synthesis of chlorophyll and protein as well as carbohydrates and the uptake of phosphate and nitrogen.	[74]
Nickel (Ni)	• It reduces protein content. • It causes chlorosis and necrosis in different plant species. • Impairment of nutrient balance and resulted in disorder of cell membrane functions. • It decreases the chlorophyll contents and growth.	[75]
Arsenic	• It inhibited photosynthesis, hence impeding the plant's growth.	[76]
Zinc (Zn)	• It reduces chlorophyll content. • It reduces protein content. • It also causes chlorosis, a purplish-red color appearance in the leaves, and enhances the shedding of older leaves. • Browning and decaying of roots could be observed. • It results in retarded growth and causes senescence.	[77]
Copper (Cu)	• It inhibits the synthesis of chlorophyll and protein as well as carbohydrates and the uptake of phosphate and nitrogen. • Withering of roots, chlorosis, necrosis, and lower leaves gets decayed. • It significantly slowed or stopped the growth.	[74, 78]

Metals	Effects	
Mercury (Hg)	• It decreases the chlorophyll content of the aquatic plant. • It binds to water channel proteins, thus inducing leaf stomata to close and physically obstructing water flow in plants. • It leads to the disruption of biomembrane lipids and cellular metabolism in plants • It causes developmental and behavioral abnormalities, impaired reproduction and survival, and in some cases direct mortality.	[79, 80]
Lead (Pb)	• It reduces the uptake of essential elements such as magnesium and iron by plants. • Thereby, powerfully inhibits the vital enzyme of chlorophyll biosynthesis, α-amino levulinate dehydrogenase, and nitrate reductase. • Strongly limits the development and sprouting of seedlings. • It inhibits the growth of roots and aerial plant parts (even in low amounts).	[81]
Cobalt (Co)	• It restricted the concentration of Fe, chlorophyll, protein, and catalase activity. • Significantly decreased water potential and transpiration rate. • It affected the translocation of P, S, Mn, Zn, and Cu from roots to tops.	[82]
Manganese (Mn)	• It causes a reduction of photosynthetic rate, and necrotic brown spotting on leaves, petioles, and stems.	[82]
Chromium (Cr)	• It severely affects the physiological and biochemical processes of the plant; causing a reduction in root growth, enzymes (catalase, peroxidase, and cytochrome oxidase), nutrient imbalance, leaf chlorosis, inhibition of seed germination, and depressed biomass.	[83]

Table 4.
Toxicity effects of excessive amounts of metals on aquatic plants.

7. Man as a final consumer

Man is a heterotrophic and therefore also rely on aquatic plants and animals for feeding. Aquatic animals play an essential role in feeding the world's growing population. Out of the world population, 3.3 billion people (approximately 42%) depend on aquatic animals for their protein resources. It is estimated to have 20% daily intake among other protein sources that humans consume [84]. Through the food chain, there is a transfer of heavy metals from prey into the predators and finally to man (**Figure 2**). Some of the predators are larger species with a longer life span and can be able to accumulate larger amounts of heavy metals as long as they live. According to Vajargah [39], they lived up to 25 years or even more. If it is not the whole body of the animal that will be consumed (such as crayfish, prawn, small fish, and crab), among the body parts of the aquatic animal that stored heavy metals which human beings generally consume is the muscle tissue or fillet. Although liver, kidney, and bone also stored heavy metals, Mohammadi et al. [85], Forouhar et al. [86], and Sattari et al. [87] opined that some people do not usually consume them. It is a fact that heavy metals are of serious threat to human health and cannot be easily removed from the human body when consumed, thereby accumulating over time. Papp et al. [88] revealed that individuals can be exposed to synergistic health risks of Hg with other

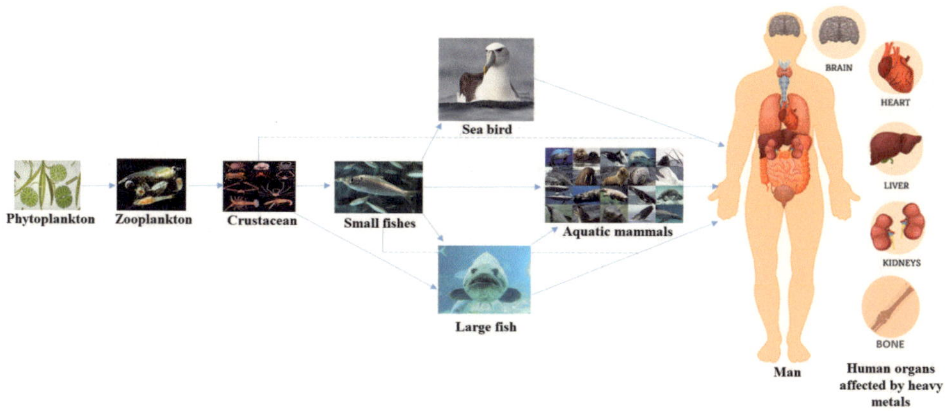

Figure 2.
Bioaccumulation of heavy metals and affected organs in human beings (Source: By the author).

metals like Al, Cu, Pb, Cd, and Mn when shark meat is consumed. Such accumulation and combination can result in acute, chronic, or long-term hazardous effects such as physiological, structural changes, teratogenic, damaging organs (brain, liver, kidney, lungs, and blood composition), and poisoning effects, once the physiological load of the human body is exceeded [33, 89].

8. Bioindicator of heavy metals

Bioindicators are living organisms or species that are used to evaluate the health condition of the natural ecosystem in the environment [90]. Citing from literature, plants, animals, and microorganisms have been identified to serve as an indicator of pollution in an ecosystem. Such as fungi, bacteria, algae, Zooplankton, lichen, earthworm, frog and toads, insects, date mussles as well as parasites [90–92]. Their availability, abundance, toxin and enzymes produced, appearance, physiology, and cellular alterations have served as parameters to presume contamination in the environment. Any bioindicators that reflect the aforementioned parameters are called effect bioindicators while those that show no deleterious outcomes even after effectively accumulating heavy metals or other chemicals into their body system more than the available concentration in their surroundings are called accumulation bioindicators (sentinels) [93]. For instance, no negative effects on shark health parameters have been reported yet on heavy metal intakes, but they are vulnerable to bioaccumulation of heavy metals.

8.1 Insects as bioindicator

Heavy metal bioaccumulation on the aquatic insect body, their abundance and diversity can serve as bioindication of heavy metal contamination of an aquatic ecosystem. Some of them have a high tolerance to environmental pollutants (*Chironomidae* - chironomids and *Hydropsychidae* - caddisflies) while others are relatively less tolerant. Mohd Rasdi *et al*. [94], Starr and Wallace [23] also identified the order Ephemeroptera (mayflies), Plecoptera (stoneflies), and Trichoptera (caddies flies) to be very sensitive to pollution and only inhabit an aquatic ecosystem with good water quality. Similarly, Odonata are important insects for assessing the quality of aquatic habitats, particularly in evaluating the overall health of water bodies. They undergo a larval stage in water and live in adult form on land near freshwater, which makes them a potent tool for assessment. Dragonfly larvae have a vast distribution within waters, an extended lifespan, a predatory nature, and a role as prey in aquatic ecosystems. Consequently, they have the ability to acquire and transport substantial quantities of heavy metals within the food chain. An index quantifying the proportion of heavy metal present in the aquatic population can be established by collecting a sufficient number of dragonfly larvae samples. Adults may contain less metal because, during the molting process to adulthood, the larvae shed their exoskeletons (exuviae) and lose the metals that were present in them. Dragonflies have been identified to be the most sensitive to habitat disturbance, as their presence confirms such water is free of pollution [95]. Many researchers have studied and confirmed the potentiality of dragonflies and damselflies as a bioindicator of environmental metallic elements [96–98]. Moreover, instead of using chemicals to check for heavy metals pollution, insects are readily available, accessible, cheap, and provide a more accurate understanding and useful ways to bio-monitor aquatic systems [99].

8.2 Fish as bioindicator

Fish consume metals dissolved or usable and can therefore function in a water ecosystem as a reliable metal contamination indicator. Atlantic mackerel (*Scomber scombrus*), due to its feeding activity and bottom-feeding habits, was considered an excellent research organism for heavy metal contamination [100]. The most susceptible to contaminants is commonly regarded as fish embryos and larvae, so they are widely used as bioindicators for water quality assessment [101].

8.3 Plant as bioindicator

Plants in the aquatic environment, accumulate dissolved metals through their roots, stems, and leaves into their body system. So, they are also a good bioindicator for metals monitoring in an aquatic ecosystem. *Ceratophyllum demersium* has been identified by many researchers to be a useful plant for biomonitoring of heavy metals in a water environment [102–104]. Likewise, *P. crispus*, *Salvinia natans, Typha latifolia* L., *Lemna sp*, *Hydrilla verticillata*, *Elodea canadensis*, *Hyrocotyle ranocloides*, *Azolla filiculoides*, *Pistia stratoites*, *Spirodela polyrhhiza*, *Wolffia globose*, *Potamogeton pectinatus*, Najas armata, and Eichhornia crassipes are also suitable as bioindicator of heavy metal pollution [69, 70, 80, 105].

8.4 Parasites as bioindicator

Recently, research has focused on using parasites to check the health status of an aquatic system. As parasites react differently to a variety of stressors in the environment, their presence or absence in a host signals the contamination of the host environment [106–108]. In fish, trematodes, nematodes, cestodes, and acanthocephalans have been studied to indicate heavy metal contamination in an aquatic environment [107, 109]. Keke et al. [47] reviewed that the presence of aforementioned helminth parasites helps their host to survive the metal toxicity by taking up the higher concentration of the heavy metal absorbed by their host; serving as a metal sink. The research of Tellez and Merchant [41] conformed to this, as the parasites in the alligator stomach, lung and intestine accumulated a higher amount of As, Cu, Se, and Zn out of the other metals investigated than the amount found in its liver. Keke et al. [47] further established that zinc has the closest relationship with most of the fish parasites, and therefore, the parasites could be used to indicate metal pollution and as well as zinc detection in the water signal parasitic infestation in fish. Therefore, research has to look more at the relationship that exists between heavy metals and some parasites, the effects of heavy metals on host parasites, and the resistant mechanism against metal toxicity.

9. Remediation of heavy metals in aquatic

Remediation is the removal or the cleanup of contaminants from the environment. Different methods have been introduced and tested to remove heavy metals from the water body. The efficiency of each method differs from one another and also the application of these methods in a wide-open aquatic system, and without having a negative effect on the dwellers is another research area that is of interest. Some of these methods are:

Phytoremediation, is the use of plants to remove, detoxify, or immobilize heavy metals from the environment. Such plant species must be fast-growing, highly resistant, and have a high capacity to bioaccumulate heavy metals. There are associated microbes that may likely work hand in hand with such a plant for the cleanup. This method is a low-cost solution, environmentally friendly, and useful for a wide range of metals [110]. The limitation is that the process may take a longer time, may not be efficient where there is heavy contamination, or becomes an invasive species as suggested by Ansari et al. [111].

Bacteria, fungi, and algae have been used in several researches to demonstrate the remediation ability of microorganisms. Microorganisms are everywhere interacting with the environment either positively or negatively. Some of the beneficial consequences of microbes are their involvement in biodegradation, elemental biogeochemical cycles of metal transformations, and full mineralization of heavy metals [112, 113]. Their ability to remediate depends on water pH, temperature, oxygen, bioavailability and concentration of metal, presence of other ions, osmotic pressure, and electron acceptors [114].

Nanobioremediation, is a bionanotechnology that uses nanoparticles to enhance microbial activity in order to remove toxic metals from the environment. This has been reported that nanobioremediation is very effective for a wide range of heavy metals removal than conventional methods [115]. Genomics is another explorable aspect for the remediation of heavy metals in the aquatic environment.

Conventional methods like adsorption, electro-dialysis, chemical precipitation, ion exchange, chemical oxidation or reduction reactions evaporative recovery, reverse osmosis, and sludge filtration [116] are also in use for the removal of heavy metals in an environment but are expensive and sometimes impracticable in local conditions.

10. Conclusion

Different kinds of dwellers in aquatic ecosystems can be microorganisms, plants, insects, and animals. From this review, biodiversity in an aquatic environment depends on the type of aquatic ecosystem. In addition, many factors regulate the presence and distribution of dwellers in an aquatic environment. Meanwhile, alterations in the environment and human-induced actions are causing an elevated discharge of some metals from their natural storage locations back into the water body. It was revealed that aquatic organisms serve as storage for heavy metals through bioaccumulation, and the accumulation of heavy metals in organisms depends on body parts, size, age, environment, and other factors. A combination of heavy metals may pose more risk than individual effects. Heavy metals have a deleterious impact on the vital growth processes of aquatic species, and the effects worsen as the concentration of pollutants increases with time. Contamination factors (CF), the geoaccumulation index method, the enrichment factor, the ecological risk index (RI), and so on are notable ecological risk assessment methods used in evaluating the likelihood and degree of the effects of exposure to environmental stressors. Man, depending on aquatic dwellers as a source of food, is significantly impacted by the synergetic health risk of this accumulation, thereby leading to acute or chronic health issues, poisoning, and damaging vital body organs in man. Given the significant role that aquatic species play in the food chain, the relatively small amounts of heavy metals in these organisms, even if they are below permissible maximum limits, need to be taken into consideration. Further study is required due to the possibility of heavy metal deposition and transfer through food chains in aquatic environments, as it will shed light on potential risks to human health.

Among the many bioindicators discussed, insects, particularly dragonflies and damselflies, are quick and useful indicators of whether or not aquatic habitats are contaminated with heavy metals. In addition to conventional approaches, plants and microorganisms serve as very efficient biological agents for eliminating heavy metals from severely contaminated settings. However, the choice of using either plants or microorganisms depends on the specific type of contaminant and its concentration in the environment. It is crucial to consistently assess the levels of contaminants in aquatic organisms in order to gain a deeper comprehension and forecast the impact of evolving aquatic ecosystems on human health and the ecosystem. Therefore, controlling heavy metal contamination in the environment should be prioritized by the government and other environmental organizations. Future research should prioritize the investigation of factors that contribute to enhancing in situ bioremediation methods. Additionally, it is crucial to examine the feasibility and flexibility of these methods in many challenging settings, including unfavorable and heavily contaminated environments with multiple heavy metals. Future studies could potentially further understand the combined toxicity and the underlying mechanisms of multi-heavy metal contamination.

Acknowledgements

The author sincerely appreciates Mrs. Temilade Adesina's support for this research.

Conflict of interest

There is no conflict of interest.

Author details

Femi Peter Adesina
Department of Biology, Federal University of Technology Akure, Ondo State, Nigeria

*Address all correspondence to: femi.adesina@outlook.com

References

[1] Ali H, Khan E. Environmental chemistry in the twenty-first century. Environmental Chemistry Letters. 2017;**15**(2):329-346

[2] Ali H, Khan E, Sajad MA. Phytoremediation of heavy metals-concepts and applications. Chemosphere. 2013;**91**(7):869-881

[3] Ali H, Khan E, Ilahi I. Environmental chemistry and ecotoxicology of hazardous heavy metals: Environmental persistence, toxicity, and bioaccumulation. Journal of Chemistry. 2019;**6730305**:1-14

[4] Shah AI. Heavy metal impact on aquatic life and human health – An over view, IAIA17 conference proceedings | IA's contribution in addressing climate change. In: 37th Annual Conference of the International Association for Impact Assessment. USA: IAIA Publisher; 2017. pp. 4-7

[5] Tao W, Li H, Peng X, Zhang W, Lou Q, Gong J, et al. Characteristics of heavy metals in seawater and sediments from Daya bay (South China): Environmental fates, source apportionment and ecological risks. Sustainability. 2021;**13**:10237

[6] European Environmental Agency. Contamination of European Seas Continues Despite Some Positive Progress. Available from: https://www.eea.europa.eu/highlights/contamination-of-european-seas-continues, 2019

[7] Akinnifesi O, Adesina F, Ogunwole G, Abiya S. Occurrence and impact of heavy metals in some water, land, floral and Fauna resources across four states in southwestern Nigeria. In: Nazal MK, Zhao H, editors. Heavy Metals - their Environmental Impacts and Mitigation. London, UK: IntechOpen, 5 Princes Gate Court; 2021. pp. 1-19

[8] Addo-Bediako A, Nukeri S, Kekana M. Heavy metal and metalloid contamination in the sediments of the Spekboom River, South Africa. Applied Water Science. 2021;**1**(133):1-9

[9] Al-Ghanim KA, Mahboob S, Seemab S, et al. Monitoring of trace metals in tissues of Wallago Attu (lanchi) from the Indus River as an indicator of environmental pollution. Saudi Journal of Biological Sciences. 2016;**23**(1):72-78

[10] Rajeshkumar S, Li X. Bioaccumulation of heavy metals in fish species from the Meiliang Bay, Taihu Lake, China. Toxicology Reports. 2018;**2018**(5):288-295

[11] Marshall SJ. Hydrology, Reference Module in Earth Systems and Environmental Sciences. Amsterdam, Netherlands: Elsevier; 2013

[12] Southichak B, Nakano K, Nomura M, Chiba N, Nishimura O. Phragmites Australis: A novel biosorbent for the removal of heavy metals from aqueous solution. Water Research. 2006;**40**(12):2295-2302

[13] Yunus K, Zuraidah MA, John A. A review on the accumulation of heavy metals in coastal sediment of peninsular Malaysia. Ecofeminism and Climate Change. 2020;**1**(1):21-35

[14] Gage MS, Spivak A, Paradise CJ. Effects of land use and disturbance on benthic insects in headwater streams draining small watersheds north of Charlotte, NC. Southeastern Naturalist. 2004;**3**(2):345-358

[15] Crisci-Bispo VL, Bispo PC, Froehlich CG. Ephemeroptera, plecoptera and trichoptera assemblages in two Atlantic rainforest streams, Southeastern Brazil. Revista Brasileira de Zoologia. 2007;**24**(2):312-318

[16] Joshi P, Pande V, Joshi P. Microbial diversity of aquatic ecosystem and its industrial potential. Journal of Bacteriology and Mycology is an Open Access. 2016;**3**(1):177-179

[17] Tong Y, Lin G, Ke X, et al. Comparison of microbial community between two shallow freshwater lakes in middle Yangtze basin, East China. Chemosphere. 2005;**60**(1):85-92

[18] O'Hare MT, Baattrup-Pedersen A, Baumgarte I, Freeman A, Gunn IDM, Lázár AN, et al. Responses of aquatic plants to eutrophication in Rivers: A revised conceptual model. Frontiers in Plant Science. 2018;**9**:451

[19] Suren AM, Smart GM, Smith RA, Brown SLR. Drag coefficients of stream bryophytes: Experimental determinations and ecological significance. Freshwater Biology. 2000;**45**:309-317

[20] Gross EM, Johnson RL, Hairston NG. Experimental evidence for changes in submersed macrophyte species composition caused by the herbivore Acentria ephemerella (Lepidoptera). Oecologia. 2001;**127**:105-114

[21] Wolters J-W, Verdonschot RCM, Schoelynck J, Verdonschot PFM, Meire P. The role of macrophyte structural complexity and water flow velocity in determining the epiphytic macroinvertebrate community composition in a lowland stream. Hydrobiologia. 2018;**806**:157-173

[22] Derka T, Zamora-Muñoz C, de Figueroa JMT. Aquatic insects. In: Rull V, Vegas-Vilarrúbia T, Huber O, Señaris C, editors. Biodiversity of Pantepui. Amsterdam, Netherlands: Academic Press; 2019. pp. 167-192

[23] Starr SM, Wallace JR. Ecology and biology of aquatic insects. Insects. 2021;**12**(1):51

[24] Dijkstra K-DB, Monaghan MT, Pauls SU. Freshwater Biodiversity and Aquatic Insect Diversification. Annual Review of Entomology. 2014;**59**(1):143-163

[25] Huryn AD, Wallace JB. Habitat, life history, secondary production, and behavioral adaptations of aquatic insects. In: Merritt RW, Cummins KW, Berg MB, editors. An Introduction of Aquatic Insects of North America. 5th ed. Dubuque, IA, USA: Kendall Hunt Publishing Company; 2019. pp. 65-116

[26] Williams DD, Williams SS. Aquatic insects and their potential to contribute to the diet of the globally expanding human population. Insects. 2017;**8**(3):72

[27] Lundberg J, Kottelat M, Smith GR, Stiassny ML, Gill AC. So many fishes, so little time: An overview of recent Ichthyological discovery in continental waters. Annals of the Missouri Botanical Garden. 2000;**87**(1):26-62

[28] Irfan S, Alatawi AMM. Aquatic ecosystem and biodiversity: A review. Open Journal of Ecology. 2019;**9**:1-13

[29] Yahya AN, Mohamed SK, Mohamed AG. Environmental pollution by heavy metals in the aquatic ecosystems of Egypt. Open Access Journal of Toxicology. 2018;**3**(1): 555603

[30] Saroop S, Tamchos S. 4- monitoring and impact assessment approaches for heavy metals. In: Kumar V, Sharma A,

Cerdà A, editors. Heavy Metals in the Environment. Amsterdam, Netherlands: Elsevier; 2021. pp. 57-86

[31] Sawe SF, Shilla DA, Machiwa JF. Assessment of ecological risk of heavy metals using probabilistic risk assessment model (AQUARISK) in surface sediments from Wami estuary, Tanzania. BioMed Research International. 2021;**6635903**:1-9

[32] Long Y, Luo YJ, Xiao J, et al. Research advances in effects of heavy metals stress on fish in Chinese. Journal of Southern Agriculture. 2016;**47**(9):1608-1614

[33] Ya-jun H, Wei L, Zhen-fei Y, Ying-chen B, Cheng-lian F, Zu-xin X, et al. Progress in the research of the toxicity effect mechanisms of heavy metals on freshwater organisms and their water quality criteria in China. Journal of Chemistry. 2020;**9010348**:1-12

[34] Baby J, Raj JS, Biby ET, Sankarganesh P, Jeevitha MV, Ajisha SU, et al. Toxic effect of heavy metals on aquatic environment. International Journal of Biological and Chemical Sciences. 2010;**4**(4):939-952

[35] Griscom SB, Fisher NS. Bioavailability of sediment-bound metals to marine bivalve mollusks: An overview. Estuaries. 2004;**27**(5):826-838

[36] Fu J, Zhao C, Luo Y, Liu C, Kyzas GZ, Luo Y, et al. Heavy metals in surface sediments of the Jialu River, China: Their relations to environmental factors. Journal of Hazardous Materials. 2014;**270**:102-109

[37] Bonnail E, Sarmiento AM, DelValls TA, Nieto JM, Riba I. Assessment of metal contamination, bioavailability, toxicity and bioaccumulation in extreme metallic environments (Iberian Pyrite Belt) using Corbicula fluminea. Science of the Total Environment. 2016;**544**:1031-1044

[38] Roosa S, Prygiel E, Lesven L, Wattiez R, Gillan D, Ferrari BJ, et al. On the bioavailability of trace metals in surface sediments: A combined geochemical and biological approach. Environmental Science and Pollution Research. 2016;**23**(11):10679-10692

[39] Vajargah MF. A review on the effects of heavy metals on aquatic animals. Journal of Biomedical Research and Environmental Sciences. 2021;**30, 2(9)**: 865-869

[40] Du Preez M, Govender D, Bouwman H. Heavy metals in muscle tissue of healthy crocodiles from the Kruger National Park, South Africa. African Journal of Ecology. 2016;**54**(4):1-6

[41] Tellez M, Merchant M. Biomonitoring heavy metal pollution using an aquatic apex predator, the American alligator, and its parasites. PLoS One. 2015;**10**(11):e0142522

[42] Kim SW, Han SJ, Kim Y, Jun JW, Giri SS, Chi C, et al. Heavy metal accumulation in and food safety of shark meat from Jeju island, Republic of Korea. PLoS One. 2019;**14**(3):e0212410

[43] Merly L, Lange L, Meÿer M, Hewitt AM, Koen P, Fischer C, et al. Blood plasma levels of heavy metals and trace elements in white sharks (Carcharodon carcharias) and potential health consequences. Marine Pollution Bulletin. 2019;**142**:85-92

[44] Shipley ON, Lee CS, Fisher NS, Sternlicht JK, Kattan S, Staaterman ER, et al. Metal concentrations in coastal sharks from the Bahamas with a focus on the Caribbean reef shark. Science Report. 2021;**11**(218):1-11

[45] Wise JP Jr, Wise J, Wise CF, Wise SS, Zhu C, Browning CL, et al. Metal levels in whales from the Gulf of Maine: A one environmental health approach. Chemosphere. 2019;**216**:653-660

[46] Martínez-López E, Peñalver J, Escriña A, Lara L, Gens MJ, Dolores EM, Alcaraz A, García-Fernández AJ. Trace metals in striped dolphins (Stenella coeruleoalba) stranded along the Murcia coastline, Mediterranean Sea, during the period 2009-2015. Chemosphere, 2019;229, 580-588.

[47] Keke UN, Mgbemena AS, Arimoro FO, Omalu ICJ. Biomonitoring of effects and accumulations of heavy metals insults using some helminth parasites of fish as bio-indicators in an Afrotropical stream. Frontiers of Environmental Science. 2020;**8**:576080

[48] Gilbert BM, Hussain E, Jirsa F, Avenant-Oldewage A. Evaluation of trace element and metal accumulation and edibility risk associated with consumption of Labeo umbratus from the Vaal dam, South Africa. International Journal Environmental Research and Public Health. 2017;**14**(7):678

[49] Alam M, Rohani F, Hossain S. Heavy metals accumulation in some important fish species cultured in commercial fish farm of Natore, Bangladesh and possible health risk evaluation. Emerging Contaminants. 2023;**9**(4):100254

[50] Lazăr NN, Simionov IA, Petrea ȘM, Iticescu C, Georgescu PL, Dima F, et al. The influence of climate changes on heavy metals accumulation in Alosa immaculata from the Danube River basin. Marine Pollution Bulletin. 2024;**200**:116145

[51] Zachritz AM, O'Reilly KE, Smith DL, Cooper MJ, Schlaht KM, Lamberti GA. Bioaccumulation of mercury in Lake Michigan painted turtles (Chrysemys picta). Environmental Monitoring and Assessment. 2023;**196**(1):75

[52] Nyeste K, Zulkipli N, Uzochukwu IE, Somogyi D, Nagy L, Czeglédi I, et al. Assessment of trace and macroelement accumulation in cyprinid juveniles as bioindicators of aquatic pollution: Effects of diets and habitat preferences. Scientific Reports. 2024;**14**(1):11288

[53] Le Saux A, David E, Betoulle S, Bultelle F, Rocher B, Barjhoux I, et al. New insights into cellular impacts of metals in aquatic animals. Environments. 2020;7:46

[54] Khayatzadeh J, Abbasi E. The effects of heavy metals on aquatic animals. In: The 1 St International Applied Geological Congress. Iran: Department of Geology, Islamic Azad University-Mashad Branch; 2010. pp. 26-28-688-694

[55] Tolins M, Ruchirawat M, Landrigan P. The developmental neurotoxicity of arsenic: Cognitive and behavioral consequences of early life exposure. Annals of Global Health. 2014;**80**(4):303-314

[56] Fowler BA. Handbook on the Toxicology of Metals. 4th ed. Cambridge, MA, USA: Academic Press; 2015

[57] Stomiñska I, Jezierska B. The effect of heavy metals on post embryonic development of Common CARP, *Cyprinus carpio* L. Archives of Polish Fisheries. 2000;**8**(1):119-128

[58] Hansen JA, Welsh PG, Lipton J, Suedkamp MJ. The effects of long-termcadmium exposure on the growth and survival of juvenile bull trout (Salvelinus confleuentus). Aquatic Toxicology. 2002;**58**:165-174

[59] Garcia-Santos S, Fontainhas-Fernandes A, Wilson JM. Cadmium

tolerance in the Nile tilapia (Oreochromis niloticus) following acute exposure: Assessment of some ionoregulatory parameters. Environmental Toxicology. 2006;**21**:33-46

[60] Sobha K, Poornima A, Harini P, Veeraiah K. A study on biochemical changes in the fresh water fish, Catla catla (Hamilton) exposed to the heavy metal toxicant cadmium chloride. Kathmandu University Journal of Science, Engineering and Technology. 2007;**1**:1-11

[61] Perera PACT, Kodithuwakku SP, Sundarabarathy TV, Edirisinghe U. Bioaccumulation of cadmium in freshwater fish: An environmental perspective. Insight Ecology. 2015;**4**(1):1-12

[62] Wang Y, Fang P. Analysis of the impact of heavy metal on the Chinese aquaculture and the ecological hazard. In: International Forum on Energy, Environment and Sustainable Development (IFEESD 2016). Amsterdam, Nederland: Atlantis Press;2016. pp. 42-47

[63] Ghosh S, Pal P, Pati K. Bioaccumulation of cadmium in fish and human health risk assessment. International Journal of Chemical Studies. 2019;**7**(3):721-724

[64] Fasmina MFA, Jayamanne SC, Liyanage NPP, Perera JLCS, De Silva DPN. Determination of the toxicity cause by trace metals on zebrafish (Danio rerio) embryo. Journal of Aquaculture and Fish Health. 2021;**10**(2):176-185

[65] Vijayavel K, Gopalakrishnan S, Thiagarajan R, Thilagam H. Immunotoxic effects of nickel in the mud crab Scylla serrate. Fish and Shellfish Immunology. 2009;**26**(1):133-139

[66] Farag AM, May T, Marty GD, et al. The effect of chronic chromium exposure on the health of Chinook salmon (Oncorhynchus tshawytscha). Journal of Aquatic Toxicology. 2006;**76**(3/4):246-257

[67] Zhou Q, Gu Y, Yue X, Mao G, Wang Y, Su H, et al. Combined toxicity and underlying mechanisms of a mixture of eight heavy metals. Molecular Medicine Reports. 2016;**15**(2):859-866

[68] Li M, Liu W, Slaveykova V. Effects of mixtures of engineered nanoparticles and metallic pollutants on aquatic organisms. Environments. 2020;**7**:27

[69] Bai L, Liu X-L, Hu J, Li J, Wang Z-L, Han G, et al. Heavy metal accumulation in Common aquatic plants in Rivers and lakes in the Taihu Basin. International Journal of Environmental Research and Public Health. 2018;**15**(2857):1-12

[70] Xiuling L, Henglun S, Yongjun Z, Weixing C, Changwei H, Chen S. Distribution and potential ecological risk of heavy metals in water, sediments, and aquatic Macrophytes: A case study of the junction of four Rivers in Linyi City, China. International Journal of Environmental Research and Public Health. 2019;**16**(2861):1-13

[71] Sagarika N. The effect of different environmental pollutants on metabolic activities in aquatic plants. International Journal of Scientific and Engineering Research. 2018;**9**(2):1978-1993

[72] Fornazier RF, Ferreira RR, Vitoria AP, et al. Effects of cadmium on antioxidative enzyme activities in sugarcane. Biologia Plantarum. 2002;**45**:91-97

[73] Mohanpuria P, Rana NK, Yadav SK. Cadmium induced oxidative stress influence on glutathione metabolic

genes of Camellia sinensis (L.) O Kuntze. Environmental Toxicology. 2007;**22**:368-374

[74] Xing W, Huang W, Liu G. Effect of excess iron and copper on physiology of aquatic plant Spirodela polyrrhiza (L.) Schleid. Environmental Toxicology. 2010;**25**:103-112

[75] Jagetiya B, Soni A, Yadav S. Effect of nickel on plant water relations and growth in green gram. Indian Journal of Plant Physiology. 2013;**18**(4):372-376

[76] Singh VP, Srivatava PK, Prasad SM. Nitric oxide alleviates arsenic-induced toxic effects in ridged luffa seedlings. Plant Physiology and Biochemistry. 2013;**71**:155-163

[77] Chanu LB, Gupta A. Toxicity of zinc on growth of an aquatic macrophyte, ipomoea aquatica forsk. Current World Environment. 2016;**11**(1):218-227

[78] Rolli NM, Hujaratti RB, Giddanavar HS, Mulagund GS, Taranath TC. Toxicity effect of copper on aquatic Macrophyte (Pistia Stratiotes L.). International Journal of Current Research and Review. 2017;**9**(15):14-20

[79] Beauvais-Flück R, Slaveykova VI, Skyllberg U, Cosio C. Molecular effects, speciation, and competition of inorganic and methyl mercury in the aquatic plant Elodea nuttallii. Environmental Science and Technology. 2018;**52**(15):8876-8884

[80] Yang J, Li G, Bishopp A, Heenatigala PPM, Hu S, Chen Y, et al. A comparison of growth on mercuric chloride for three Lemnaceae species reveals differences in growth dynamics that effect their suitability for use in either monitoring or remediating ecosystems contaminated with mercury. Frontiers in Chemistry. 2018;**6**(112):1-12

[81] Nas FS, Ali M. The effect of lead on plants in terms of growing and biochemical parameters: A review. MOJ Ecology and Environmental Sciences. 2018;**3**(4):265-268

[82] Ahmood QA, Al-Jawasim MH. Effects of heavy metals on physiological status of plants. Plant Archives. 2019;**19**(2):2865-2871

[83] Suljević D, Fočak M, Alijagic A. Assessing chromium toxicity across aquatic and terrestrial environments: A cross-species review. Drug and Chemical Toxicology. 2024;**47**(6):1312-1324

[84] FAO (Food and Agriculture Organization of the United Nations). The State of World Fisheries and Aquaculture 2016. Contributing to Food Security and Nutrition for all. Rome: FAO; 2016. p. 200

[85] Mohammadi GM, Sanati AM, Bozorgpanah KZ. Investigation of total petroleum hydrocarbons (TPH) and index metals (Ni & V) on Gammarus's tissue (Pontogammarus maeoticus) and coastal sediments of the Caspian Sea: Guilan Province. Journal of Aquatic Ecology. 2018;**7**(4):9-19

[86] Forouhar VM, Imanpoor MR, Shabani A, Hedayati A, Faggio C. Effect of long-term exposure of silver nanoparticles on growth indices, hematological and biochemical parameters and gonad histology of male goldfish (Carassius auratus gibelio). Microscopy Research and Technique. 2019;**82**(7):1224-1230

[87] Sattari M, Majidi S, Imanpour Namin J, Bibak M, Forouhar Vajargah M, Nasrollahzadeh A. Investigating the relationship between some element concentrations in liver and muscle of Vimba persa and growth indices during different seasons in the

southwest coasts of the Caspian Sea. Journal of Aquaculture Development. 2020;**14**(3):43-53

[88] Papp A, Pecze L, Szabo A, Vezer T. Effects on the central and peripheral nervous activity in rats elicited by acute administration of lead, mercury and manganese, and their combinations. Journal of Applied Toxicology. 2006;**26**:374-380

[89] Rose M, Fernandes A, Mortimer D, Baskaran C. Contamination of fish in UK fresh water systems: Risk assessment for human consumption. Chemosphere. 2015;**122**(2):183-189

[90] Parmar TK, Deepak R, Agrawal YK. Bioindicators: The natural indicator of environmental pollution. Frontiers in Life Science. 2016;**9**(2):110-118

[91] Miedico O, Ferrara A, Tarallo M, Pompa C, Bisceglia D, Chiaravalle AE. Hazardous and essential trace elements profile in the different soft tissues of Lithophaga lithophaga (Linnaeus, 1758) from southern Adriatic Sea (Italy). Toxicology and Environmengtal Chemistry. 2016;**98**(8):877-885

[92] Zaghloul A, Saber M, Gadow S, Awad F. Biological indicators for pollution detection in terrestrial and aquatic ecosystems. Bulletin of the National Research Centre. 2020;**44**(127):1-11

[93] Gerhardt A. Bioindicator Species and their Use in Biomonitoring. Environmental Monitoring I (47 Pages). Encyclopedia of Life Support Systems. UNESCO Ed. Oxford (UK): Eolss Publisher; 2002

[94] Mohd Rasdi Z, Fauziah I, Ismail R, Mohd Hafezan S, Fairuz K, Hazmi AD, et al. Diversity of aquatic insects in Keniam River, National Park, Pahang, Malaysia. Asian Journal of Agriculture and Rural Development. 2012;**2**:312-328

[95] Zia A, Rafi MA, Hussain Z, Naeem M. Occurrence of Odonata in northern area of Pakistan with seven new records. Halteres. 2009;**1**(1):48-56

[96] Villalobos-Jiménez G, Dunn AM, Hassall C. Dragonflies and damselflies (Odonata) in urban ecosystems: A review. European Journal of Entomology. 2016;**113**:217-232

[97] Jacob S, Thomas AP, Manju EK. Odonata (dragonflies and damselflies) as bio indicators of water quality. International Journal of Innovative Research in Science, Engineering and Technology. 2017;**6**(9):19464-19474

[98] Lesch V, Bouwman H. (2018). Adult dragonflies are indicators of environmental metallic elements. Chemosphere. 2018;**209**:654-665

[99] Abdullah MZ, Hazwani NMJ, Abas MT. The use of ipomoea pes-caprae plant species to monitor metal pollutions of the coastal soil. Jurnal Teknologi. 2016;**78**:1-6

[100] Abdul Rashid W, Wan VL, Harun AA. Accumulation and depuration of heavy metals in the hard clam (Meretrix meretrix) under laboratory conditions. Tropical Life Sciences Research. 2009;**20**(1):17-24

[101] Gabr HA, Gab-Alla A. Effect of transplantation on heavy metal concentrations in commercial clams of Lake Timsah, Suez Canal. Egypt: Oceanology. 2008;**50**(1):83-93

[102] Polechońska L, Klink A. Estimation of background concentrations of macro- and trace elements in an aquatic plant as a basis for the passive biomonitoring of pollution. The Science of the Total Environment. 2023;**899**:1-8

[103] Du Y, Wu Q, Kong D, Shi Y, Huang X, Luo D, et al. Accumulation and translocation of heavy metals in water hyacinth: Maximising the use of green resources to remediate sites impacted by e-waste recycling activities. Ecological Indicators. 2020;**1**(115):106384

[104] Khudhair IH, Lazim II, Naqeeb NA, Maktoof AA. Estimation of some trace metals in water, sediments and two species of aquatic plants in the Al-Garaf River at Al-Rafa District-southern Iraq. Journal of Chemical Health Risks. 2021;**11**(2):227-236

[105] Unadkat K, Parikh P. A review on heavy metal absorption capacity of aquatic plants: Sources, impact and remediation technique. International Journal of Allied Practice, Research and Review. 2017;**4**(12):23-30

[106] Vidal-Martínez V, Wunderlich A. Parasites as bioindicators of environmental degradation in Latin America: A meta-analysis. Journal of Helminthology. 2017;**91**(2):165-173

[107] Hassan A, Moharram S, El Helaly H. Role of parasitic helminths in bioremediating some heavy metal accumulation in the tissues of Lethrinus mahsena. Turkish Journal of Fisheries and Aquatic Sciences. 2018;**18**:435-443

[108] Mehana EE, Khafaga AF, Elblehi SS, Abd El-Hack ME, Naiel MA, Bin-Jumah M, et al. Biomonitoring of heavy metal pollution using acanthocephalans parasite in ecosystem: An updated overview. Animals. 2020;**10**:811

[109] Al-hasawi ZM. Environmental parasitology: Intestinal helminth parasites of the siganid fish Siganus rivulatus as bioindicators for trace metal pollution in the Red Sea. Parasite. 2019;**26**:12

[110] Mahmoud RH, Hamza AHM. Phytoremediation application: Plants as biosorbent for metal removal in soil and water. In: Ansari AA et al., editors. Phytoremediation. New York, US: Springer International Publishing AG; 2017. pp. 405-422

[111] Ansari AA, Naeem M, Gill SS, AlZuaibr FM. Phytoremediation of contaminated waters: An eco-friendly technology based on aquatic macrophytes application. The Egyptian Journal of Aquatic Research. 2020;**46(4**:371-376

[112] Thenmozhi R, Arumugam K, Nagasathya A, Thajuddin N, Paneerselvam A. Studies on mycoremediation of used engine oil contaminated soil samples. Advances in Applied Science Research. 2013;**4**(2):110-118

[113] Kapahi M, Sachdeva S. Bioremediation options for heavy metal pollution. Journal of Health and Pollution. 2019;**9**(24):1-20

[114] Ayangbenro AS, Babalola OO. A new strategy for heavy metal polluted environments: A review of microbial biosorbents. International Journal of Environmental Research and Public Health. 2017;**14**(1):94

[115] Dixit R, Wasiullah MD, Pandiyan K, Singh UB, Sahu A, Shukla R, et al. Bioremediation of heavy metals from soil and aquatic environment: An overview of principles and criteria of fundamental processes. Sustainability. 2015;7:2189-2212

[116] Siddiquee S, Rovina K, Azad S, Naher L, Suryani S. Heavy metal contaminants removal from wastewater using the potential filamentous fungi biomass: A review. Journal of Microbial and Biochemical Technology. 2015;7:384-393

Chapter 7

Phytoplankton Diversity and Heavy Metal Accumulation: A Case of Suneka Wastewater Treatment Plant, Kenya

Rayori Douglas Mosoti, Nyabaro Obed, Getabu Albert, Omondi Reuben, Orina Paul and Omondi Argwings

Abstract

Knowledge of the nature and composition of wastewater is critical in wastewater treatment, reuse, and disposal. Suneka wastewater treatment plant (Suneka WWTP) treats wastewater from Kisii municipality. The treated effluent is discharged into river Riana. The Suneka WWTP does not have adequate capacity to treat all the wastewater from the municipality fully. The discharge of partially or untreated wastewater into the Riana River, particularly during system breakdown, is of great concern due to the potential health risks it poses to the environment, humans, and animals. This chapter discusses phytoplankton community structure in the Suneka WWTP and their roles in wastewater treatment, especially in heavy metal accumulation. Phytoplankton species identified belonged to Bacillariophyceae, Chlorophyceae, Cyanophyceae, Euglenaphyceae, Zygnematophyceae, and Dinophyceae. The total phytoplankton biovolume recorded was 680.99 mm^3L^{-1}. The phytoplankton also contributed to wastewater polishing by converting nutrients into their biomass and removing heavy metals from the wastewater column through bioaccumulation.

Keywords: wastewater, phytoplankton diversity, heavy metals, accumulation, Suneka wastewater treatment plant

1. Introduction

Wastewater is of poor quality as it is rich in pathogenic microorganisms [1, 2], heavy metals [3–5], organic and inorganic chemicals, and toxic substances [6–8]. Therefore, wastewater is unsuitable for domestic, agricultural, and industrial uses and pollutes the environment [9–11]. The effects associated with release of untreated or partially treated wastewater from domestic and industrial sources include the degradation of aquatic ecosystems and pollution of surface waters resulting in outbreaks of food poisoning, and water-borne diseases [1, 9, 12].

Phytoplankton are free-floating microscopic plants in water, and they are the primary producers providing food to aquatic organisms [13, 14]. The species commonly found in most aquatic environments include Cyanophytes (blue-green algae), Bacillariophytes (diatoms), Chlorophytes (green algae), and Pyrrophytes (desmids). Fauna and flora community structures in aquatic environments are normally disrupted with pollution [15]. Recent studies in Kisumu Bay, polluted with sewage discharge, showed that cyanophyte predominate [15]. Also, a study in treated wastewater maturation ponds showed that physico-chemical parameters affect the dominance of Cyanophyta, Chlorophyta, and Euglenaphyta. At low N:P ratio ≤ 10, Cyanophyta predominate [16]. Therefore, phytoplankton community structure can be used as a bio-indicator of the health status of wastewater during the treatment process in the ponds [14, 16, 17].

In this study, the phytoplankton community structure of Suneka WWTP is discussed including their role in wastewater polishing especially heavy metals accumulation, evaluated. Comprehensively, the objective of this research was to investigate the diversity of phytoplankton and the accumulation of heavy metals in the context of the Suneka WWTP in Kenya. The study aimed to explore the variety of phytoplankton species present in the wastewater treatment environment and to understand how these organisms interact with and potentially accumulate heavy metals. By examining both the biodiversity of phytoplankton and the levels of heavy metals within these organisms, the research seeks to elucidate the ecological impacts of wastewater treatment processes and the potential implications on environmental and public health.

2. Materials and methods

2.1 Study area

The study was conducted in Suneka WWTP in Kisii County, Kenya. Physically, it is located at Suneka at latitude 0039′30“S and Longitude 34042’30”E. The plan is a lagoon system with functional ponds through which wastewater passes for polishing. The effluent is discharged into River Riana, which eventually discharges into River Kuja, a large river in Southwest Kenya, emptying its water into Lake Victoria (**Figure 1**). The WWTP's design capacity is 15,000 m^3/day following renovation from the previous design, which had a capacity of 8000 m^3/day. The intent of renovation was to improve the wastewater treatment capacity and its efficiency due to increased domestic, institutional, and agricultural wastewater resulting from an increased population within the municipality. The current design has additional ponds with enhanced capacity. The WWTP serves Kisii Municipality and its environs.

2.2 Sampling

Field sampling was done during morning hours between 8:00 am and 10:00 am once every month during May–August 2021. Nine sampling points were identified, including the Influent (inlet point when wastewater enters the WWTP), two anaerobic ponds, one facultative, and two maturation ponds. The effluent before discharge into the river besides to three points in the river, that is, at the confluent and 100 meters up and downstream at the effluent discharge point into River Riana (**Figure 1**).

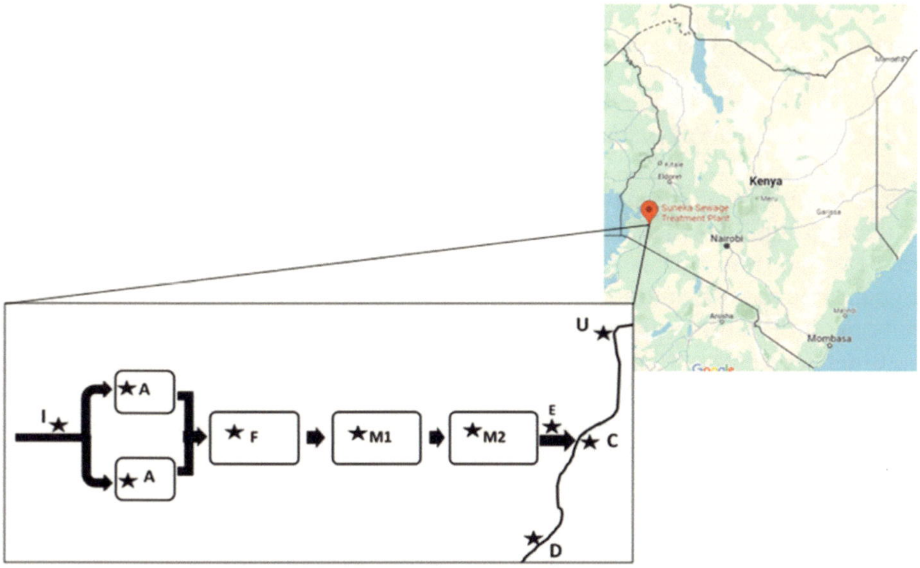

Figure 1.
The study area map. Where: The sketch shows the Suneka WWTP design; the star indicates the sampling points. The alphabet letters indicates: I-Influent; A-Anaerobic pond; F-Facultative pond; M1-Maturation pond 1; M2-Maturation pond 2; E-Effluent; U-Upstream; C-Confluent; D-Downstream.

2.2.1 Phytoplankton sampling procedure

Field sampling was done during morning hours between 8:00 am and 10:00 am once every month during May to August 2021. Nine sampling points were identified, including the Influent (inlet point when wastewater enters the WWTP), two anaerobic ponds, one facultative, and two maturation ponds. The effluent before discharge into the river besides to three points in the river, that is, at the confluent and 100 meters up and downstream at the effluent discharge point into river Riana (**Figure 1**).

2.2.2 Phytoplankton identification and enumeration

Phytoplankton samples were collected at sub-surface level in triplicate by filtering 20 liters of wastewater through 20 μm mesh size plankton net and then the net contents were poured into pre-cleaned well-labeled 500 ml plastic bottles at each sampling station. The collected samples were immediately preserved by adding 2–3 drops of 1% acidic Lugol's iodine solution, and were then kept in a cool box and transported to Kenya Marine and Fisheries Research Institute (KMFRI), Kisumu Centre laboratory for analysis. For heavy metals analysis, the phytoplankton samples collected were preserved by adding 1 ml of concentrated nitric acid. The samples were then clearly labeled, packed, put into an ice-cool box, and transported to the laboratory for analysis. Quality control in laboratory analyses involves various measures to ensure accuracy, proper calibration, and appropriate handling of samples. Accuracy was maintained using certified reference materials, standard additions, and participation in inter-laboratory proficiency testing. Calibration of instruments included regular use of calibration standards, adherence to manufacturer guidelines, and routine maintenance checks. Proper sample handling involved strict protocols for collection, labeling, storage, and processing to prevent contamination and degradation, along with chain-of-custody documentation.

2.2.3 Determination of phytoplankton biomass

The phytoplankton biomass was determined as biovolume (mm^3L^{-1}). For each taxonomic group identified, the cell dimensions such as cell length, width, and depth were measured from at least 20 specimen samples, which were randomly selected. The geometric approximation method was used to calculate biovolume. Algal cell density per unit of the projection area of the colony was estimated at a higher magnification (400X) except for Microcystis colonies, which were measured at a lower magnification of 40X. To estimate the depth of each colony, we used the fine adjustment of the microscope to focus on the top and bottom of each colony. Biovolume was calculated using the formula below [14]:

$$Biovolume\left(\mu g / L\right) = Geometric\ shape \times Number\ of\ counts \times Factor \qquad (1)$$

where Factor is the dimensions of the counting chamber and the microscope at magnification level, that is, 400X.

2.2.4 Phytoplankton diversity indices

To determine the phytoplankton diversity and abundance across the different sampling stations, four diversity indices were computed that is Shannon-Wiener (H′), Margalef's Index (d), Dominance (D), and species evenness) by following formulas according to [18, 19].

Shannon-Weiner diversity index in a given habitat is the uncertainty of identity of unknown individual calculated using the formula below;

$$\text{Shannon} - \text{Wiener Index } \boldsymbol{H'} = \sum\nolimits_i \boldsymbol{PiLniPi} \qquad (2)$$

where Pi is the proportion (n/N) of all the phytoplankton which belong to the ith species, Ln is the natural log and $\sum$ is the sum of the calculation.

In the Shannon-Wiener Diversity Index computation, it is assumed that all species are represented in a sample and are randomly selected. Also, it accounts for both the abundance and evenness of the species present. After calculation, the value obtained normally ranges between 0 and 4. If the index value obtained after calculation is high, then it indicates greater number complexity that is a diverse number of species within a community in the sampling site.

Species richness is the number of species in a species list and is calculated using the following equation [14];

Margalef's Index (d) determined as;

$$d = \frac{\mathrm{s} - 1}{Ln(\mathrm{N})} \qquad (3)$$

Where S is the total number of species, Ln is the Natural log and N is the total number of individuals.

Species evenness is the closeness in numbers of each species in a given habitat. Calculated as follows [14];

Evenness (E) was given as;

$$E = \frac{H'}{\log S} \tag{4}$$

where H' is the Shannon-Wiener diversity index and S is the total number of species.

2.2.5 Heavy metal analysis

The heavy metals that were analyzed include zinc (Zn), lead (Pb), copper (Cu), and cadmium (Cd).

In the laboratory, each phytoplankton sample was filtered through a pre-weighed 42 Whatman GF/C filter paper. The filter papers and the samples were then dried in a horizontal flow oven (Model: Wisd-SWOF 50, 250VAC, 750 W) to a constant weight. The weight of the dried filter paper with sample was then determined using an electronic analytical balance with an accuracy of 0.1 mg (Model: Shimadzu-ATX224). To obtain the phytoplankton's dry weight, the filter paper's initial weight was subtracted from the final weight of the dried filter paper with the sample. The dried filter papers with the samples were then ground into a fine powder using a mortar and pestle. Each obtained powder was then put into separate pre-cleaned beakers.

Wet digestion was conducted using Aquaregia (a mixture of HNO_3/HCl in the ratio 3:1). The digested samples were then filtered using Whatman No. 42 filter paper. The obtained filtrate was transferred into a separate 100 ml volumetric flask. The solution was diluted to the mark with distilled water, and then the samples were transferred into separate plastic bottles, which were clearly labeled and awaiting heavy metal analysis. The blank was prepared identically using plain filter paper following similar steps above, and distilled water was used in the place of a sample. Heavy metal concentration analysis was done using the atomic absorption spectrophotometer model AA 7000 Shimadzu, Japan.

To validate the method used for heavy metal analysis and ascertain the accuracy of the AAS analytical procedure, samples with unknown heavy metal concentrations were spiked with standards of known concentrations, and percentage recovery was determined for the respective heavy metals. Briefly, a sample with an unknown metal concentration was spiked with a known standard metal concentration and then digested. Then, the amount of spiked metal recovered after digestion of the spiked sample was used to calculate the percentage recovery using the formula below:

$$\textit{Percentage recovery} = \left(\frac{\text{Conc.of spiked sample} - \text{conc.of unspiked sample}}{\text{Conc.of spike added}} \right) \times 100 \tag{5}$$

If the calculated values were within 80–120%, they indicated good accuracy for the analysis procedure [20].

2.3 Physico-chemical parameters sampling and analyses

2.3.1 Physical parameters

Wastewater temperature (°C), pH, electrical conductivity (μScm^{-1}), dissolved oxygen concentrations (DO) (mgL^{-1}), and total dissolved solids (TDS) (mgL^{-1}) were measured *in situ* using a calibrated portable professional series (YSI) multi-parameter meter model

35C at sub-surface level. After each measurement, the probes were rinsed thoroughly with deionized water. The readings were taken in triplicates at each sampling station.

2.3.2 Total suspended solids (TSS)

For total suspended solids (TSS) and nutrients (namely nitrites, nitrates, ammonium-nitrogen, total nitrogen, soluble reactive phosphorous, total phosphorous, and Silicates, wastewater samples were collected at sub-surface level in triplicate using pre-washed 500 ml polypropylene plastic bottles rinsed with double distilled water. The filled bottles with wastewater were labeled and immediately kept in a cooler boxer at 4°C and transported to the laboratory for analysis.

For TSS analysis, the filters used to determine TSS were pre-weighed using analytical balance (Model: Shimadzu-ATX224). The filter paper was then transferred to the top of filtration flask, followed by the screwing in of the filtration cap. Wastewater sample measuring 100 ml was put into the filtration flask, followed by switching on the vacuum pump to initiate the filtration process. After filtration process, the filter papers were carefully removed from filtration apparatus using forceps, transferred to a glass weighing dish as a support and dried for 1 hour at 110°C to constant weight in drying oven. This was followed by removing the filter from the petridish and reweighing it. The TSS was calculated using the following formula:

$$TSS\left(\frac{mg}{\mathrm{L}}\right)=\frac{(\mathrm{Residue}+\mathrm{Filter})(mg)-\mathrm{Filter}(mg)}{\mathrm{Sample\ filtered}\,(mL)}\times x\,1000\left(\frac{mL}{L}\right) \tag{6}$$

2.3.3 Chemical parameters

Nutrient analysis was done using the spectrophotometric method to determine water and wastewater as described in APHA (2014). The analyzed nutrients included total nitrogen (TN), and total phosphorous (TP).

2.4 Data analyses

Microsoft Excel version 2010 was used to organize the phytoplankton and physico-chemical data obtained from the different sampling stations. Paleontological Statistics (PAST) software was used to determine phytoplankton diversity indices. Spatial variations in the physico-chemical parameters were determined by two-way analysis of variance (ANOVA) at a pre-determined alpha value of 0.05 to test for significant differences. Where the variations in means were significant, post hoc analysis was done using the Tukey pairwise comparisons under SPSS version 22 to establish where the differences existed between the sampling stations. To handle potential confounding factors, random sampling ensured representative data, and multiple samples per station provided reliable estimates, minimizing selection bias and accounting for natural variability.

3. Results

3.1 Phytoplankton diversity and species composition

A summary of phytoplankton species recorded in the Suneka WWTP are presented in **Table 1**. The family Bacillariophyceae had the highest number of species

DOI: http://dx.doi.org/10.5772/intechopen.1006276

(37), consisting of 33% by species composition, followed by Chlorophyceae, which was represented by 29 species consisting of 26% by species composition. The family Cyanophyceae was represented by 22 species, which led to a 20% species composition. Other taxonomic families included Zygnemophyceae, Euglenaphyceae, and Dinophyceae represented by 11 (10%), 10 (9%), and 3 (3%) species, respectively (**Table 1**).

Chlorophyceae	*Cyanophyceae*	*Bacillariophyceae*
Ankistrodesmus falcatus	*Anabaena flos-aquae*	*Amphora ovaris*
Botryococcus braunii	*Anabaenopsis tanganyikae*	*Amphora* sp.
Coelastrum microporum	*Aphanocapsa pulchra*	*Aulacoseira ambigua*
Crucigenia sp	*Aphanocapsa rivularis*	*Aulacoseira nyasensis*
Dictyosphaerium reniforme	*Aphanothece* sp	*Aulacoseira schroidera*
Kirchnella contorta	*Chodatella armata*	*Chodatella subsalsa*
Kirchnella lunaris	*Chroococcus dispersus*	*Chodatella armata*
Monoraphidium sp	*Chroococcus limnetica*	*Cyclotella kutzinghiana*
Oocystis parva	*Chroococcus turgidus*	*Cymbella* sp
Pediastrum boryanum	*Cylindrospermopsis africana*	*Cymbella cistula*
Pediastrum duplex	*Merismopedia tennuissima*	*Diatoma elongatum*
Scenedesmus curvatus	*Microcystis aeruginosa*	*Diatoma hemiale*
Scenedesmus acuminatus	*Microcystis wasenbergii*	*Diatoma vulgare*
Scenedesmus obliquus	*Oscillatoria tenuis*	*Navicula* sp
Schroidera setigera	*Plankolyngbya tallingii*	*Navicula contenta*
Tetraedron arthromisforme	*Planktolyngbya circumcreta*	*Navicula exicilis*
Tetraedron trigonum	*Planktolyngbya limnetica*	*Nitzschia dissipata*
	Planktolyngbya talingii	*Nitzschia lacustris*
Dinophyceae	*Romeria ankensis*	*Nitzschia palea*
Ceratinium branchycerous	*Romeria elegans*	*Nitzschia recta*
Glenodinium pernardii		*Nitzschia sub acicularis*
Glenodinium pulvastitus		*Stephanodiscus astraca*
	Zygnematophyceae	*Stephanodiscus* sp
Euglenaphyceae	*Coelomoron merostoides*	*Surirella affins*
Euglena acus	*Cosmarium* sp	*Surirella ovalis*
Euglena virids	*Closterium navicula*	*Surirella* sp
Euglenaphytalena acus	*Cosmarium menenghiana*	*Synedra cunningtonii*
Euglenaphytalena virids	*Crucigenia menenghiana*	*Synedra ulna*
Phacus longicauda	*Cosmarium connatum*	
Phacus sp	*Cosmarium cunningtonii*	
Strombomonas sp.	*Straurastrum dickiei*	
Trachelomonas armata		
Trachelomonas volvocina		

Table 1.
A list of phytoplankton taxa found in the current Suneka WWTP.

Shannon-Wiener diversity index (H′) ranged from 1.18 at the effluent station to 3.31 at the downstream sampling station. The maximum dominant index (D) of 0.63 was recorded at the effluent station while the downstream sampling station had the least (0.05). In terms of Margalef's diversity, that is, species richness (d), the effluent was richer (with a value of 5.09), while the facultative pond had the least (3.96). On the other hand, the Evenness (E), ranged from 0.07 to 0.61 (**Table 2**).

3.2 Phytoplankton abundance

The total phytoplankton biovolume for the WWTP was 680.99 mm^3L^{-1} with a mean of 113.5 ± 127.8 mm^3L^{-1}. The family Chlorophyceae was contributing to 45.7% while Dinophyceae contributed the least with 0.4% of the total phytoplankton biovolume. The family Euglenaphyceae contributed to 33.9%, followed by Bacillariophyceae (12.9%), then Zygnematophyceae with 4.2% of the total phytoplankton biovolume. The family Cyanophyceae contributed with 3.0% of the total phytoplankton biovolume (**Table 3**).

3.3 Spatial variation

The results obtained on the total phytoplankton biovolume depicted variation between the sampling stations in the WWTP pond series. The facultative pond had the highest total phytoplankton biovolume by composition with 21.01%, followed by anaerobic pond with 8.47%. The effluent had a slightly higher total phytoplankton biovolume with 6.12% compared with the influent sampling station (4.33%) (**Table 4**). The systematic increase in the total algal biovolume from the influent to the facultative pond indicates availability of nutrients as a result of biological breakdown of nutrients, which in turn promotes algal growth.

The results of each family's composition in the WWTP sampling stations are shown in **Figure 2**. Bacillariophyceae dominated the maturation of pond 1 and downstream sampling stations, indicating better water quality resulting from wastewater polishing. Chlorophyceae dominated the upstream sampling station, while *Cyanophyceae* recorded a low total phytoplankton biovolume in most of the sampling stations. Similarly, *Dinophyceae* was only recorded in facultative pond, maturation pond 2, effluent, and downstream sampling stations with low total biovolume. *Euglenaphyceae* family recorded the highest biovolume at the facultative pond compared to other sampling stations.

Spatial changes in the composition of the different algal taxa in the WWTP pond series are evidence of progressive wastewater polishing in the WWTP pond series. Also, nutrients are available in the water column for algae to utilize for growth following the degradation of organic matter and other substances in wastewater. Moreover, changes in the environmental conditions within the WWTP pond series encouraged the growth and dominance of different algal taxa. In the WWTP ponds, algal blooms with different colors were observed on the water surface but were more pronounced in the facultative pond. Algal blooms have been associated with the presence of algal toxins. Moreover, Cyanophytes, especially the *Anabaena* sp. and *Microcystis* sp., attest to this fact and release algal toxins that are toxic to animals, consequently of great concern.

	Influent	Anaerobic pond	Facultative pond	Maturation pond 1	Maturation pond 2	Effluent	Confluent	Upstream	Downstream
Taxa (s)	28	47	32	37	43	44	34	34	45
Individuals	677	1659	2497	470	2587	4689	723	645	934
Shannon_H	2.69	2.96	1.74	2.29	2.61	1.18	3.04	3.02	3.31
Dominance_D	0.09	0.10	0.30	0.15	0.14	0.63	0.07	0.07	0.05
Margalef (Species richness)	4.14	6.21	3.96	4.26	5.35	5.09	5.01	5.10	6.43
Evenness_e^H/S	0.53	0.41	0.18	0.27	0.32	0.07	0.61	0.60	0.61

Table 2.
The phytoplankton diversity indices of the Suneka WWTP.

Taxonomic group	Phytoplankton biovolume (mm^3L^{-1})	Percentage biovolume (%)
Bacillariophyceae	87.56	12.7
Chlorophyceae	310.96	45.7
Cyanophyceae	20.44	3.0
Dinophyceae	2.57	0.4
Euglenaphyceae	230.90	33.9
Zygnematophyceae	28.55	4.2
Grand Total	**680.99**	**100**

Table 3.
Phytoplankton biovolume of water samples from the Suneka WWTP.

Sampling stations	Phytoplankton Biovolume (mm^3L^{-1})	Percentage (%)
Influent	29.49	4.33
Anaerobic pond	57.71	8.47
Facultative pond	143.04	21.01
Maturation pond 1	32.25	4.74
Maturation pond 2	34.39	5.05
Effluent	41.67	6.12
Confluent	29.84	4.38
Upstream	278.09	40.84
Downstream	34.52	5.07

Table 4.
Spatial variations of phytoplankton biovolume of water samples from the Suneka WWTP.

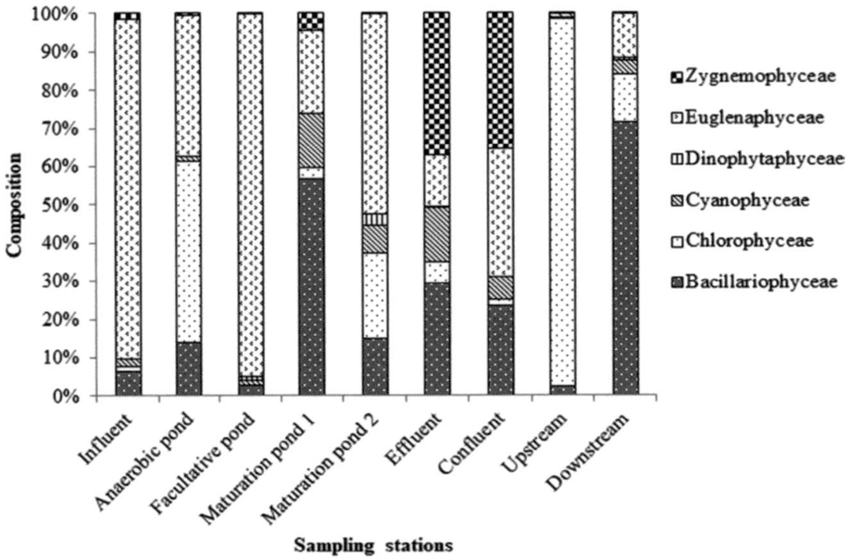

Figure 2.
Relative abundance of phytoplankton taxa in sampling stations in the Suneka WWTP.

3.4 Heavy metal concentrations in phytoplankton

Figure 3 shows the mean (± SE) spatial variation of heavy metal concentrations (in ppm) in phytoplankton samples collected from the WWTP including three sampling stations along river Riana.

Throughout the study period, Cd concentrations were below the detection limit (<BDL) in all phytoplankton samples analyzed.

The mean Cu concentration in phytoplankton from the sampling stations ranged from 0.25 ± 0.07 ppm to 0.38 ± 0.05 ppm. The phytoplankton collected from the upstream sampling station had the highest mean Cu concentration while the anaerobic pond had the least mean concentration. One-way ANOVA test showed that the mean Cu concentrations in phytoplankton were not significantly different among the sampling stations ($F_{(8, 27)} = 0.648$; $p = 0.731$).

For Zn, the average phytoplankton concentration range was between 0.33 ± 0.04 ppm and 0.21 ± 0.03 ppm. The confluent sampling station's concentration was below the detection limit (<BDL) in the phytoplankton samples. One-way ANOVA test for the Zn mean concentration difference in phytoplankton samples was insignificant among the sampling stations ($F_{(7, 8)} = 1.489$; $p = 0.293$).

The mean Pb concentration ranged from 1.17 ± 0.01 ppm to 0.27 ± 0.004 ppm. The downstream had the highest Pb mean concentration (1.17 ± 0.01 ppm) while the influent had the least mean Pb concentration (0.27 ± 0.00 ppm). Moreover, Pb mean concentration was below the detection limit in the influent sampling station. One-way ANOVA test showed that mean Pb concentrations were significantly different among the sampling stations ($F_{(7, 8)} = 21.731$; $p = 0.000$). *Post hoc* Tukey Pairwise Comparisons revealed that the mean Pb concentration for phytoplankton samples from the anaerobic station was lower by 0.45 ppm, 0.41 ppm, and 0.5 ppm than those in the effluent, upstream, and downstream sampling stations, respectively (**Table 5**).

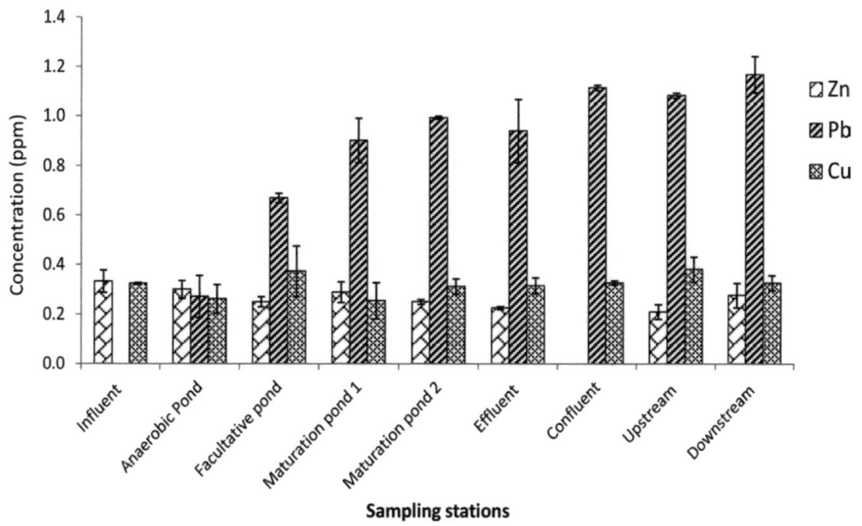

Figure 3.
Spatial variation of heavy metal concentrations (in ppm) in phytoplankton samples from Suneka WWTP.

Parameters	Sampling station								
	Influent	Anaerobic pond	Facultative pond	Maturation pond 1	Maturation pond 2	Effluent	Confluent	Upstream	Downstream
pH	7.33 ± 0.01	6.9 ± 0.05	7.77 ± 0.13	7.75 ± 0.07	7.72 ± 0.06	7.77 ± 0.06	7.64 ± 0.07	7.45 ± 0.08	7.27 ± 0.1
EC (μScm^{-1})	1097.75 ± 128.92^{a}	752.75 ± 83.95^{b}	704.92 ± 48.92^{b}	676.83 ± 46.79^{b}	658.25 ± 46.92^{b}	665.58 ± 41.17^{b}	295.67 ± 18.67^{c}	115.75 ± 2.27^{d}	146.67 ± 10.23^{d}
Temp (°C)	22.1 ± 0.34^{c}	25.2 ± 0.21^{b}	25.2 ± 0.4^{b}	26.1 ± 0.21^{b}	27.4 ± 0.3^{a}	26.1 ± 0.16^{b}	22.6 ± 0.3^{c}	22.6 ± 0.5^{c}	21.9 ± 0.3^{c}
DO (mgL^{-1})	0.2 ± 0.04^{f}	0.6 ± 0.09^{e}	3.5 ± 1.08^{a}	$3.0 \pm 0.81^{b,c}$	2.2 ± 0.65^{d}	2.7 ± 0.76^{c}	$3.3 \pm 0.71^{a,b}$	3.3 ± 0.59^{a}	$3.3 \pm 0.67^{a,b}$
TSS (mgL^{-1})	65.34 ± 7.4^{c}	57.88 ± 4.3^{d}	45.78 ± 7.9^{e}	65.38 ± 9.9^{c}	53.19 ± 5.5^{e}	77.16 ± 6.0^{a}	35.22 ± 2.6^{f}	73.33 ± 14.9^{b}	79.78 ± 13.2^{a}
TDS (mgL^{-1})	$438.3 \pm 79.5^{a,b}$	400.3 ± 57.0^{a}	$267.9 \pm 46.9^{b,c}$	260.3 ± 45.7^{c}	212.1 ± 51.6^{c}	259 ± 42.5^{c}	59.9 ± 17.8^{d}	107.5 ± 1.9^{d}	67.1 ± 6.7^{d}
TN (μgL^{-1})	$236.2 \pm 28.84^{c,d}$	185.1 ± 30.05^{e}	276.4 ± 45.32^{b}	249.6 ± 69.91^{c}	199.5 ± 30.11^{e}	390.7 ± 64.88^{a}	223.3 ± 88.45^{d}	226.6 ± 39.7^{d}	110.1 ± 21.93^{f}
TP (μgL^{-1})	801.0 ± 300.12^{e}	2742.0 ± 206.26^{a}	$2411.0 \pm 399.64^{a,b}$	$1844.0 \pm 352.98^{c,d}$	1628.0 ± 335.74^{d}	2557.0 ± 172.55a	$2272.0 \pm 154.14^{a,b}$	$2378.0 \pm 415.28^{a,b,c}$	$1943.0 \pm 358.41^{b,c,d}$

Where: Different superscript letters signify that the means differ significantly ($p < 0.05$).

Table 5.
Mean (±SE) spatial variation of physico-chemical parameters in the sampled stations in Suneka WWTP.

3.5 Physico-chemical parameters

The mean pH values recorded ranged from weakly acidic to alkaline level that is from 6.58 to 8.40. The influent sampling station had a neutral mean PH of 7.33 ± 0.08, which was lower than the effluent station (7.77 ± 0.06), indicating an improvement in wastewater polishing. Among the wastewater stabilizing ponds, the facultative pond had the highest mean pH of 7.77 ± 0.13. There was a general increase in the mean pH as the wastewater underwent polishing between the influent and effluent, a further indication of an improvement in wastewater quality.

The mean (± SE) conductivity value recorded was 568.24 μScm^{-1} ± 34.97, and the recorded values ranged between 99.0 and 2134.0 μScm^{-1}, with the influent sampling station recording the highest mean value of 1097.75 μScm^{-1} ± 128.92 while the upstream point recorded the least mean value of 115.75 μScm^{-1} ± 2.27. These results indicate a considerable declining trend in the mean conductivity from the influent to effluent as the wastewater underwent polishing. Two-factor ANOVA showed that differences in the mean conductivity at different sampling stations were statistically significant ($F_{(8, 108)} = 210.15$; $p = 0.000$). *Post hoc* Tukey Pairwise Comparisons revealed that the conductivity mean for the influent (1097.75 ± 128.92 μScm^{-1}) station was significantly higher than that of the anaerobic station (752.75 ± 83.95 μScm^{-1}). However, anaerobic pond sampling station mean conductivity was not significantly different from that of facultative (704.92 ± 48.92 μScm^{-1}), maturation pond 1 (676.83 ± 46.79 μScm^{-1}), maturation pond 2 (658.25 ± 46.92 μScm^{-1}), and effluent (665.58 ± 41.17 μScm^{-1}) stations.

For temperature, the mean value recorded was 24.36°C ± 0.21 with a minimum value of 20.0°C and maximum value of 28.5°C. The maturation pond 2 had the highest mean temperature of 27.36 ± 0.3°C closely followed by maturation pond 1 with 26.1 ± 0.21°C then by effluent sampling station with 26.09 ± 0.16°C. Like pH, there was an increase in the mean temperature from the influent to the effluent as wastewater underwent polishing through the WWTP. Two-factor ANOVA showed that the mean temperature among the sampling stations differed significantly ($F_{(8, 108)} = 73.21$; $p = 0.000$). *Post hoc* Tukey Pairwise Comparisons revealed that the mean temperature of the maturation pond 2 (27.4 ± 0.3°C) was significantly higher than that of the other sampling stations. The influent station's mean temperature did not differ significantly from that of the confluent, upstream, and downstream sampling stations. Similarly, the mean temperature of the anaerobic station did not differ significantly from that of the facultative, maturation pond 1, and effluent sampling stations.

The recorded mean (± SE) DO was 2.45 mgL^{-1} ± 0.24 with a minimum value of 0.1 mgL^{-1} and maximum value of 10.0 mgL^{-1}.). In terms of trend, in general, there was an increase in the mean of dissolved oxygen concentration between the influent through the WWTP system to the effluent sampling station similar with the trends for pH, and temperature. Two way ANOVA indicated there were significant differences between the sampling stations in the mean dissolved oxygen concentrations ($F_{(8, 108)} = 227.8$; $p = 0.000$). *Post hoc* Tukey Pairwise Comparisons revealed that the mean dissolved oxygen of the influent (0.23 ± 0.2 mgL^{-1}) was significantly lower compared with those of other sampling stations.

The mean (± SE) TSS concentration recorded for the sampling stations was 61.45 ± 3.14 mgL^{-1} with a minimum and maximum values of 7.76 and 132.0 mgL^{-1} respectively. The highest mean of TSS was recorded in the downstream sampling station with 79.78 ± 13.2 mgL^{-1}. The confluent sampling station had the lowest mean of 35.22 ± 2.62 mgL^{-1}. In terms of trend, TSS mean concentrations fluctuated with no significant trend between the inlet and effluent. Two way ANOVA indicated significant differences in the mean TSS values between the sampling stations ($F_{(8, 108)} = 551.5$; $p = 0.000$). *Post hoc* Tukey Pairwise Comparisons revealed that the mean TSS of the influent station (65.34 ± 7.4 mgL^{-1}) was significantly lower from that of the effluent (77.16 ± 6.0 mgL^{-1}), but the latter mean TSS was not significantly different from that of the downstream sampling station (79.78 ± 13.2 mgL^{-1}).

The mean (± SE) TDS recorded was 230.3 ± 18.35 mgL^{-1} with minimum and maximum values of 18.9 and 10.68 mgL^{-1} respectively. The influent and anaerobic sampling stations had the highest mean values of 438.3 ± 79.5 and 400.3 ± 57.0 mgL^{-1}, respectively. The downstream and upstream stations had the lowest mean values of 67.1 ± 6.7 and 59.9 ± 1.9 mgL^{-1}, respectively. In terms of trend, there was a steady decline in mean TDS concentration between the influent and effluent sampling stations, indicating wastewater polishing. Two-way ANOVA indicated significant differences between the sampling stations ($F_{(8, 108)} = 32.48$; $p = 0.000$). *Post hoc* Tukey Pairwise Comparisons revealed that the mean of the influent station (438.3 ± 79.5 mgL^{-1}) was higher and significantly different from that of the effluent (259.0 ± 42.5 mgL^{-1}). The latter had no significant difference with that of maturation ponds 1 and 2.

The mean TN concentration recorded was 233.1 ± 17.83 μgL^{-1} with minimum and maximum values of 22.0 and 745.2 μgL^{-1}. The effluent sampling station had the highest mean TN concentration of 390.7 ± 64.88 μgL^{-1} while the downstream sampling station had the lowest mean of 110.1 ± 21.93 μgL^{-1}. In terms of trend, the mean TN concentration showed no spatial trend between the influent and effluent sampling stations. Two-way ANOVA showed that TN significantly differed among the sampled stations ($F_{(8, 108)} = 344.62$; $p = 0.000$). *Post hoc* Tukey Pairwise Comparisons revealed that the mean TN concentration of influent station (236.2 ± 28.84 μgL^{-1}) differed significantly from the effluent station (390.7 ± 64.88 μgL^{-1}), while the upstream station (226.6 ± 39.7 μgL^{-1}) differed significantly with the downstream (110.1 ± 21.93 μgL^{-1}) station.

The mean TP concentration recorded was 2064 ± 114.10 μgL^{-1} with minimum and maximum values of 20.69 and 3710.20 μgL^{-1}. The anaerobic station had the highest mean TP concentration of 2742 ± 206.26 μgL^{-1} while the influent had the least mean of 801 ± 300.12 μgL^{-1} which was lower compared to the effluent station which had mean of 2557 ± 172.55 μgL^{-1}. In terms of trend, there was fluctuation in mean TP concentrations with no significant trend spatially similar to the mean trends for TSS, silicates, nitrate-nitrogen, and TN. Two-way ANOVA showed that TP concentrations significantly differed among the sampling stations ($F_{(8, 108)} = 32.23$; $p = 0.000$). *Post hoc* Tukey Pairwise Comparisons revealed that the mean TP concentration of influent station (801 ± 300.12 μgL^{-1}) differed significantly with the effluent station (2557 ± 172.55 μgL^{-1}).

4. Discussions

4.1 Phytoplankton

4.1.1 Phytoplankton diversity and species composition

Phytoplankton are aquatic plants and primary producers providing food to other aquatic organisms [13]. However, their community structure is affected by water quality; as a result, they have been used as bio-indicators in water quality monitoring. The cyanobacteria *Microcystis* sp. and the Diatom *Nitzchia* sp. are good indicators of polluted water. *Chlamydomonas*, *Euglena*, *Scenedesmus*, and *Microcystis* are good indicators of eutrophic waters. Wastewater is of poor quality as it is rich in pollutants, and phytoplankton plays an important role in wastewater treatment through the utilization of nutrients and the conversion of them into biomass [9, 16, 21]. In addition, they bioconcentrate and bioaccumulate heavy metals, thus contributing to their removal from the water column [22]. The current study showed a considerable spatial variation in the phytoplankton abundance and diversity between the sampling stations. These variations were linked to physical, chemical, and biological environmental changes as the wastewater underwent treatment.

During this study, a total of 112 phytoplankton species were recorded, unlike previous study carried out at the plant by Rayori et al. [17], who recorded a total of 124 phytoplankton species. Nevertheless, the identified species in both studies belonged to six taxonomic groups: *Euglenaphyceae*, *Bacillariophyceae*, *Dinophyceae*, *Cyanophyceae*, *Chlorophyceae*, and *Zygnematophyceae* despite changes in the wastewater capacity of the Suneka WWTP. Moreover, the families *Bacillariophyceae*, *Chlorophyceae*, and *Cyanophyceae* generally dominated the phytoplankton species by composition in both studies. The differences in the total number of species recorded between the studies can be attributed to the difference in the wastewater treatment plant designs due to the renovation and expansion of the WWTP, environmental conditions, and seasonality. Also, human activities during renovation disturbed the establishment of phytoplankton species in the latter design. These findings corroborate the findings obtained by Babu et al. [23] in L. victoria and Omondi [14] in L. Naivasha where both showed that algal community structures are affected by varying physico-chemical parameters.

The *Euglenaphyceae* was dominant in the influent, facultative, and the effluent while the *Bacillariophyceae* dominated in the anaerobic pond. The dominance of *Euglenaphyceae* can be attributed to organic load from the municipal wastewater. On the other hand, the cyanophytes were moderate in all stations, which could be attributed to their tolerance to variable changes in the environmental conditions. For instance, this was seen in *Microcystis aeruginosa* and *Anabaena* spp. which were found in the sampling stations. The low diversity was attributed to extreme environmental conditions observed in the sampling stations, in line with similar studies in polluted aquatic environments [23, 24].

The different sampling stations recorded variations in phytoplankton diversity indices. In the current study, the Shannon-Wiener diversity index was higher than in the previous design, indicating an improvement in biodiversity. The low diversity of species in the initial design can be associated with the fact that the well establishment species competed with other species leading to dominance of some as depicted by the dominance indices. The phytoplankton species diversity index (H′) during this study was generally low, indicating low diversity but with a few taxa dominating. Species Evenness (E), values ranged between 0.07 and 0.61, depicting a significant difference and an improvement of the species evenness in the WWTP. Generally, the low evenness values recorded in the WWTP that a few species dominated in the sampling stations, can be linked to harsh conditions in the wastewater treatment lagoons; hence the species were not evenly distributed in the sampling stations.

4.1.2 Phytoplankton biomass and distribution

In terms of total phytoplankton biovolume percentage, the *Euglenaphyceae* and *Dinophyceae* were dominant before the WWTP was renovated [17] while *Chlorophyceae* and *Euglenaphyceae* were the most dominant after renovation of the

WWTP. Therefore, it appears that phytoflagellate and zooflagellates were the most dominant taxa in the WWTP. These are indicators of poor-quality water. This dominance can be attributed to favorable environmental conditions favoring optimum survival of the two taxa. This seemingly led to competitive exclusion of other species. Moreover, there were changes in the composition of different algal taxa in the lagoon pond series, indication of progressive wastewater polishing in the WWTPs. Evidence that the WWTPs polished the effluent was discerned by comparing the algal composition within the WWTP pond series with that of upstream sampling station along river Riana, which in this case served as the control and indicated that *Bacillariophyceae* were the dominate taxa there. *Bacillariophyceae* are indicators of good water quality. The increasing representativeness of *Bacillariophyceae* along the pond treatment series toward the effluent attests to the fact that ponds were treating the wastewater. This also indicates the progressive degradation of organic matter in wastewater thus releasing nutrients into the water column. Other breakdowns of organic substances lead to variation in the environmental conditions within the water treatment pond series. The environmental variation encourages dominance of different algal taxa in the different ponds that is from anaerobic, facultative, and maturation ponds.

The anaerobic pond had the highest total phytoplankton biovolume by composition of *Dinophyceae*, with the dominant species being *Ceratium branchyceros*. This was attributed to extended mixing periods and resident time in the anaerobic pond. In the facultative pond, *Euglenaphyceae* and *Dinophyceae* groups had a higher biovolume. *Euglenaphyceae*, *Euglena virids*, and *Euglenaphyta lena acus* recorded the highest biovolume while for *Dinophyceae*, only *Ceratium branchyceros* species were recorded. This could be attributed to the availability of nutrients that were dislodged from the organic load, hence making them available for utilization, and reduced turbidity, which enhances transparency and leads to higher photosynthetic activity in the pond. The least phytoplankton total biovolume was recorded in the downstream sampling station. This can be attributed to the washing down of the species by water currents, increased turbidity leading to elevated temperature, and an aspect of competition leading to mutual exclusion. This could further be due to the dilution effect of the large volume of water in River Riana, which changed the physical and chemical parameters such as nutrient concentrations and transparency, which therefore could encourage a change in species composition.

The highest phytoplankton total biovolume was recorded in the facultative pond sampling station. The algal composition was dominated by *Euglenaphyceae* by biovolume, among other groups, of which *Euglena acus* species was the most abundant. Moreover, *Euglenaphyceae* dominated the influent, anaerobic pond, and maturation pond 2, of which *Euglena acus* species accounted for the highest biovolume. The dominance of *Euglenaphyceae* (*Euglena acus*) can be attributed to higher organic and inorganic loads in the sampling ponds. The differences in the total phytoplankton biomass and species distribution in the sampling stations were attributed to the variable unfavorable conditions in the lagoons and operational and design differences of the WWTP [14, 23, 24]. This corroborates the study findings by Babu et al. [23], which showed significant levels of Euglenaphyta in Kisumu Bay in which receives a large volume of semi-treated sewage wastewater from both the traditional wastewater treatment trickling filter plant adjacent to Kasat River, and the WWTP at Nyalenda. Due to this, the water quality in Kisumu Bay has rapidly declined in the last three decades, between 1990 and the present, and the physical and chemical environment there is very close to that existing in the WWTPs. The algal composition in the inner Kisumu Bay is dominated by Cyanophyceae, mainly by *Anabaena* sp., and *Microcystis aeruginosa* species, and presently, *Euglenaphyceae* have started to appear there. In general, the results obtained from this study indicate that the phytoplankton community structure was influenced by the variation in the physico-chemical parameters in line with findings in other studies [14, 23–25].

During the study, algal blooms with different colors throughout the ponds on the water surface were observed. These algal blooms are an indicator of the presence of algal toxins in the WWTP. The predominant algae species observed in this study such as *Microcystis aeruginosa* and *Anabaena* sp. together with other species from other taxa are known to produce potent algal toxins known as microcystins. The produced algal toxins can cause serious illness or death to humans, wildlife, livestock, and other forms of life in aquatic ecosystems. Babu et al. [23], in their study at Kisumu Bay in Lake Victoria, observed algal toxins were in high concentration where algal blooms were present. Therefore, there is a need for other studies to be conducted in the Kisii municipality WWTP to assess the presence, concentrations, and types of algal toxins.

4.1.3 Phytoplankton heavy metals accumulation

This study revealed the presence of Cu, Pb, and Zn, but Cd was not recorded throughout the study period. Moreover, the study revealed accumulation of heavy metals in phytoplankton which might lead to biomagnification of heavy metals in the food web, posing health hazards [26].

In phytoplankton samples, the order of dominance of the assessed heavy metals followed Pb > Cu > Zn. This could be associated with the phytoplankton's preferential uptake and utilization of the respective constituents. These findings are similar to other studies showing significant variations in metal levels in water column, sediments, and plankton. The bioaccumulation of heavy metals by phytoplankton reduced the heavy metal concentration in wastewater column similar to other studies [22].

4.2 Physico-chemical parameters

4.2.1 pH

The recorded mean pH values of the effluent were significantly higher than that of the influent; that is, the influent pH was more acidic than that of effluent which was more alkaline. The low pH value was attributed to the high organic load in the sewage wastewater, and the process of bio-treatment led to pH increase in the effluent. For instance, dense algal growth increases pH due to the assimilation of nitrates and carbon dioxide. This holds as during the study, there was increase in algal biomass in the lagoons. These findings are in agreement with observations by Ronoh [27] in a study conducted at Moi University WWTP and Wanjohi et al. [8] in his study conducted at the University of Eldoret WWTP, where they both observed an increase in pH between the influent and the effluent. Effluent discharge with extreme pH ranges can harm aquatic organisms as most biochemical reactions are catalyzed by enzymes that act within small pH ranges.

4.2.2 Conductivity

The mean conductivity values for the influent were higher than those of the effluent; that is, it depicted a general declining trend from the influent point to the discharge point. This was attributed to decline in dissolved salts and heavy metals concentrations responsible for conductance as the wastewater underwent treatment. Also, the higher conductivity value recorded in the influent can be attributed to high inorganic load as the water enters the ponds. On the other hand, the low conductivity value recorded at the effluent can be attributed to the utilization of dissolved nutrients by phytoplankton found in the lagoons. This is evidenced by the increasing trends of algal biovolume from the inlet to maturation ponds. These findings corroborate well with those obtained from the University of Eldoret WWTP [8]. The upstream sampling point on River Riana recorded a lower mean conductivity value during the entire sampling period than the downstream sampling point. This can be attributed to minimal human activity upstream while the downstream can be attributed to the treatment plant, which acts as a source of pollution. This corroborates well with the studies done by Omondi [14] and Babu et al. [23] in L. Naivasha and L. Victoria, respectively, whereby they observed polluted sampling sites in the respective lakes recorded higher conductivity.

4.2.3 Temperature

The recorded mean temperature of discharged effluent into Riana River was within the ranges suitable for wastewater treatment and within the optimal ranges of microbial activities responsible for organic and inorganic substance breakdown. In addition, the recorded temperature ranges were optimal for algal growth; hence, lagoons had diverse communities of algae. In turn, the algae convert the nutrients into biomass, thus polishing the wastewater in the lagoons. The temperature recorded for WWTP, showed a slight increase in the mean temperature from the influent to effluent sampling stations. These changes were attributed to harsh environmental conditions (in the anaerobic ponds), sampling time, and the dimensions of the stabilization ponds with shallow mean depth of the ponds. Moreover, the increasing trends of algal biovolume in the lagoons contributed to increased suspended solids that absorb solar energy and retain their heat for longer periods.

The increase in temperature in wastewater in the sampling stations also influenced the solubility of compounds, which was evidenced by the spatial variation of heavy metal concentrations in wastewater during the study. Ronoh [27], in his study at Moi University wastewater treatment plant, and Wanjohi et al. [8], in the University of Eldoret wastewater treatment plant both observed decline in mean temperature between the influent and the effluent sampling points. The high temperatures were attributed to exothermic reactions during breakdown of organic and inorganic matter in the wastewater.

Also, domestic wastewater has a higher temperature than that of maturation and tertiary ponds because large institutions within Kisii such as hospitals, schools and polytechniques could be discharging effluents with high temperature due to their cooking ventures. Also, domestic households with domestic hot water instant showers contribute to effluent with high temperatures at the influent at the WWTP. High temperature is suitable for organic matter decomposition in the ponds as they are shallow.

4.2.4 Dissolved oxygen (DO)

Dissolved oxygen percentage increase in wastewater indicated that the quality of wastewater improved as it underwent polishing through the lagoons before being discharged as expected. Anaerobic ponds had least mean DO. The significantly low levels of DO in the influent and anaerobic pond were attributed to the decomposition of sewage rich in organic matter by microbes that required oxygen. The increase of DO concentrations in the facultative pond was linked to photosynthetic algal activities. The high level of DO in the effluent was linked to the reduction of organic matter in the maturation ponds and exposure of wastewater in the ponds to atmospheric oxygen resulting from mixing brought about by strong winds blowing above the ponds. With reference to Riana River, for both initial and current designs, higher DO mean levels recorded at the upstream sampling station can be attributed to lower temperature levels and lower organic matter while the lower levels at the downstream stations can be attributed to higher organic loads from the treatment plant, and higher sediments; thus absorbing solar radiation leading to higher temperature resulting low DO concentrations. The current findings corroborate the findings of Ronoh [27] in the Moi University sewage treatment plant and other similar studies [23, 28]. Monthly, the mean DO showed a significant difference between the sampling months for both wastewater treatment plant designs. These differences were attributed to seasonality, where water temperature plays a role in DO solubility [14, 29, 30]. Also, low DO concentrations can be attributed to variable levels of organic matter and differential flow rates of sewage water from the municipality.

4.2.5 Total suspended solids (TSS)

The recorded mean TSS concentrations for the influent were relatively higher than in the effluent. The reduction was attributed to removal of suspended solids in wastewater column by algae and microorganisms and sedimentation of settleable solids forming sludge. On the other hand, the availability of nutrients in the lagoons promoted algae growth, which, in turn, contributed to increased suspended solids in the lagoons, including the effluent discharged. The upstream sampling station had lower mean TSS concentrations than the downstream sampling station, which was attributed to lower suspended particulate matter. However, the higher mean TSS concentrations at the downstream sampling station can be attributed to the treatment plant's organic load discharge into the river. The study findings corroborate well with findings by Ronoh [27], who noted in their study that there was a significant difference in mean TSS values of samples collected from influent compared with effluent from the sewage treatment plant, and the outlet did not meet NEMA and Moi University effluent discharge standards. Also, surface runoff contribute to increased amount of inorganic and organic matter in rivers, leading to increased TSS levels [31].

4.2.6 Total dissolved solids (TDS)

Total dissolved solids steadily declined in mean concentration between the influent and effluent stations in the WWTP. This reduction indicates significant polishing of wastewater by the WWTP. High mean TDS concentrations recorded at the influent sampling stations were linked to higher organic and inorganic materials load from municipal wastes. In the anaerobic pond, progressive decomposition of organic and inorganic matter in wastewater resulted in simpler, harmless soluble substances (TDS). Reduction in the mean TDS levels through the WWTP pond series toward the effluent sampling station was associated with phytoplankton nutrient uptake and utilization and settlement of both organic and inorganic particulate matter forming sludge. This is in line with previous studies showing that settling the inorganic and organic substances in the stabilization ponds results in TDS concentration reduction while the substances form part of sludge [27]. The higher rates of evaporation reduced the volume of wastewater in the WWTP, thus concentrating dissolved solids in wastewater into smaller volumes. This study's findings are consistent with other findings [27, 30].

4.2.7 Nutrient concentrations

The Suneka WWTP did not attenuate TN between the influent and the effluent. The influent had lower mean TN concentrations than the effluent, and these study findings were contrary to Ronoh et al.'s [27] in Moi University WWTP. The low levels of TN in the influent were attributed to raw sewage rich in lodged nitrogen compounds in organic material yet to be broken down. The upstream sampling station recorded a higher mean TN (223.3 ± 39.70 μgL^{-1}) than that of the downstream sampling station (110.1 ± 21.93 μgL^{-1}), an indication that the river had other sources of TN in addition to the WWTP. The discharged effluent means for TN concentrations did not exceed the allowable limits by EPA standards. Therefore, improvement should be geared toward increasing the number of ponds so that the WWTP can be effective in TN removal. TN is important in eutrophication and water quality assessment.

The mean value of TP in the effluent (2557.0 ± 172.55 μgL^{-1}) was higher than that of the influent (801.0 ± 300.12 μgL^{-1}), recording a 219% increase in mean TP concentration and this implied inefficiency of TP removal from wastewater during treatment. Consequently, the discharged effluent mean TP concentrations exceeded NEMA maximum set standards. Increase in TP concentrations in the WWTP pond series can be associated with conversion of the organic and inorganic phosphorous in the sediment through biogeochemical processes.

In general, these results indicate that the wastewater treatment plant was a point source of nutrient enrichment (eutrophication) to river Riana for TP due to its incomplete removal. The continued discharge of TP into river Riana will lead to eutrophication, eventually affecting species diversity and river water quality [27, 32]. To improve TP removal efficiency, algal cells harvesting can be done, water retention period increased to enhance sedimentation.

5. Conclusions and recommendations

5.1 Conclusions

Suneka WWTP is crucial in polishing wastewater from Kisii municipality and its environs. Based on the results obtained from the current study, the following conclusions have been drawn:

1. This study revealed spatial variation in terms of diversity and distribution in phytoplankton community structure. In terms of taxonomic groups, the phytoplankton species that were identified belonged to six broad taxonomic groups which include: *Bacillariophyceae*, *Chlorophyceae*, *Cyanophyceae*, *Euglenaphyceae*, *Zygnematophyceae*, and *Dinophyceae*. Moreover, the phytoplankton contributed to wastewater polishing by converting nutrients into their biomass. In addition, they played a significant role in heavy metals removal from the wastewater column through bioaccumulation.

2. The population densities and species composition of the phytoplankton communities in the lagoon were influenced by the physico-chemical parameters.

3. There was an improvement in the levels of selected physico-chemical parameters after wastewater treatment by Suneka WWTP, indication of wastewater polishing.

5.2 Recommendations and future studies

The collected data forms a baseline for future studies. Gaps were identified based on the current study findings for further research. Therefore, the following recommendations have been drawn:

1. There is a need for further research in the Suneka WWTP to identify plankton community structures using other methods like molecular techniques in addition to microscopic techniques using standard identification keys. Moreover, algal toxins found in the WWTP need to be identified and characterized following the observation of cyanobacteria blooms known to produce toxins.

2. During this study, only four heavy metals were analyzed. Therefore, this study recommends that more studies be conducted to include other heavy metals not covered during this study.

3. To address the issue of nutrient enrichment in the effluent-receiving waters, this study proposes construction of a wetland with appropriate macrophytes for further polishing of effluent. Moreover, plants will also contribute to the removal of heavy metals through phytoremediation, but the extent to which they can mitigate the pollutants needs to be determined.

4. Investigate the seasonal variations in phytoplankton community structure and their impact on wastewater polishing efficiency. Conducting sampling throughout different seasons will provide insights into how environmental factors such as temperature and precipitation influence phytoplankton composition and their ability to contribute to wastewater treatment.

5. Assess the long-term ecological impacts of the Suneka WWTP effluent on the receiving water bodies. This should include monitoring changes in water quality parameters downstream of the discharge point over an extended period. Long-term studies can reveal trends in water quality improvement or degradation, helping to refine wastewater treatment processes and mitigate potential environmental impacts.

Acknowledgements

The authors are grateful to the African Development Bank (AfDB) for the financial support through a project research grant that Kisii University hosted. We also acknowledge the National Commission for Science Technology and Innovation (NACOSTI) for the permission to carry out this research and Kisii Town Water and Sewerage Company (GWSCO), Kisii County, for free access to the wastewater treatment plant. We sincerely thank Kenya Marine Fisheries and Research Institute (KMFRI) for its laboratory facilities.

Conflict of interest

The authors declare no conflict of interest.

Author details

Rayori Douglas Mosoti[1*], Nyabaro Obed[2], Getabu Albert[3], Omondi Reuben[3], Orina Paul[4] and Omondi Argwings[5]

1 Department of Biological Sciences, School of Pure and Applied Sciences, Kisii University, Kisii, Kenya

2 Department of Chemistry, School of Pure and Applied Sciences, Kisii University, Kisii, Kenya

3 Faculty of Agriculture and Natural Resources Management, Department of Environment, Natural Resources and Aquatic Sciences, Kisii University, Kisii, Kenya

4 Kenya Marine and Fisheries Research Institute (KMFRI), Kegati Aquaculture Research Centre, Kisii, Kenya

5 Department of Agriculture and Environmental Studies, Sigalagala National Polytechnic, Kakamega, Kenya

*Address all correspondence to: drayori@kisiiuniversity.ac.ke

References

[1] Karimi RD. The bacterial flora of tilapia (*Oleochromis Niloticus*) and catfish (*Clarias gariepinus*) from Earthen ponds in Sagana fish farm and Masinga dam [thesis]. Kenya: Kenyatta University; 2015

[2] Song E, Nyerere A, Magoma G, Gunturu R. Occurrence of highly resistant microorganisms in Ruai wastewater treatment plant and Dandora dumpsite in Nairobi County, Kenya. Advances in Microbiology. 2019;**9**:479-494

[3] Douglas R, Albert G, Reuben O, Paul O, Hellen N, Boniface G. Assessment of heavy metal concentrations (Cu, Cd, Pb, and Zn) in wastewater from Gusii treatment plant in Kisii County, Kenya. Pan Africa Science Journal. 2022;**1**(2):122-138

[4] Sayo S, Kiratu JM, Nyamato GS. Heavy metal concentrations in soil and vegetables irrigated with sewage effluent: A case study of Embu sewage treatment plant, Kenya. Scientific African. 2020;**8**:e00337

[5] Kinuthia GK, Ngure V, Beti D, Lugalia R, Wangila A. Levels of heavy metals in wastewater and soil samples from open drainage channels in Nairobi, Kenya: Community health implication. Scientific Reports. 2020;**10**:1-13

[6] Atwebembeire J, Andama M, Yatuha J, Lejju JB, Rugunda GK, Bazira J. The physico-chemical quality of effluents of selected sewage treatment plants draining into river Rwizi, Mbarara Municipality, Uganda. Journal of Water Resource and Protection. 2019;**3**:20-36

[7] Jin Z, Zhang X, Li J, Yang F, Kong D, Wei R. Impact of wastewater treatment plant effluent on an urban river. Journal of Freshwater Ecology. 2017;**32**(1):697-710

[8] Wanjohi L, Mwasi S, Mwamburi L, Isaboke J. Efficiency of University of Eldoret Wastewater Treatment Plant, Kenya. Africa Environmental Review Journal. 2019;**3**(2):99-109

[9] UNESCO. Wastewater the Untapped Resource. The United Nations World Water Development Report. Ecohydrology and Hydrobiology; 2017

[10] Girard JJ, Girard JJ. Feasibility of wastewater reuse for fish production in small communities in a developing world setting [thesis]. USA: University of South Florida; 2011

[11] Igbinosa EO, Okoh AI. Impact of discharge wastewater effluents on the physico-chemical qualities of a receiving watershed in a typical rural community. International Journal of Environmental Science and Technology. 2009;**6**(2):175-182

[12] Emily CL, Mainya NO, Mosoti RD, Evans KO. Assessment of Chemosit river pollution with urbanization of Chemosit Centre, Kericho County, Kenya. Pan Africa Science Journal. 2023;**2**(3):1-12

[13] Emmanuel IC, Onyema BE. The plankton and fishes of a tropical creek in South-Western Nigeria. Turkish Journal of Fisheries and Aquatic Sciences. 2007;**7**:105-113

[14] Omondi AO. Diversity, Distribution and Biomass of Phytoplankton in Lake Naivasha, Kenya. Kenya: Kisii University; 2019

[15] Were-Kogogo P, Adhiambo J. Status of phytoplankton community of Kisumu

Bay, Winam Gulf, Lake Victoria, Kenya. International Journal of Engineering Science. 2017;**6**(5):22-28

[16] Pastich MT, Gavazza EA, Florencio S, Kato L. Structure and dynamics of the phytoplankton community within a maturation pond in a semiarid region, Brazil. Journal of Biology. 2016;**76**(1):144-153

[17] Rayori D, Getabu A, Omondi R, Orina P, Gisacho B, Omondi A. Phytoplankton diversity in Gusii wastewater treatment plant in Kisii County, Kenya. International Journal of Fisheries and Aquatic Studies. 2021;**9**(3):299-306

[18] Ogbeigbu AE. Biostatistics; a Practical Approach to Research and Data Handling. Benin City Nigeria: Minex publishing Company Limited; 2005. pp. 153-155

[19] Eyo VO, Ekpo PB, Andem AB, Okorafor KA. Ecology and diversity of phytoplankton in the great Kwa River, Cross River State, Nigeria. International Journal of Fisheries and Aquatic Studies. 2013;**1**(2):1-7

[20] Agoro AM, Adeniji OA, Adefisoye AM, Okoh AO. Heavy metals in wastewater and sewage sludge from selected municipal treatment plants in eastern Cape Province, South Africa. Journal of Water. 2020;**12**(2746):2-19

[21] Gani M, ALfassane MA, Khondker MA. Limnology of wastewater treatment lagoons at Pagla, Narayanganj. Bangladesh Journal of Botany. 2011;**40**(1):35-40

[22] Tessema K, Lemma B, Fetahi T, Kebede E. Accumulation of heavy metals in the physical and biological systems of Lake Koka, Ethiopia: Implications for potential health risks. Lakes Reserv. 2020;**25**:314-325

[23] Babu J et al. Water quality, phytoplankton composition and microcystin concentrations in Kisumu Bay (Kenya) of Lake Victoria after a prolonged water hyacinth infestation period. Lakes & Reservoirs. 2021;**26**:1-23

[24] Islam SMD, Huda ME. Water pollution by industrial effluent and phytoplankton diversity of Shitalakhya River, Bangladesh. Journal of Scientific Research. 2016;**8**(2):191-198

[25] Adelakun KM, Mu'azu MM, Amali RP, Omotayo OL. Diversity of phytoplankton communities in a tropical river basin, Nigeria. Hydro Nepal Journal. 2016;**19**:52-56

[26] Chinnaraja V, Santhanam P, Balaji B, Dinesh S, Jothiraj K. An investigation on heavy metals accumulation in water, sediment and small marine food chain (plankton and fish) from Coromandel Coast, southeast coast of India. Indian Journal of Natural Sciences. 2011;**1**(8):532-540

[27] Ronoh AK. Efficacy of sewage treatment plant in Moi University, Uasin Gishu County, Kenya [thesis]. Kenya: Moi University; 2017

[28] Zobeidi AA, Bebba A. Seasonal variations of physical, chemical parameters in a wastewater treatment plant by aerated lagoons at southern-east of Algeria. Research Journal of Pharmaceutical, Biological and Chemical Sciences. 2015;**6**(3):1097-1102

[29] Waithaka AW. Wastewater management in cotton wet processing in Thika cloth mills in Kiambu County, Kenya [thesis]. Kenya: Kenyatta University; 2017

[30] Otieno OS. Physico-chemical and bacteriological quality of water from five rural catchment areas of lake Victoria

Basin in Kenya [thesis]. Nairobi: Kenyatta University; 2015

[31] Nikuze MJ, Niyomukiza J, Nshimiyimana A, Kwizera J. Assessment of the efficiency of the wastewater treatment plant: A case of Gacuriro Vision City. IOP Conference Series: Earth and Environmental Science. 2020;**448**(12046):1-10

[32] Musungu PC, Jondiko IO, Lalah J, Ongeri D, Chepkui R, Kiema F. The extent of nutrient removal by wastewater treatment plants along the Nyalenda Wigwa stream and the river Kisat (Kenya). Ecohydrology and Hydrobiology. 2013;**13**(4):236-240

Chapter 8

Heavy Metal Uptake by Pearl Millet and Sorghum Grown on Contaminated Soils from Letseng and Kao Village in Lesotho

Motsamai Tjakata and Fisseha Itanna

Abstract

The aim of the study was to determine the level of heavy metals in sorghum and Pearl millet and plant responses to heavy metals in contaminated soils close to Letseng and Kao diamond mining sites in Lesotho. A number of plant parameters were measured, namely, height, chlorophyll content, plant morphology, and heavy metal contents in plants to find out how heavy metals affect plant growth and the food chain around Kao and Letseng. The samples were collected from three sites, two of which are from contaminated sites (Letseng and Kao village) and one from a nearby uncontaminated control site (Mokhotlong Agricultural Research Station) in Lesotho. A wet digestion method was used to determine the composition of each crop using strong acids and oxidizing agents. Plant heavy metals were later determined using an AAS 500 model. The results indicated that there are higher concentrations of heavy metals in Pearl millet and sorghum in Letseng Village from pot experiment in the order of Pb > Cr > Cd > Ni > Fe > Mn > Cu > Co > Zn. Around the Kao Village, three metals were in the toxic range, in the order of Cd > Pb > Cr. Sorghum absorbs more heavy metals than Pearl millet in all study sites from pot experiments in glass houses, indicating that the health of humans and animals in the community is at risk. Results of this study indicate that the heavy metal risk around the mining sites is alarming and hence appropriate measures need to be taken to rectify the situation.

Keywords: heavy metals, plant morphology, wet digestion, contaminated soils, health threat, plant response

1. Introduction

Metal toxicity decreases the uptake and translocation of nutrients and water and allows oxidative damage, thus restricting plant growth [1]. The concentration of heavy metals can differ in plants and soils depending on the seasons. For instance, concentrations of Cd, Cu, and Mn in soils are positively correlated with concentrations of Cd, Cu, and Mn in plants only during summer [2]. Zinc and nickel are essential metals [3], but increased concentrations pose a risk to plants and their roots, whereby there will be stunted growth and oxidative damage in plants [4]. An increased level of heavy metals in plants could cause the following symptoms in plants; chlorosis, necrosis, damage to plant organs, and plant phenotype changes [5].

Heavy metal accumulation and dispersion in plants vary based on plant species, element species, chemistry and bioavailability, redox, pH, CEC, dissolved oxygen, temperature, and root secretion [6]. Irrigating Pearl millet with contaminated water increases the heavy metal concentration in the plants mostly when applied directly, and the effects on the crop could be easily seen [7].

High concentrations of heavy metals have an impact on plant growth, photosynthesis, and phytoaccumulation efficiency. Root elongation for absorption of heavy metals can be inhibited by heavy metals such as Mn due to its meristematic cell division potential [8]. The level and amount of heavy metal accumulation in plants rely on soil type, pH, humidity, and micronutrient content as well as the time of crop harvest. For example, in alkaline soils, leaching and bioavailability of heavy metals to plants are lower. The presence of organic matter can also restrict metal uptake from soil solution [9]. The mobility of heavy metals is decreased when soil pH is above 5. Thus, plant growth is weakened, and deficiencies of Fe and Mn will be there [10].

High metal contamination in plants decreases leaf production rate and reduces plant mass, and there is a high chance of delayed phenological development [11]. The late and decreased reproduction may have large effects at population, community, and ecosystem levels, even if the effect of low metal levels on plant growth may be small and attributed to increased evolution of metal tolerance. Flowering phenology is considered very active in response to metals. Advanced seed ripening due to heavy metals causes leaf life span to be decreased at the highest and the lowest degree of metals [12].

High levels of Cd (>3 mg/kg) in soils affect plant height, stem diameter, plant biomass, root volume, root surface area, and root biomass, as well as soil enzyme activities and microbial biomass [13]. Cd availability exponentially reduces the shoot of the plant. Cd has several effects on plant photosynthesis [14]. The aim of this study is to find out the levels of heavy metals (mild, toxic, or sufficient) in Pearl millet and sorghum and determine plant response to the extent of heavy metals from contaminated soils around Letseng and Kao Village in Lesotho.

2. Materials and methods

2.1 Study site

The study was conducted in a pot experiment in Roma village on contaminated soils collected from Kao village (29°00'56"S and 28°37'19″E at 2378 m elevation) and Letseng village near Letseng Diamond Mining (29°00'45" S and 28°52'39″ E at 2885 m elevation) and Mokhotlong Agricultural Research Station (29°17'14" S and 29°04'45″ E at 2153 m elevation) as reference point. The soils from the contaminated sites were collected along wastewater drainage lines at the mines' outlets. Kao village is affected by the wastewater from the Kimberlite Diamond mine, while Letseng mining is the largest mining site in Lesotho, with a cold climate and frequent snowfall. The research aimed to understand the impact of contaminants on agricultural fields and environment (**Figure 1**).

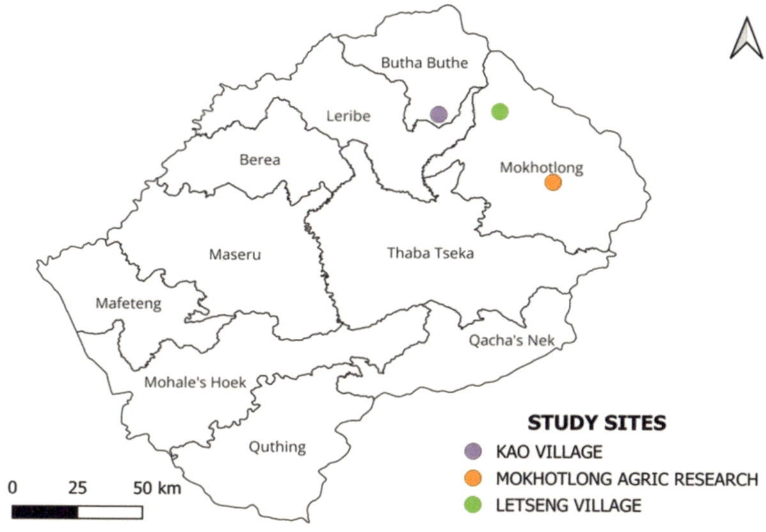

Figure 1.
Map of Lesotho showing three study sites of interest. Letseng Village and Kao village are of interest and focus area since there are mining activities nearby.

2.2 Pot experiment

The pot experiment was conducted to observe the effect of heavy metals on the crop plants (Pearl millet and sorghum) from soils obtained from Mokhotlong Agricultural Research, Letseng, and Kao in the glass house at the National University of Lesotho. Three replications of sorghum and three replications of Pearl millet were grown in pots filled with soils collected from different spots at all three sites. Soils from the Mokhotlong Research Station were the control samples, while soils from the two villages sites (Letseng Village and Kao Village) represent the contaminated soils. Plants were watered regularly to maintain the required optimum moisture, and the growing plants were monitored daily to assess plant changes in leaves and overall growth.

2.3 Plant analyses

Plant samples of Pearl millet and sorghum from the pot experiment were analyzed in the Soil Science Laboratory of NUL. For heavy metal determination in plants, plants were ground using a leaf powder grinding machine. A 0.5 g of ground plant samples were weighed in Erlenmeyer flask for analysis. A combination of different reagents such as H_2SO_4, HCl, and H_2O_2 were added in glass tubes for solubilization of minerals. The samples were then evaporated on a hot plate at 320°C to reduce contamination of the sample with substance in air, vessel walls, and the local environment, and the solution was cooled at room temperature. Lastly, deionized water was added. AAS 500 model was used to read all heavy metals in plant tissues.

2.4 Analysis of data

Data was analyzed using SPSS software. Linear regression analysis was computed to assess the relationship between individual heavy metal concentrations in the Pearl

Name of metal	FAO/CODEX [15, 16]	WHO [17]
Zn	60	30–100
Ni	1.5	10
Cu	30	10
Mn	5	15–100
Cd	1	0.02
Pb	2	2
Cr	1	1.3
Fe	48	500–640
Co		50

Table 1.
Allowable heavy metal levels in crops.

millet and sorghum samples and in the soil. Means, standard deviation, and coefficient of variation (CV %) were determined for each treatment:

$$CV = \frac{\text{standard deviation}}{\text{mean}} \quad (1)$$

Significant differences between treatments from the mining sites and control were estimated, where p values less than 0.05 were considered as significant ($P < 0.05$), allowing only a 5% chance error in the results (**Table 1**).

3. Results

3.1 Plant analyses of sorghum and pearl millet

All plants sampled show that the amount of copper in the plants is less than the permissible limit (**Table 2**); therefore, there is no contamination of copper in the three different study sites. The amount of zinc from all plants is also less than the allowable limit, and it may even be deficient for plants (**Table 2**). At all sites, the amount of iron in crop samples was very low, implying a deficiency of iron in all soils (**Table 2**). Lead in the crop samples, on the other hand, exceeded the allowable limits for plants. Crops contained high amounts of chromium as well (**Table 2**).

The permissible level of cobalt is 50 mg/kg; therefore, plants are not contaminated with cobalt. Cadmium is high in the crops grown on soils from contaminated sites (**Table 2**), while it is not detectable in the crop samples grown on soils from MAR, the control site.

4. Discussion

4.1 Concentration of essential metals in pearl millet and sorghum

In **Table 2**, the statistical values show the overall performance or response of sorghum and Pearl millet to heavy metals, while **Figures 2–4** show how sorghum

		Cu	Zn	Fe	Pb	Ni	Co	Cr	Mn	Cd
		mg/kg	**mg/kg**	**mg/kg**	**mg/kg**	**mg/kg**	**mg/kg**	**mg/kg**	**mg/kg**	**mg/kg**
Method detection limit		0.008	0.089	0.320	0.069	0.105	0.064	0.072	0.042	0.016
Sorghum for Letseng	Mean	8.43	26.23	105.3	7.00	6.43	14.33	13.63	7.73	2.86
	Std. Dev	5.42	1.85	56.87	2.85	1.23	0.80	1.53	1.81	0.90
Pearl millet for Letseng	Mean	2.03	23.20	85.83	8.53	6.43	13.20	10.06	4.53	2.10
	Std. Dev	0.80	2.17	12.55	1.10	0.66	0.81	0.92	1.10	0.10
Sorghum for Kao	Mean	1.88	28.88	181.6	5.86	8.88	12.46	15.84	21.42	1.62
	Std. Dev	1.54	5.05	123.6	0.58	1.60	0.65	6.58	22.60	0.25
Sorghum for MAR	Mean	2.17	26.67	217.4	8.05	7.22	12.80	2.80	8.90	0.00
	Std. Dev	0.59	1.57	76.39	2.88	1.46	0.92	0.66	3.43	0.00
Pearl millet for Kao	Mean	0.80	29.44	180.0	8.22	8.56	11.98	7.72	9.16	1.22
	Std. Dev	0.43	2.04	88.03	0.90	0.74	0.62	1.90	4.61	0.19
Pearl millet for MAR	Mean	1.75	26.00	149.5	8.70	8.07	10.15	2.95	7.55	0.00
	Std Dev	0.79	2.42	47.66	0.32	0.79	3.68	1.40	1.48	0.00

Table 2.
Heavy metal contents in sorghum and Pearl millet grown on soils from the three study sites under a pot experiment.

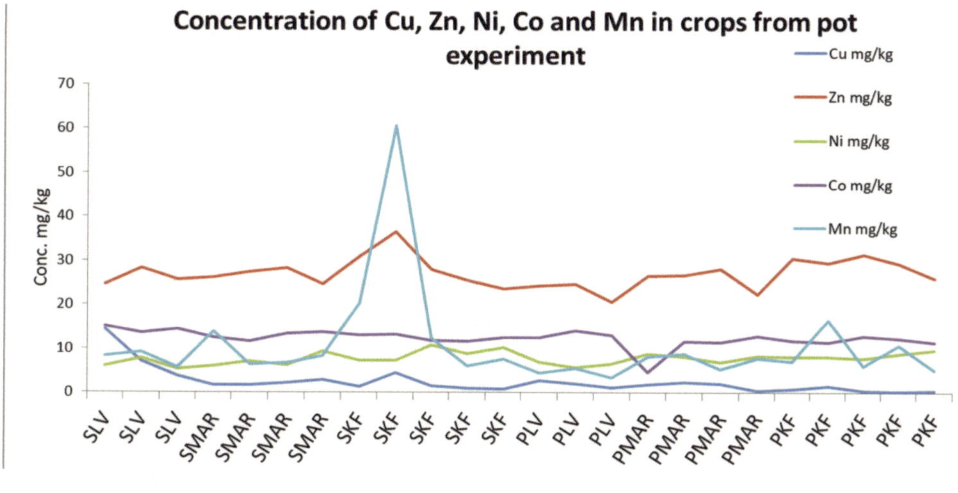

Figure 2.
Concentrations of essential metals in the crops studied (LV, KF, and MAR represent Letseng Village, Kao Farmlands, and Mokhotlong Agricultural Research, respectively) while S and P represent Sorghum and Pearl millet.

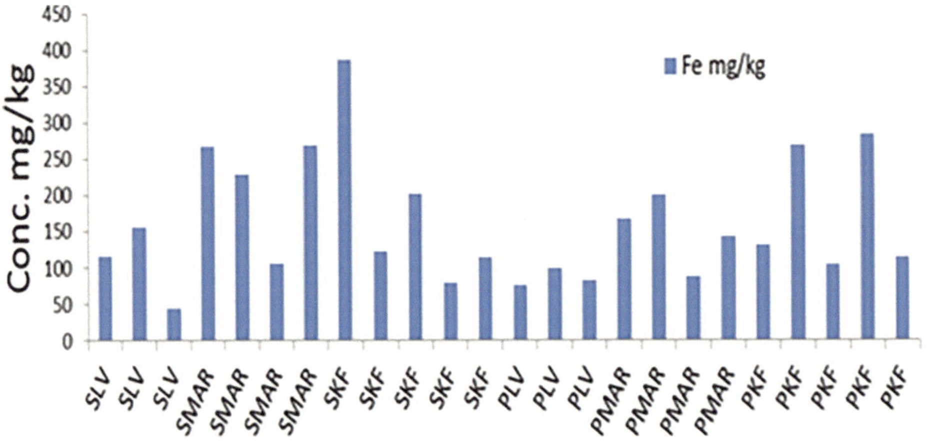

Figure 3.
Concentration of iron found in different crops sampled per three sites (LV, KF, and MAR represent Letseng Village, Kao Farmlands, and Mokhotlong Agricultural Research, respectively) while S and P represent Sorghum and Pearl millet.

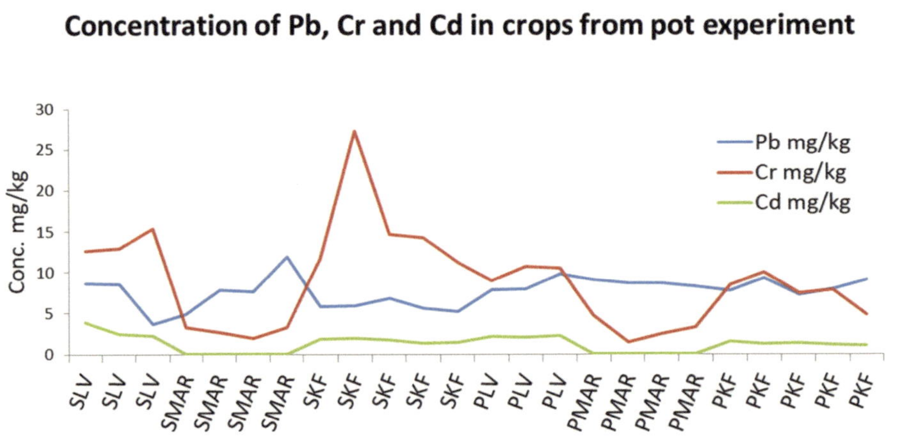

Figure 4.
Concentrations of toxic heavy metals in the crops studied (LV, KF, and MAR represent Letseng Village, Kao Farmlands, and Mokhotlong Agricultural Research, respectively) while S and P are representing Sorghum and Pearl millet.

and Pearl millet respond to heavy metals at different spots where soils were collected and pot experiments were done. None of the crops tested from the pot experiment exceeded the allowable limit of Cu in plants (**Table 2**). Cu levels in sorghum grown on soils from the Letseng Village, Kao Village, and MAR have similar amounts of Cu in plants, which is close to the permissible level. Copper assists in controlling plant diseases. The ingestion of copper through the food chain in plants can bring chronic exposure [18, 19].

All crops tested had Zn concentrations below the recommended concentration of Zn in plants (**Table 2**). The plant molecular regulators are very sensitive in controlling the adaptation of plants when Zn is in excess, and plants also adapt to a shortage of zinc supply by enhancing the zinc uptake capacity [20]. Zinc affinity becomes strong with soil pH. Thus, the adsorption also increases with increasing Zn to become more available to plants [21].

Zn is essential in the physiology and biochemistry of plants. However, excessive amounts could cause oxidative damage by allowing the levels of reactive radicals [22].

Sorghum has a high potential of absorbing more essential nutrients than Pearl millet, as can be seen from **Figure 2**.

Nickel was found to be lower than the maximum permissible level in the plants, thus imposing no risk to any plants grown in the environments. Nickel strongly plays an essential role in metabolic reactions in plants. In cases where nickel exceeds permissible values in plants, its effects are reduced or inhibited growth, chlorosis, necrosis, and wilting. The plant height on soils from the control site (MAR) performed well compared to Letseng Village and Kao Village.

All sites and plants tested showed that cobalt is below the range expected in healthy plants based on the permissible level of cobalt in the plant, which is 15–25 mg/kg in the plants [23]. The level of cobalt is low in soils from Kao Village and Letseng Village. Cobalt in the roots and leaves cannot disrupt a range of metabolic processes due to less competitive interactions with other micronutrient cations [24].

The level of manganese in plants is way too low in the studied sites. The highest Mn content in the crops was 21.4 mg/kg for sorghum in the Kao village farm and low (4.53 mg/kg) in the Pearl millet sample that was taken from a pot experiment on Letseng Village soil. Manganese is very important in chloroplast formation, photosynthesis, and nitrogen metabolism. When manganese is low in the plant, the plant growth and quality are hindered. It is essential mostly in the catalyst of the oxygen-evolving complex (OEC) of photosystem II (PSII); thus, the amount of Mn should be increased in soil for plant root uptake [25].

Iron is high in sorghum because MAR, Kao Village, and Letseng Village soils have 217, 180, and 105 mg/kg in the plant, respectively. Iron plays many roles in plant enzymes such as cytochromes, mainly in the transport chain through electrons. It is needed in higher amounts by plants, and its deficiency will cause plants to trample [26].

4.2 Concentration of toxic heavy metals in sorghum and Pearl millet

It appears that the sorghum and Pearl millet experiments under a pot trial have both similar lead (>2 mg/kg) concentrations in the plants, and it should be noted that the permissible concentration of Pb in the plant is 2 mg/kg according to WHO [17]. Letseng had soil contamination of lead that could have a significant contribution in affecting the crop directly, since the Pb level was above the allowable limit in soils.

It should be noted that lead in **Figure 4**, unlike other heavy metals, is easily absorbed by plants, and this has a very big effect on plant growth. Plants, to protect themselves, should have strong cell wall to prevent the accumulation of lead inside the cell [27].

Chromium in the soils of the Kao Village and Letseng Village was high, and its concentration in the plants also shows that the plants are contaminated with chromium (**Table 2**). Chromium is a serious threat to plants because, at certain stages, it has totally killed the plants besides restricting the germination of seeds by triggering protease activity [28]. Chromium in the environment should be controlled totally.

Pearl millet and sorghum had high amounts of Cd, way above the allowable 0.02 mg/kg of Cd in their tissues, unlike other heavy metals discussed (**Table 2**). Cd toxicity in crop plants has the potential to decrease the uptake and translocation of nutrients and water. There are also other effects that cadmium toxicity could cause on crops such as higher oxidative damage, plant metabolism interruption, and plant morphology and physiology restrictions [29].

Cadmium in water is highly soluble, and this makes it even more harmful to plants, even at lower concentrations [30]. The symptoms that have been observed in the plant samples from Kao village as well as those from the pot experiment are stunting and chlorosis of plants, indicating the effect of Cd toxicity. The main reason for stunting in growth and chlorosis is due to Cd restricting the plant roots to absorb nutrients that they require for growth [31].

4.3 Plant height and chlorophyll content in Pearl millet and sorghum

The bar chart in **Figure 5**, clearly gives an indication of how contaminants have affected the plants in the pot experiment. Crops grown on the soils of the control site MAR outperformed the crops grown on soils of the contaminated sites during the three-month growth period. Both crops tested on soils of the mining sites performed badly because their plant height was increasing but very slowly because of the low chlorophyll measured since was very low to allow the plant to grow to the optimum. Heavy metals in the soil compete with other needed nutrients; therefore, the growth of the plants is affected, as this could clearly be seen with the soils from Kao and Letseng Villages. With plants grown on soils from MAR, the growth is high and rapid due to the fact that there were no contaminants in the soils.

Heavy metals present in plants in larger quantities are considered to be toxic and could also affect human life by entering the food chain. The plant height of plants grown in the soils of Kao Village is very small, which is indicative of the severe

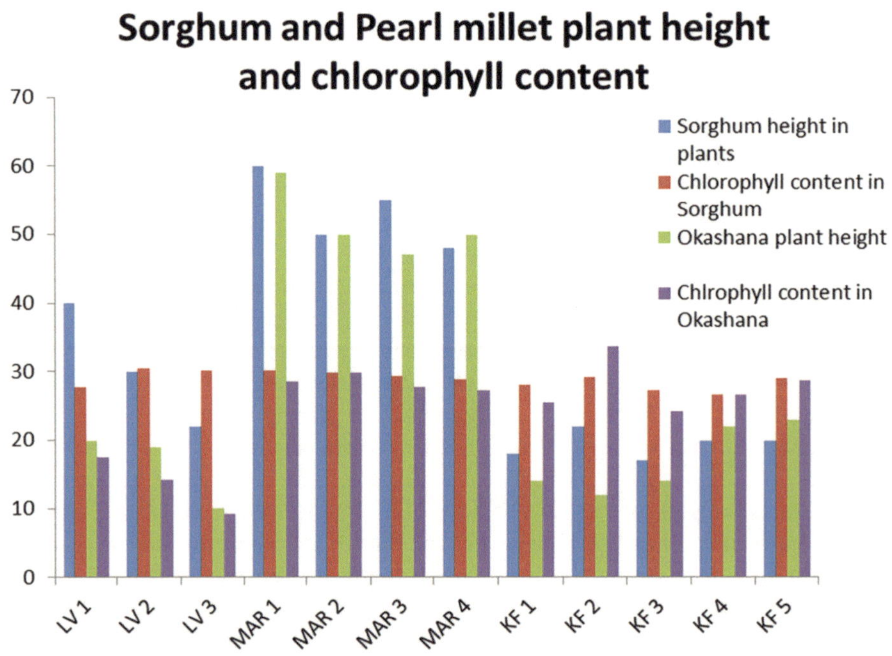

Figure 5.
Plant height and Chlorophyll analyses in pot experiment for 3 sited LV, KF and MAR represents Letseng village, Kao Farmlands and Mokhotlong agricultural research respectively.

contamination of cadmium and other heavy metals that affect plant growth and photosynthesis. With crops grown in a pot experiment on soils of the Letseng Village, chromium, lead, and cadmium are apparently the ones restricting plant growth and height. It should be noted that at low concentrations, non-essential heavy metals truly affect plant growth, ecosystems, microorganisms as well as human life [32].

These non-essential elements affect plant growth because plants are inhibited from absorbing the nutrients in the soil, and growth is highly affected at that moment. In both Letseng and Kao Diamond Mining, the chlorophyll content is very low, implying that the synthesis of glucose and oxygen is very low, which results in the plants not growing to the needed optimum growth due to the effects of heavy metal concentrations in the soil [33]. The soil content of these metals (Cr, Cd, and Pb) was also in the toxic range, hence why the plant contents are also high.

5. Conclusion

Sorghum and Pearl millet have absorbed significant levels of heavy metals from the soils into their roots. Kao Village is highly contaminated with chromium, cadmium, and lead, and stunted growth and chlorosis are observed in the plants.

There are higher concentrations of heavy metals in Pearl millet and sorghum in Letseng Village from the pot experiment in the order of Pb > Cr > Cd > Ni > Fe > M n > Cu > Co > Zn. Around the Kao Village, three metals were in the toxic range in the order of Cd > Pb > Cr. Sorghum absorbed more heavy metals than Pearl millet in all study sites from pot experiments in glass houses. The health of humans and animals consuming sorghum could be at risk.

Author details

Motsamai Tjakata[1*] and Fisseha Itanna[2]

1 Department of Soil Science and Resource Conservation, National University of Lesotho (NUL), Lesotho

2 School of Interdisciplinary Research and Graduate Studies (SIRGS), University of South Africa (UNISA), South Africa

*Address all correspondence to: motsamai.tjakata@yahoo.com

References

[1] White PJ, Pongrac P. Heavy-metal toxicity in plants. In: *Plant Stress Physiology*. Cabi: Wallingford UK; 2017. pp. 300-331

[2] Sharma RK, Agrawal M, Marshall F. Heavy metal contamination of soil and vegetables in suburban areas of Varanasi, India. Ecotoxicology and Environmental Safety. 2007;*66*(2):258-266

[3] Itanna F, Coulman B. Phyto-extraction of copper, iron, manganese, and zinc from environmentally contaminated sites in Ethiopia, with three grass species. Communications in Soil Science and Plant Analysis. 2003;**34**(1-2):111-124

[4] Tahar K, Keltoum B. Effects of heavy metals pollution in soil and plant in the industrial area, West Algeria. Journal of the Korean Chemical Society. 2011;**55**(6):1018-1023

[5] Wang Y, Zhong B, Shafi M, Ma J, Guo J, Wu J, et al. Effects of biochar on growth and heavy metals accumulation of moso bamboo (Phyllostachy pubescens), soil physical properties and heavy metals solubility in soil. Chemosphere. 2019;**219**:510-516

[6] Cheng Shui Ping CS. Heavy metals in plants and phytoremediation. Environmental Science and Pollution Research. 2003;**10**(5):335-340

[7] Ankush, Prakash R, Singh V, Diwedi A, Popat RC, Kumari S, et al. Sewage sludge impacts on yields, nutrients and heavy metals contents in pearl millet–wheat system grown under saline environment. International Journal of Plant Production. 2021;*15*:93-105

[8] Benzarti S, Mohri S, Ono Y. Plant response to heavy metal toxicity: Comparative study between the hyperaccumulator Thlaspi caerulescens (ecotype Ganges) and nonaccumulator plants: Lettuce, radish, and alfalfa. Environmental Toxicology: An International Journal. 2008;**23**(1):607-616

[9] Kumari S, Mishra A. Heavy metals contamination. In: Soil Contamination-Threats and Sustainable Solutions. London, UK: IntechOpen; 2021

[10] Do Nascimento ARVJ, Cunha GKG, do Nascimento CW.A., and da Cunha K.P.V. Assessing soil quality and heavy metal contamination on scheelite mining sites in a tropical semi-arid setting. Water Air, and Soil Pollution. 2021;**232**(9):1-15

[11] Peralta-Videa JR, De la Rosa G, Gonzalez JH, Gardea-Torresdey JL. Effects of the growth stage on the heavy metal tolerance of alfalfa plants. Advances in Environmental Research. 2004;*8*(3-4):679-685

[12] Ryser P, Sauder WR. Effects of heavy-metal-contaminated soil on growth, phenology and biomass turnover of Hieracium piloselloides. Environmental Pollution. 2006;*140*(1):52-61

[13] Pinto-Poblete A, Retamal-Salgado J, López MD, Zapata N, Sierra-Almeida A, Schoebitz M. Combined effect of microplastics and Cd alters the enzymatic activity of soil and the productivity of strawberry plants. Plants. 2022;**11**(4):536

[14] Krantev A, Yordanova R, Janda T, Szalai G, Popova L. Treatment with salicylic acid decreases the effect of cadmium on photosynthesis in maize

plants. Journal of Plant Physiology. 2008;**165**(9):920-931

[15] Joint, F.A.O, World Health Organization and WHO Expert Committee on Food Additives. Evaluation of Certain Food Additives and Contaminants: Seventy-Fourth [74th] Report of the Joint FAO/WHO Expert Committee on Food Additives. Rome, Italy: World Health Organization; 2011

[16] FAO/CODEX. Global Assessment of Soil Pollution: Report. Rome, Italy; 2022

[17] World Health Organization (WHO). Permissible Limits of Heavy Metals in Soil and Plants. Geneva, Switzerland: WHO; 2011

[18] Coelho FC, Squitti R, Ventriglia M, Cerchiaro G, Daher JP, Rocha JG, et al. Agricultural use of copper and its link to Alzheimer's disease. Biomolecules. 2020;*10*(6):897

[19] La Torre A, Iovino V, Caradonia F. Copper in plant protection: Current situation and prospects. Phytopathologia Mediterranea. 2018;*57*(2):201-236

[20] Assunção AG, Schat H, Aarts MG. Regulation of the adaptation to zinc deficiency in plants. Plant Signaling and Behavior. 2010;*5*(12):1553-1555

[21] Recena R, García-López AM, Delgado A. Zinc uptake by plants as affected by fertilization with Zn sulfate, phosphorus availability, and soil properties. Agronomy. 2021;*11*(2):390

[22] Natasha N, Shahid M, Bibi I, Iqbal J, Khalid S, Murtaza B, et al. Zinc in soil-plant-human system: A data-analysis review. Science of the Total Environment. 2022;*808*:152024

[23] Hu X, Wei X, Ling J, Chen J. Cobalt: An essential micronutrient for plant growth? Frontiers in Plant Science. 2021;*12*:768523

[24] Liu J, Reid RJ, Smith FA. The mechanism of cobalt toxicity in mung beans. Physiologia Plantarum. 2000;*110*(1):104-110

[25] Schmidt SB, Jensen PE, Husted S. Manganese deficiency in plants: The impact on photosystem II. Trends in Plant Science. 2016;*21*(7):622-632

[26] Rout GR, Sahoo S. Role of iron in plant growth and metabolism. Reviews in Agricultural Science. 2015;*3*:1-24

[27] Nas FS, Ali M. The effect of lead on plants in terms of growing and biochemical parameters: A review. MOJ Ecology & Environmental Science. 2018;*3*(4):265-268

[28] Saud S, Wang D, Fahad S, Javed T, Jaremko M, Abdelsalam NR, et al. The impact of chromium ion stress on plant growth, developmental physiology, and molecular regulation. Frontiers in Plant Science. 2022;*13*:994785

[29] Haider FU, Liqun C, Coulter JA, Cheema SA, Wu J, Zhang R, et al. Cadmium toxicity in plants: Impacts and remediation strategies. EcotoF. 2021;*211*:111887

[30] Kumar A, Maiti SK. Availability of chromium, nickel and other associated heavy metals of ultramafic and serpentine soil/rock and in plants. International Journal of Emerging Technology and Advanced Engineering. 2013;**3**(2):256-268

[31] Dağhan H. The effect of heavy metal treatments on uptake of nitrogen, phosphorus and potassium in transgenic and non-transgenic tobacco plants. Journal of Agricultural Sciences. 2013;**19**(2):129-139

[32] Bharti R, Sharma R. Effect of heavy metals: An overview. Materials Today: Proceedings. 2022;**51**:880-885

[33] Zhang H, Xu Z, Guo K, Huo Y, He G, Sun H, et al. Toxic effects of heavy metal Cd and Zn on chlorophyll, carotenoid metabolism and photosynthetic function in tobacco leaves revealed by physiological and proteomics analysis. Ecotoxicology and Environmental Safety. 2020;**202**:110856

Chapter 9

Perspective Chapter: Heavy Metals-Mediated Chemical Contamination in Foods, Associated Health Risks, and Remediation Techniques

Hamid Ullah, Abdul Majeed Khan, Sadeeq Ullah, Muhammad Asif, Muhammad Imtiaz and Bibi Shazia

Abstract

Environmental contamination is one of the great challenges worldwide. It is exponentially increasing through natural and non-natural sources, particularly through anthropogenic activities. Pollutants such as heavy metals, SO_2, CO, nitrogen oxides, biological contaminants, ozone, etc., are serious threats to the environment. Among others, the heavy metals exploration through mining, their natural addition to the various vegetables/foods from the soil, and their presence in air and water are recognized as the riskiest factors contributing to environmental contamination. The presence of heavy metals in the environment, particularly in foodstuffs poses highly toxic effects on human health which compel the scientist to identify the levels of these heavy metals in the foodstuffs as well as to design green strategies to overcome the health-related challenges with the used of the heavy metals contaminated foods. The present chapter focuses on the heavy metals presence in different foods through dietary intake, sources of these metals, the associated risks, and reported heavy metals remediation strategies for foods including physical, chemical, bioremediation, and others.

Keywords: environment contamination types, heavy metals in foodstuffs, heavy metals associated risks, remediation strategies, human health

1. Introduction

1.1 Background of the chapter

Contamination is the basic term used when we discuss about the environment and environmental science. Although the present chapter focuses on the contamination caused by heavy metals in foods, sources of these heavy metals, the associated health risks, and remediation strategies, however an understanding of the basic terms such as environment, environmental science, contaminations, and its types are highly necessary to describe the topic in the more understandable way. Thus, the mentioned terms will be discussed in the initial sections followed by the topic under focus.

1.2 Environment and environmental science

The environment generally refers to the surroundings or conditions in which a person, animals, or plants live, operate, and influence each other. More explicitly an environment is recognized as the circumstances, objects, physical, chemical, biotic factors, and conditions in one's surroundings that influence an organism, or an ecological community that ultimately determine its survival [1]. The physical and chemical factors include but not limited to light, temperature, color, hydrology (such as rainfall, soil moisture, flow rates, and sea level), pH, nutrients, level of oxygen, all toxic gases and compounds, organic and inorganic pollutants and others [2]. For example, light directly influences the process of photosynthesis in plants which is essential for their survival. The biotic factor includes all biological conditions/components/contaminants of biological and microbial origin, e.g. animals, microbes, plants, human beings, and others. These conditions/components interact in various ways and predict the level of survival and overall ecological balance [3]. The ecological community includes a group of species, living and interacting in the same area thereby determining the survival of each other. Surrounding conditions (climate or presence of other species or human activity) is also an important factor in defining the environment as it includes all external conditions affecting organisms.

Environmental science deals with the physical, chemical, and biological interaction of living organisms with their environment. In other words, environmental science is an interdisciplinary field that studies the interactions between the natural environment and human activities. It encompasses various scientific disciplines, including biology, chemistry, geology, and physics, to understand the processes that shape the Earth's ecosystems, the impact of human actions on the environment, and the methods for mitigating environmental issues. In addition, this branch of science discusses how the natural phenomena take place and the changes that occur in the climate, as well as the impact of human/living being activities on these natural phenomena creating various environmental problems. Furthermore, environmental sciences deal with all the challenges related to the environment and find and suggest solutions to how to overcome the different environmental problems in a sustainable manner [4].

1.3 Environmental contamination and its types

There are various contaminants present in the environment that badly affect the human health by causing or contributing to different diseases [2, 3, 5–6]. Although the different types of pollutants are responsible for more than one type of contamination but still based on these different types of pollutants/contaminants, the environmental contamination can be [7], but not limited to the following 7 major types:

1.3.1 Biological contamination

Biological pollution is caused by the presence or introduction of adequate quantities of non-indigenous or invasive species, such as pathogenic bacterial strains, viruses, algae, protozoa, fungi, insect pupae, animal dander and their saliva, house dust, mites, cockroaches, pollen, and others. These biological contaminants pollute water, air, food, etc., which ultimately cause various diseases such as typhoid, cholera, polio, hepatitis, and others, and subsequently affect humans' activities [7–8]. This kind of contamination could be managed via adopting various approaches, including bioremediation through the use of indigenous microbes, chemicals, changes in growth conditions for the microorganisms, and others [8].

1.3.2 Air contamination

Air contamination refers to the introduction of harmful substances such as volatile organic substances, particulate matter, and toxic gases like CO_2, CH_4, CO, SO_2, NO, O_3, and others. The sources of these toxic gases are diverse including power-producing stations, gas stations, automobiles, combustion engines, burning of fossil fuels, furnaces, agricultural activities, volcanic and soil eruptions, and others. Air pollution carries a serious threat not only to climate change but also directly affects human health since the exposure of the individual to air pollution has caused suffering from various diseases such as respiratory diseases, cancer, cardiovascular diseases, and others [7, 9].

1.3.3 Radioactive pollution

Radioactive contamination generally refers to the release of radiation into the environment, i.e., air, soil, and water, and ultimately contaminate it. It originates usually from the mining of radioactive substances, their uses and deposition, nuclear tests, nuclear accidents, managing nuclear wastes, diagnostic tests using radiation, nuclear medicines, and others [7, 10, 11]. Common radioactive substances include uranium, radon gas, cesium, and others. As a result of the radioactive pollution, the whole ecosystem is badly affected and various health issues are observed in the surrounding inhabits, ranging from simple illness till mortality [10–14].

1.3.4 Water contamination

Water is one of the essential requirements of living beings. Water contamination is one of the most highlighted problem worldwide among the different type of contamination. It refers to the presence of chemicals beyond the acceptable limit and pathogens in the water systems, including rivers, oceans, storage tanks, and others, deteriorating these systems and ultimately causing health issues in humans and other inhabitants. The microbial contaminants include pathogenic bacteria, viruses, microscopic protozoa, and worms which are spread by human and animal wastes. The chemical (organic and inorganic) or microorganisms are added unintentionally to the water sources both drinking and underground water from the natural mineral deposits or from industrial processes and their effluents. The chemical contaminants include heavy metals, paints, detergents, solvents, batteries, medicines, disinfectants, pesticides, oil and petrochemicals, and others. The UN report 2003 (UN WWAP 2003) reveals that 2 million tons of sewage, industrial effluent, and agricultural wastes are discharged into the world's water each day. Radiological contamination of water is also one the types of water contamination [7, 12, 15–20].

1.3.5 Soil contamination

Soil contamination is among the major types of contamination which is caused by the natural or non-natural introduction/presence of the toxic xenobiotic chemicals/synthesized substances. The different types of soil contaminants include paints, industrial byproducts, pesticides, herbicides, medical waste, animal waste, heavy metals/metalloids, microplastics, antibiotic resistance genes, increase in salt contents, acidic nature of the soil, and others. These contaminants degrade the soil quality. Literature reveals that the increased levels of multiple soil contaminants are associated with changes in microbial traits including genes associated with environmental stress resistance, nutrient cycling, crop growth, and pathogenesis. Furthermore, these contaminants accumulate in crops, fruits, and vegetables and ultimately cause serious threat to the human's health [7, 21, 22]. Various strategies have been reported to be considered to minimize the level of the aforementioned containments in solid or in other words to manage soil contaminations. These strategies include identification of the sources of contaminants, reduction of heavy metal phytoavailability in soil with liming or other immobilizing materials, cultivation of low accumulating crops, use of clean water, fertilizer management such as utilization of natural fertilizers, bioremediation, and others. Other remediation strategies to overcome soil contamination include, the use of physicochemical processes/physical barrier application via capping water surface to limit the infiltration of the uncontaminated water, Separation/concentration processes in which the pollutants are treated with other substances to reduce the related hazards, Chemical treatments that reduce the bioavailability/mobility of contaminants upon reaction with specific reagents, polymer encapsulation, phytoextraction (extraction of both metallic and organic constituents through direct uptake by plants) and phytovolatalization processes (involve specialized enzymes that can transform and volatilize contaminants in the plant-microbe–soil system), soil amelioration/amendments via lime, phosphate, and organic matter e.g., the addition of lime reduce the bioavailability of heavy metals [22, 23].

1.3.6 Contamination by medical wastes

The contamination caused by medical waste is a serious threat to the environment, which is directly affecting human beings and other living organisms, particularly in developing countries where biomedical waste management is very poor. The riskiest of medical waste includes injections, syringes, needles, blades, scalpels, and others. The hospitals, clinics, and diagnostic centers generate medical waste, and improper deposition, causes various fatal diseases. It is reported that 16,000 hepatitis C infections, 66,000 hepatitis B infections, and about 1000 human immunodeficiency (HIV) infections are caused by mismanagement of medical waste, which led to approximately 1100 deaths and significant disability [6, 24, 25]. The medical waste contains toxic and radioactive substances. It is reported that 15% of the total amount of medical waste is hazardous and can be infectious and hazardous. Among the careful management strategies, a massive volume of medical waste is mostly treated through incineration, but this process may release toxic gases and solid waste with contents of heavy metals and other toxic substances. It is therefore other strategies such as steam autoclave disinfection, microwave disinfection, mechanical/chemical disinfection, recycling, and others, may be adopted to reduce medical waste to the maximum. However, careful background knowledge is necessary about the type/nature or composition of the medical waste before adopting any of the mentioned management strategies to avoid any incident [26, 27].

1.3.7 Chemical contamination

As previously mentioned at the start of the chapter that the vast nature of the topic "environmental contaminations, associated risk and remediation strategies" is almost impossible to cover each type of contamination. In addition, since food safety is one of the global challenges, therefore the present chapter greatly focuses on one of the major types of contamination, i.e., chemical contamination mediated via heavy metals in foods. Furthermore, the associated health risks of the contaminated foodstuff via heavy metals and remediation techniques are also a main focus of the present chapter.

Chemical contamination encompasses a vast category of contaminants that occurs either by waste of toxic chemicals, heavy metals, and others that are not intentionally added to the ecosystem and pose a serious threat to plants, animals, humans, and all living things. Origins of chemical contaminants are diverse, namely soil, environment, disinfection byproducts, personal care products, packaging material, air, and water. Food and water contamination by chemicals is one of the serious global issues. The chemicals that contaminate food include but are not limited to metals/metalloids, aromatic compounds, persistent organic pollutants, perfluorinated compounds, pharmaceuticals, undesired edible colors, radioactive elements, electronic waste, plastics, and others [20, 28–31].

The contamination of food happened due to multiple reasons, such as contamination during long-chain processing of foods, baking, roasting, heating, fermentation, or hydrolysis, transportation of foods, deliberate mixing of additives to the foods for improving the shelf life of a food product, packaging of the foods, and others. Chemical contaminants enter the food chain naturally as well as via pathogens. Chemical contaminants of foods typically include environmental contaminants, food processing contaminants, unapproved food additives, migrants from packaging materials, etc. Among the environmental contaminants, the toxic heavy metals entered foods via anthropogenic activities and/or naturally through water, air, or soil.

The remediation strategies to mitigate the level of these containments and their hazardous impact on humans' health may vary, which depends on the sources as well as the nature of the containments. These strategies include bioremediation, chemical oxidation, electrokinetic remediation, precipitation, flocculation, adsorption, and others [32].

2. Heavy metals-derived contamination

Heavy metals are naturally occurring elements that have high atomic weight, high density (>5 g/cm^3), even more than water, usually having an atomic weight over 63.5, and the majority are toxic at ppb levels. Examples of heavy metals include lead (Pb), iron (Fe), mercury (Hg), cadmium (Cd), zinc (Zn), arsenic (Ar), copper (Cu), chromium (Cr), and others. Heavy metals have the characteristics of environmental persistence and bioaccumulation and are added to the environment in various ways (through multiple industrial, domestic, agricultural, medical, and technological applications), thus causing "heavy metals derived contamination" [33]. It is a type of chemical contamination in which the undesired presence of inorganic pollutants, "heavy metals," exceeds the permissible limit in the environment, i.e., in the air, water, soil, food, and others. As a result, an undesirable alteration in the natural environment occurs that affects both flora and fauna badly. The heavy metals may be considered

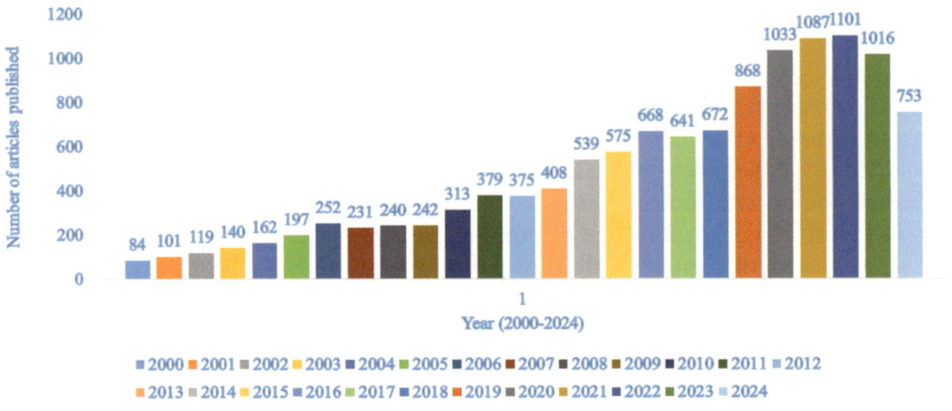

Figure 1.
Number of articles published regarding heavy metals toxicity to human health.

as one of the major types of environmental pollutants that originate from natural or anthropogenic activities [33, 34]. Heavy metals pollute the environment in various aspects, such as directly accumulating and badly affecting soil, air, water, food, sediments, and biota (plants, animals, and aquatic life), thus degrading the overall quality of these components and resultantly making the environment unpleasant for the living systems. Numerous research have been done regarding the heavy metals toxic effect on humans, and the PubMed [34] survey regarding this is shown in **Figure 1**.

The graph given in **Figure 1** was generated by entering the combination of the terms "heavy metal toxicity effect on human health" to gather the literature data published from 2000 to 2024, which reflects the main concept of the chapter.

2.1 Heavy metals in foods

Various biological contaminants (bacteria, viruses, fungi, and others), chemical contaminants, and physiological factors are involved in contaminating foodstuffs, i.e., cake, bread, beans, fish, meat, fruits, milk, herbal drinks, alcoholic and non-alcoholic beverages, etc. Among the chemical contaminants, heavy metals in foods (of both plant and animal origin) are one of the major causes of food contamination, which leads to foodborne diseases, and thus food safety becomes a global issue [34, 35]. The most toxic among the heavy metals that occur in foodstuffs are Pb, Cd, As, and Hg.

2.1.1 Factors influencing the heavy metal contents in foods, their sources, and associated risks

Various factors influence the presence and concentration of these toxic metals in foods, which range from soil type and environmental conditions during the growth of the fruits providing plants and post-harvesting stages such as handling, processing, packaging, cooking/preparation, and others. These factors are responsible for either an increase or decrease in the toxic heavy metal contents in the respective food, e.g., washing after post-harvesting may reduce the metal contents but can be increased if the water itself has heavy metal contaminants. Cooking may reduce metal contents in foods, while processing equipment and packaging may contribute to the heavy metal contents of foods [35, 36].

2.1.2 Heavy metal possibility in foods, their sources, and associated risks

Heavy metals are found in different foods in the form of different species. The most toxic one is arsenic, which may be present in foods in the form of arsenocholine, in organic arsenic, arsenite, arsenate, dimethylarsinate, dimethylarsenate, methylarsonate, methylarsonite, arsenobetaine, arsenosugar, trimethylarsonio, propionate, and trimethylarsine oxid [35, 37–40].

Heavy metals may be classified into essential and non-essential elements on the basis of their demand by the plant-animal-human system. For example, Iron (Fe), Molybdenum (Mo), Copper (Cu), Manganese (Mn), and Zinc (Zn) are important for plants, animals, and human beings. Selenium (Se) and Cobalt (Co) are important to animal-human beings. Nickel (Ni) and Chromium (Cr) are important to plants, human beings. Lead (Pb), Mercury (Hg), Cadmium (Cd), and Arsenic (As) are not needed by any one organism and serve as toxic metals for all the living organisms. Despite the fact that some of the mentioned essential heavy metals are necessary for the animals, and plants, their presence more than permissible limits is highly toxic to the growth of the living beings [41].

The detailed information about the types of heavy metals present in the food, their possible sources, and associated toxicity are shown in the following **Table 1**.

Heavy metals	Foods having respective metals	Possible sources of the respective heavy metals from which it is added to the foods	Potential associate health risks	References
Arsenic	Rice (mostly arsenic is found in highest concentration in rice), fish, meat, poultry, dairy products, cereal grains like wheat, vegetables, fruits, and certain mushrooms.	In crop-based foods and fruits, it is added from soil and the contaminated water that is used for cultivation. Also, it is added from pesticides (e.g., calcium arsenite and copper acetoarsenite-Paris Green) and from herbicides (e.g., methylarsenic acid and dimethylarsenic acid). Seafood directly absorbs it from the contaminated seawater and in seafoods, it is found in the form of Arsencholine and Trimethylarsine oxide. In dairy products, poultry, and other meat, it is more possibly added from the contaminated water used by the animals and poultry.	The first most toxic among the heavy metals causes genetic toxicity, reproductive toxicity, cellular toxicity, chronic kidney diseases, diabetes, obesity, and others.	[35, 37–41]
Lead	Mostly in plant-based foods.	Exposure to lead can occur through contaminated air, water, dust, food, or consumer products that contain inorganic lead. Organic Pb (Tetra-ethyl Pb) is predominantly found in leaded gasoline.	The second most toxic among the heavy metals, it decreases the growth rate of children and affects the nervous system and metabolism.	[35, 42]

Heavy metals	Foods having respective metals	Possible sources of the respective heavy metals from which it is added to the foods	Potential associate health risks	References
Cadmium	The major food groups that contribute to the most Cd exposure are rice and grains, seafood, meat, and vegetables.	Major sources are anthropogenic activities such as the application of phosphate fertilizers, wastewater, sewage sludge, and manures.	Osteoporosis, kidney diseases, cardiovascular diseases, cancer, and others.	[35, 43–45]
Mercury	Marine foods, e.g., shark, swordfish, redfish, etc. Fish protein binds more than 90% of the consumed methylmercury.	Contaminated water, industrial effluents, and mercury contents in low-quality cigarettes and medicines.	Causes damage to the fetus, brain, and kidney.	[35, 46–48]
Antimony	Cereals, vegetables, fruits, meat, poultry, seafood, and dairy products.	Plant-based foods near antimony mines. Antimony, usually in the form of antimony trioxide, enters the environment mainly as a result of industrial activities and via weathering of rocks and runoff from soils.	Cardiotoxicity (49% of patients), pancreatitis, visceral leishmaniasis co-infections. It is also reported as carcinogenic.	[35, 49, 50]
Chromium	Grain products, fruits, broccoli, nuts, spices, brewer's yeast, grape juices, and others.	Chromium is a naturally occurring element. Chromium compounds are stable in the trivalent (III) state. The most common exposure routes of chromium in humans are ingestion, dermal contact, and inhalation. Chromium may get into our food chain through stainless steel, metals, and pottery when they are used as food containers.	Chromium(III) is an essential nutrient. Humans and animals' exposure to this metal in excess or even deficiency is associated with lung cancer, carcinogenicity, respiratory, cardiovascular, gastrointestinal, and musculoskeletal diseases.	[35, 51–53]

**The intensity of toxicity is associated with, (a) type of metals, as the order is As>Pb > Hg > Cd.*

Table 1.
*Most toxic heavy metals, their dietary intake, and associated health risks.**

The possible mechanisms of action of the heavy metals toxicity to the human body are that they involve the production of reactive oxygen species, the inactivation of enzymes/enzymes getting inactivated, and the development of oxidative stress in the human body. Moreover, the heavy metals generate free radicals that promote oxidative stress that subsequently damages the important biomolecules inside the living body, including proteins, lipids, and DNA [54–56]. The heavy metals such as iron, cobalt, copper, manganese, molybdenum, and zinc are required by humans. These are considered biocompatible; however, they are toxic to living organisms at higher concentrations (**Table 2**).

Heavy metals	Sources	Associated risks	References
Cobalt	Red meats, shellfish, fish, leafy green vegetables, legumes, nuts, and cereals. Foods high in vitamin B-12 are the only source of cobalt used by the body. Cobalt metal enters into the soil environment from phosphate fertilizers, industrial wastewater, and sewage sludge. This ultimately entered into foods.	Excessive exposure shows adverse effects like hearing and visual impairment, cardiovascular disease, and others.	[35, 41, 57]
Cupper	Mining, pesticides, chemical industry, and piping.	Liver and renal dysfunction, anemia, intestinal irritation, and stomach problems.	[35, 58]
Zinc	Refineries, brass manufacture, metal plating, and plumbing.	Skin irritation and nervous system disturbances.	[35, 59]
Manganese	It is an essential nutrient. Reports about dietary intake of manganese are very rare; mainly, it is added to the human body via drinking water and inhalation. However, its dietary intake has been reported in the following foods: Grains and grain-based products, vegetables, sugar plants, oilseeds and spices, fruits, peanuts, juices, meat, fish, dairy products, eggs, and others.	In high concentration, it is a potent neurotoxicant but least among all the essential heavy metals/nutrients. It shares a contribution to neurological disorders such as Parkinson's disease.	[35, 58, 60]
Iron	Iron is an essential element for human life since it participates in many functions in the human body, including oxygen transport, immunity, cell division and differentiation, and energy metabolism. Taking in excess of the heme iron (present in animal products such as meat, fish, and poultry) and nonheme iron, which is found in fruits, vegetables, dried beans, nuts, grain products, and meat. Apple has the highest content of iron.	Anemia is one of the major diseases caused by iron deficiency. It may cause cancer via free radical generation.	[35, 58, 61, 62]
Molybdenum	It is a trace element functioning as a cofactor in four enzymes, i.e., sulfite oxidase, aldehyde oxidase, xanthine oxidase, and mitochondrial amidoxime-reducing components. Legumes, grains, and nuts are the main dietary sources of molybdenum in humans.	It is an essential element having no known profound toxicity effects due to its excess or deficiency because the body regulates it efficiently.	[61–63]
Nickel	It enters into the environment from anthropogenic activity and natural sources and is ubiquitous in the environment. Monovalent, divalent, and trivalent forms of nickel occur in living organisms. The Food and Drug Administration has not established daily values for nickel to be used for food and dietary supplement labeling purposes. Chocolate, nuts, dried beans and peas, and grains are considered rich in nickel.	Allergy, cardiovascular and kidney diseases, lung fibrosis, lung and nasal cancer in humans are associated to some extent with the nickel.	[61, 64–66]

Table 2.
Biocompatible heavy metals that are toxic at high concentrations.

Elements such as aluminum, beryllium, barium, silver, gold, thallium, tellurium, tin, titanium, vanadium, and uranium have no known biological functions. These elements may be toxic at very high concentrations in the body, but not to the same extent as arsenic and other toxic metals listed in **Table 1**. The permissible limits of all the toxic heavy metals in different foods are given in the literature [35, 58, 67].

Considering the global perspectives of the heavy metals-mediated contamination in foodstuffs, recent studies revealed that the vegetables grown by using contaminated irrigation water in China were found with the highest levels of Pb and Cr among the other heavy metals. The vegetables grown in Bangladesh have shown the highest contents of Cu, Cd, and Pb. Among the European countries, the vegetables cultivated in mining-affected areas of Romania have shown high levels of heavy metals, including Pb, Cu, and Cd. Tomatoes from Romania have shown very high contents of Cu. The contamination of foods in these regions poses long-term health risks to the connected populations. The detailed data reported in the literature [68, 69] is presented in **Table 3**.

Country	Vegetable/food item	Heavy metal	Contamination level mg/kg
Bangladesh	Spinach	Cu	5.59
	Radish		4.45
	Carrot		5.35
	Brinjal		6.69
	Spinach	Pb	0.98
	Brinjal		0.83
	Cabbage		0.60
	Turnip	Cd	0.06
	Radish		0.65
	Carrot		0.06
China	Spinach	Pb	4.5
	Lettuce		5.8
	Turnip		5.2
	Maize	Cr	0.08–0.38
Spain	Lettuce	Ni^{2+}	< 0.02
Brazil	Brinjal	Pb	0.44
Romania	Lettuce	Cd^{2+}	0.97–1.52
	Tomato	Cu^{2+}	32.5–46.0
Florida, USA	Lettuce	As^{3+}	27.3
Egypt	Orange	Cu^{2+}	0.36–2.2

Table 3.
Global overview of the level of different toxic heavy metals demonstrating the contamination of different vegetables/food items.

2.2 Remediation techniques to reduce the heavy metals-based contamination in foods

There are a number of remediation strategies reported to mitigate the heavy metals in foods and subsequently ensure the safety of human health from their toxicity. Majority of the reported techniques focus on a single point in eliminating these elements from foods, which is to target the root causes/sources of these heavy metals from where they enter into foods and subsequently affect human health and the environment. A report generated from PubMed [70] about remediation strategies to mitigate the heavy metal contaminants from foods is given in **Figure 2**. The data was generated by entering a combination of the specific keywords (heavy metals remediation foods in the PubMed), which presented the number of published papers focusing on the keywords and gave an overview of the interest of the scientists during the mentioned span of time.

The various adopted remediation techniques can be classified broadly into physical, chemical, biological, and integrated approaches [71, 72].

2.2.1 Physical remediation

The physical remediation techniques in foodstuffs via targeting their source material/soil may include washing of soil, soil extraction, soil solidification, and soil stabilization of the heavy metals [73]. The physical remediation approach also includes thermal desorption. In thermal desorption, pollutants are separated from the soil by their volatilization through direct or indirect heating from the soil. In this process, steam, microwave, or infrared radiations are used to volatilize the heavy metal pollutants from the soil, such as mercury, arsenic, and others [74]. Soil replacement is also used as a heavy metals treatment method for soil. It is a short-distance replacement method for moderate quantities of extremely contaminated shallow soil, replacing the contaminated soil with uncontaminated soil to reduce the heavy metal content in the soil [72].

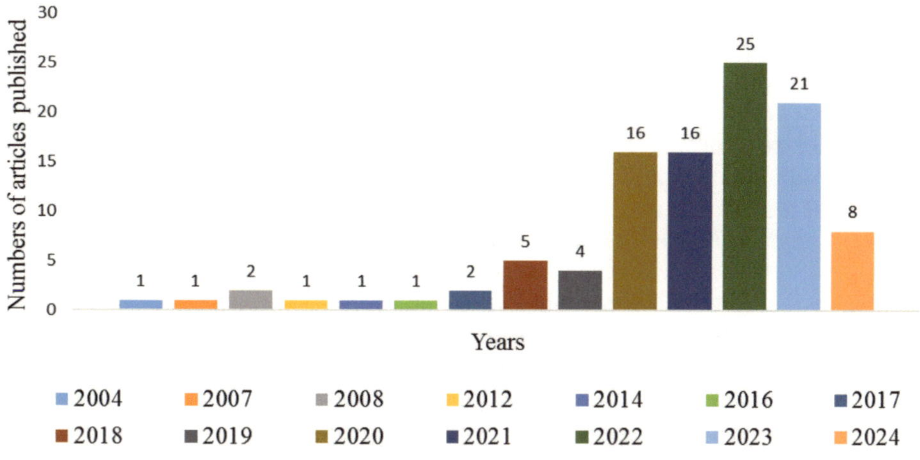

Figure 2.
Number of papers published about remediation techniques applied for heavy metals reduction from foods.

Zhou and his team categorized the physical remediation method into three types: soil substitution, sweeping of soil, and the importation of land from uncontaminated sites. Among these, the substitution of the soil technique is very suitable for land in a small area with contamination. Despite the advantages of the physical remediation techniques for heavy metals from the soil, the physical heavy metals remediation methods are associated with some disadvantages, such as being time-consuming, laborious, and not economically viable methods [73, 75].

2.2.2 Chemical remediation

Chemical remediation approach involves the decrease in mobility, availability, and toxicity of heavy metals in soil by injecting chemical remediation agents and applying the functions of adsorption, precipitation, and oxidation [72]. An example is the use of a chelating agent to improve metal accumulation and its detoxification in different plant parts without influencing plant growth. Chelators usually have functional groups that may coordinate through their lone pairs, e.g., S, N, and O atoms are commonly used as ligand atoms in chemical groups such as –SH, –S–S, $–NH_2$, =NH, –OH, $–OPO_3H$, or > C=O. Peptides are the best chelators among others. The application of the use of chelating agents in the soil site is also reported; however, this approach can also be applied directly to the human body, which results in complex formation with the respective chelator, e.g., EDTA complexes and others. The chelation complexes are formed non-toxic and aid in their excretion from the site of deposition [76]. In the last few years, nanotechnological approaches have been increasingly used in heavy metal remediation from soil, solid waste, and wastewater, which are the main sources of heavy metals-mediated contamination in foods. The nanosubstances reported for the mentioned purpose are iron oxide nanoparticles, ferrite nanoparticles, magnetic multiwall carbon nanotubes, nanoclays, nanomembranes, and others. These substances serve as direct or indirect agents for heavy metals remediation from foods or food sources [77, 78].

2.2.3 Bioremediation

Bioremediation is a broad category and is one of the most attractive to eliminate or degrade heavy metals from the soil or directly from the organism through conversion of the contaminant into less toxic substances or through reduction of the pollutant toxicity via metabolic activity of the organism. This remediation approach can be subclassified into microbial remediation, the nanobioremediation approach, the carbon-based biochar approach, phytoremediation strategies, and others [79]. In the microbial remediation approach, heavy metal-tolerable microbes are applied in the field for advanced genetic development [79, 80]. The nanobioremediation is one of the most efficient approaches to overcoming organic contamination in soils. NPs transform or absorb the HMs present in the soil, therefore minimizing their bioavailability as well as their mobility. For instance, the mobility of cadmium was reported to be decreased with the application of Fe3O4 NPs [81].

The carbon-based biochar approach is used as an ideal adsorbent applied to heavy metals-enriched sites to mitigate the heavy metals level. For effective sorption, some factors like the addition of organic additives, variation in temperature, and redox complexation are taken into consideration. Recently a new modified biochar with earthworms called "vermibiochar" has been introduced to the contaminated soil and water sites to mitigate the heavy metals concentration, and subsequently their concentration in respective foods and juices reduced [79, 81].

Phytoremediation of heavy metals is a cheaper and greener approach that is used widely to overcome the heavy metals-derived contamination. According to this approach, plants having high heavy metal accumulation capacity are grown on the sites having high contentment of heavy metals. For example, a novel arsenic-hyperaccumulating fern (*Pteris vittata*) is widely reported as a sustainable phytoremediation tool for the reclamation of arsenic-contaminated soils. The phytoremediation techniques work through the implementation of various processes such as phytoextraction, phytostabilization, phytovolatilization, rhizofiltration, and phytofiltration [79, 82].

Compared to the physicochemical method, bioremediation of toxic heavy metals in food chain contamination through soils is the most advantageous approach with respect to environmental safety, field scale application, and economical approaches. Some disadvantages include the practical and technical challenges faced during the field scale application [83].

In addition to the remediation approaches listed above, there could be others, some of which are currently being developed, to address the problem posed by the toxicity of heavy metals in foods and human health, as these toxic metals pose a major risk to the environment as a whole and human beings in particular.

3. Conclusion

Majority of the heavy metals, even in trace quantities, while the biocompatible heavy metals (iron, cobalt, copper, manganese, molybdenum, and zinc) in high concentrations can cause diverse health problems, including neuro disorders, skin problems, diabetes, chronic kidney diseases, cardiovascular diseases, cancer, and others. The research publications trend and the literature review presented in the chapter indicate the alarming situation of the issues related to the heavy metals-based contamination in foods. It can be inferred that different methods, including physical, chemical, and bioremediation methods, can be adopted to overcome the risk factor associated with the toxicity of heavy metals from foodstuffs. Among these methods, the different bioremediation methods are highly applied to address the mentioned health problem caused by heavy metals toxicity.

Despite the effective methods, the complexity of foods may be a challenge for the successful application or result-oriented application of the mentioned strategies. Different global standards regarding the permissible limits and adaptation of the advanced method due to the low economic strength of the developing countries may also be a great challenge in adopting a uniform and advanced strategy to combat the heavy metals associated risks.

In addition to the effective remediation strategies, and even after their application to address the ongoing challenges associated with the heavy metals-mediated contamination in foods, the following recommendations are made:

a. Balanced diet is needed to maintain the level of heavy metals within the permissible limits or which is required by a human body.

b. Minimize the anthropogenic activities that are responsible for heavy metals contamination in soil or directly contaminating foods.

c. More plantation is needed of those plants that have the capacity to contribute to phytoremediation.

d. Bioremediation approaches, among others, are highly recommended due to their less costly and more living-being-friendly nature.

e. Maximum awareness of the public about the heavy metals toxicity is needed.

f. Advocating for adopting uniform global standards is recommended.

g. More investment in the research and development sector in the related research area is needed, and particularly global organizations like WHO should provide financial support to developing countries for this purpose.

The present chapter discussed the foodstuff-based heavy metals associated toxicity effects on human health and remediation strategy in detail, but still, the chapter has some limitations. Although a reference (having a table) about the data about permissible limits of the heavy metals from foodstuff is given, however, the data in the form of the table within the chapter is lacking. Lack of separate tables about the comparison of the foodstuffs enriched in toxic heavy metals as well as the comparison of the remediation strategies are some limitations of the present chapter.

Author details

Hamid Ullah[1*], Abdul Majeed Khan[2], Sadeeq Ullah[1], Muhammad Asif[3], Muhammad Imtiaz[4] and Bibi Shazia[1]

1 Department of Chemistry, Balochistan University of Information Technology, Engineering and Management Sciences, Takatu Campus, Quetta, Pakistan

2 General Studies Department, Jubail Industrial College, Jubail, Industrial City, Kingdom of Saudi Arabia

3 Department of Biotechnology, BUITEMS, Takatu Campus, Quetta, Pakistan

4 Department of Chemistry, Mohi-Ud-Din Islamic University Nerian Sharif, AJ&K, Pakistan

*Address all correspondence to: hamidullah9@gmail.com

References

[1] Harper C, Snowden M. Environment and Society: Human Perspectives on Environmental Issues. London, United Kingdom: Routledge; 2017

[2] Thompson LA, Darwish WS. Environmental chemical contaminants in food: Review of a global problem. Journal of Toxicology. 2019;**2019**(1):2345283

[3] Rasheed T, Bilal M, Nabeel F, Adeel M, Iqbal HM. Environmentally-related contaminants of high concern: Potential sources and analytical modalities for detection, quantification, and treatment. Environment International. 2019;**122**:52-66

[4] Sauvé S, Bernard S, Sloan P. Environmental sciences, sustainable development and circular economy: Alternative concepts for trans-disciplinary research. Environmental Development. 2016;**17**:48-56

[5] Shetty SS, Sonkusare S, Naik PB, Madhyastha H. Environmental pollutants and their effects on human health. Heliyon. 2023;**9**:e19496

[6] Swain CK. Environmental pollution indices: A review on concentration of heavy metals in air, water, and soil near industrialization and urbanisation. Discover Environment. 2024;**2**:5. DOI: 10.1007/s44274-024-00030-8

[7] Aljamali NM, Jawad AM, Jawad A. A literature review on types of contamination (biological, chemical, medical). International Journal of International Journal of Green Chemistry. 2019;**5**(1):7-14

[8] Salami R, Kordi M, Delangiz N, Moghiseh E, Lajayer BA, Keswani C, et al. Biological contamination and the control of biological contaminants in the environment. Hazardous and Trace Materials in Soil and Plants. 2022:9-14

[9] Manisalidis I, Stavropoulou E, Stavropoulos A, Bezirtzoglou E. Environmental and health impacts of air pollution: A review. Frontiers in Public Health. 2020;**8**:14

[10] Parker LA, Moreno-Garijo A, Chilet-Rosell E, Lorente F, Lumbreras B. Gender differences in the impact of recommendations on diagnostic imaging tests: A retrospective study 2007-2021. Life. 2023;**13**(2):289

[11] Autsavapromporn N, Krandrod C, Klunklin P, Kritsananuwat R, Jaikang C, Kittidachanan K, et al. Health effects of natural environmental radiation during burning season in Chiang Mai, Thailand. Life. 2022;**12**(6):853

[12] Ahmad N, Uddin Z, Rehman JU, Bakhsh M, Ullah H. Evaluation of radon concentration and heavy metals in drinking water and their health implications to the population of Quetta, Balochistan, Pakistan. International Journal of Environmental Analytical Chemistry. 2020;**100**(1):32-41

[13] Ahmad N, Nasir T, Ur Rehman J, Ullah H, Uddin Z. Risk assessment of radon in soil collected from chromite mines of Khanozai and Muslim Bagh, Balochistan, Pakistan. Environmental Technology and Innovation. 2019;**16**: 100476

[14] Ahmad N, Nasir T, Rizwan S, Ullah H, Bakhsh M. Evaluation of 222Rn and 226Ra concentrations in cement and limestone of sheikh Buddin Hill, Pezu, Pakistan using different techniques. International Journal of

Environmental Analytical Chemistry. 2019;**99**(7):683-691

[15] Sharma S, Bhattacharya A. Drinking water contamination and treatment techniques. Applied Water Science. 2017;**7**:1043-1067

[16] Lin L, Yang H, Xu X. Effects of water pollution on human health and disease heterogeneity: A review. Frontiers in Environmental Science. 2022;**10**: 880246

[17] Shoaib M, Khan; Manzoor Iqbal Khattak; Nida Kazmi; Adnan Afridi; Munnaza Saeed; Hamid Ullah. Removal of selected heavy metals through magnetic graphitic carbon nano structures from drinking water samples of the various areas of Balochistan. Journal of Population Therapeutics and Clinical Pharmacology. 2024;**31**:1854-1891

[18] Khan MS, Khattak MI, Kazmi N, Afridi A, Saeed M, Ullah H. Synthesis and characterization of nanoparticles for the removal of heavy metals from drinking water of dams in Balochistan. Journal of Population Therapeutics and Clinical Pharmacology. 2024;**31**(5):1560-1577

[19] Wali S, Khattak MI, Khattak MI, Afridi A, Ullah H. Synthesis, characterization and application of Sibi bentonite clay-graft-poly (acrylamide/ co- acrylic acid) superabsorbent composite and its absorbency study of water. Journal of Population and Therapeutics and Clinical Pharmacology. 2024;**31**:2022-2033

[20] Nawaz F, Ullah Khan B, Ullah H, Islam ZU, Ullah Khan G, Khayam M, et al. Efficient hydroxyl radical production with sodium salts of surfactants ozonation for treatment of organic acid pollutants in wastewater. Environmental Technology and Innovation. 2021;**22**:101451

[21] Liu YR, Van der Heijden MG, Riedo J, Sanz-Lazaro C, Eldridge DJ, Bastida F, et al. Soil contamination in nearby natural areas mirrors that in urban greenspaces worldwide. Nature Communications. 2023;**14**(1):1706

[22] Zhao FJ, Ma Y, Zhu YG, Tang Z, McGrath SP. Soil contamination in China: Current status and mitigation strategies. Environmental Science and Technology. 2015;**49**(2):750-759

[23] Lombi E, Wenzel WW, Adriano DC. Soil contamination, risk reduction and remediation. Land Contamination and Reclamation. 1998;**6**(4):183-197

[24] Padmanabhan KK, Barik D. Health hazards of medical waste and its disposal. In: Energy from Toxic Organic Waste for Heat and Power Generation. Cambridge, United Kingdom: Woodhead Publishing; 2019. pp. 99-118

[25] Ghali H, Cheikh AB, Bhiri S, Bouzgarrou L, Rejeb MB, Gargouri I, et al. Health and environmental impact of hospital wastes: Systematic review. Dubai Medical Journal. 2023;**6**(2):67-80

[26] Janik-Karpinska E, Brancaleoni R, Niemcewicz M, Wojtas W, Foco M, Podogrocki M, et al. Healthcare waste—A serious problem for global health. In: Healthcare. Vol. 11, No. 2. Basel, Switzerland: MDPI; 2023. p. 242

[27] Tufail M, Khalid S. Heavy metal pollution from medical waste incineration at Islamabad and Rawalpindi, Pakistan. Microchemical Journal. 2008;**90**(1):77-81

[28] Thompson LA, Darwish WS. Environmental chemical contaminants in

food: Review of a global problem. Journal of Toxicology. 2019;**2019**(1):2345283

[29] Rasheed H, Iqbal S, Ullah H, Akbar A, Farooqi A, Syed S. Analysis of heavy metals as toxic elements in cosmetics available at local markets of Quetta, Balochistan, Pakistan. International Journal of Environmental Analytical Chemistry. 2023;**103**(1): 57-70

[30] Zahoor S, Iqbal S, Hashmi I, Akbar A, Ullah H, Panezai H, et al. Chemical and biological food safety threats associated with fresh juices consumed in capital territory Islamabad, Pakistan. Progress in Nutrition. 2021;**23**(4):e2021144

[31] Rehman JU, Ahmad N, Ullah N, Alam I, Ullah H. Health risks in different age group of nitrate in spring water used for drinking in Harnai, Balochistan, Pakistan. Ecology of Food and Nutrition. 2020;**59**(5):462-471

[32] Jovanović T, Petrović M, Kostić M, Bojić D, Bojić A. Chemical remediation technologies. Facta Universitatis, Series: Physics, Chemistry and Technology. 2021;**19**(1):1-15

[33] Singh A, Sharma A, Verma RK, Chopade RL, Pandit PP, Nagar V, et al. Heavy Metal Contamination of Water and their Toxic Effect on Living Organisms, the Toxicity of Environmental Pollutants. Vol. 1. London, UK: Intech open; 2020. pp. 1-19

[34] PubMed. Available from: https://pubmed.ncbi.nlm.nih.gov/?term=heavy+metal+toxicity+effect+on+human+health&filter=years.2000-2024&sort=date&sort_order=asc

[35] Deka AK, Kumar KJ, Basumatary S. Monitoring strategies for heavy metals in foods and beverages: Limitations for human health risks. In: Heavy Metals-Recent Advances. London, UK: IntechOpen; 2023

[36] Morgan JN. Effects of processing on heavy metal content of foods. In: Jackson LS, Knize MG, Morgan JN, editors. Impact of Processing on Food Safety. Advances in Experimental Medicine and Biology. Vol. 459. New York, NY, USA: Springer Nature; 1999

[37] Khan MI, Ahmad MF, Ahmad I, Ashfaq F, Wahab S, Alsayegh AA, et al. Arsenic exposure through dietary intake and associated health hazards in the Middle East. Nutrients. 2022;**14**(10):2136

[38] Molin M, Ulven SM, Meltzer HM, Alexander J. Arsenic in the human food chain, biotransformation and toxicology–review focusing on seafood arsenic. Journal of Trace Elements in Medicine and Biology. 2015;**31**:249-259

[39] Yoshinaga J, Narukawa T. Dietary intake and urinary excretion of methylated arsenicals of Japanese adults consuming marine foods and rice. Food Additives and Contaminants: Part A. 2021;**38**:622-629

[40] Upadhyay MK, Shukla A, Yadav P, Srivastava S. A review of arsenic in crops, vegetables, animals and food products. Food Chemistry. 2019;**276**:608-618

[41] Khan ZI, Liu W, Mubeen I, Alrefaei AF, Alharbi SN, Muhammad FG, et al. Cobalt availability in the soil plant and animal food chain: A study under a peri-urban environment. Brazilian Journal of Biology. 2023;**83**:e270256

[42] Kumar A, Kumar A, Cabral Pinto MMS, Chaturvedi AK, Shabnam AA, Subrahmanyam G, et al. Lead toxicity: Health hazards, influence on food chain, and sustainable remediation approaches. International

Journal of Environmental Research and Public Health. 2020;**17**(7):2179

[43] Chunhabundit R. Cadmium exposure and potential health risk from foods in contaminated area, Thailand. Toxicological Research. 2016;**32**(1):65-72

[44] Pokharel A, Wu F. Dietary exposure to cadmium from six common foods in the United States. Food and Chemical Toxicology. 2023;**178**:113873

[45] Qing Y, Li Y, Cai X, et al. Assessment of cadmium concentrations in foodstuffs and dietary exposure risk across China: A metadata analysis. Exposure and Health. 2023;**15**:951-961

[46] Bernhoft RA. Mercury toxicity and treatment: A review of the literature. Journal of Environmental and Public Health. 2012;**2012**:460508

[47] Carocci A, Catalano A, Lauria G, Sinicropi MS, Genchi G. 16 a review on mercury toxicity in food. Food. Toxicology. 2016;**315**:315-325

[48] Khan EA, Abbas Z. A scoping review of sources of mercury and its health effects among Pakistan's most vulnerable population. Reviews on Environmental Health. 2020;**36**(1):39-45

[49] Cheung Chung SW, Kwong KP, Yau JC, Wong WW. Dietary exposure to antimony, lead and mercury of secondary school students in Hong Kong. Food Additives and Contaminants. 2008;**25**(7):831-840

[50] Belzile N, Chen YW, Filella M. Human exposure to antimony: I. Sources and intake. Critical Reviews in Environmental Science and Technology. 2011;**41**(14):1309-1373

[51] Genchi G, Lauria G, Catalano A, Carocci A, Sinicropi MS. Prevalence of cobalt in the environment and its role in biological processes. Biology. 2023;**12**(10):1335

[52] Sazakli E. Human health effects of Oral exposure to chromium: A systematic review of the epidemiological evidence. International Journal of Environmental Research and Public Health. 2024;**21**:406

[53] Chatterjee S. Chromium toxicity and its health hazards. International Journal of Advanced Research. 2015;**3**(7): 167-172

[54] Balali-Mood M, Naseri K, Tahergorabi Z, Khazdair MR, Sadeghi M. Toxic mechanisms of five heavy metals: Mercury, lead, chromium, cadmium, and arsenic. Frontiers in Pharmacology. 2021;**12**:643972

[55] Valko MMHCM, Morris H, Cronin MTD. Metals, toxicity and oxidative stress. Current Medicinal Chemistry. 2005;**12**(10):1161-1208

[56] Jaishankar M, Tseten T, Anbalagan N, Mathew BB, Beeregowda KN. Toxicity, mechanism and health effects of some heavy metals. Interdisciplinary Toxicology. 2014;**7**(2):60-72

[57] Shin DY, Lee SM, Jang Y, Lee J, Lee CM, Cho EM, et al. Adverse human health effects of chromium by exposure route: A comprehensive review based on toxicogenomic approach. International Journal of Molecular Sciences. 2023;**24**(4):3410

[58] Singh R, Gautam N, Mishra A, Gupta R. Heavy metals and living systems: An overview. Indian Journal of Pharmacology. 2011;**43**(3):246-253

[59] Sachse B, Kolbaum AE, Ziegenhagen R, Andres S, Berg K, Dusemund B, et al. Dietary manganese

exposure in the adult population in Germany—What does it mean in relation to health risks? Molecular Nutrition and Food Research. 2019;**63**(16):1900065

[60] Goyer R. Issue Paper on the Human Health Effects of Metals. Washington, D.C: US Environmental Protection Agency; 2004

[61] Shubham K, Anukiruthika T, Dutta S, Kashyap AV, Moses JA, Anandharamakrishnan C. Iron deficiency anemia: A comprehensive review on iron absorption, bioavailability and emerging food fortification approaches. Trends in Food Science and Technology. 2020;**99**:58-75

[62] Novotny JA. Molybdenum nutriture in humans. Journal of Evidence-Based Complementary and Alternative Medicine. 2011;**16**(3):164-168

[63] Mania M, Rebeniak M, Postupolski J. Food as a source of exposure to nickel. Roczniki Państwowego Zakładu Higieny. 2019;**70**(4):393-399

[64] Khan ZI, Ahmad K, Ahmad T, Zafar A, Alrefaei AF, Ashfaq A, et al. Evaluation of nickel toxicity and potential health implications of agriculturally diversely irrigated wheat crop varieties. Arabian Journal of Chemistry. 2023;**16**(8):104934

[65] Nielsen F. Nickel. Advances in nutrition (Bethesda, Md.). 2021;**12**(1):281-282

[66] Genchi G, Carocci A, Lauria G, Sinicropi MS, Catalano A. Nickel: Human health and environmental toxicology. International Journal of Environmental Research and Public Health. 2020;**17**(3):679

[67] Codex Alimentarius Commission. General Standard for Contaminants and Toxins in Food and Feed. CODEX STAN 193-1995. Rome, Italy: Codex Alimentarius Commission; 2016

[68] Scutarașu EC, Trincă LC. Heavy metals in foods and beverages: Global situation, health risks and reduction methods. Food. 2023;**12**(18):3340

[69] Sharma A, Nagpal AK. Contamination of vegetables with heavy metals across the globe: Hampering food security goal. Journal of Food Science and Technology. 2020;**57**:391-403

[70] PubMed. Available from: https://pubmed.ncbi.nlm.nih.gov/?term=heavy+metals+remediation+foods&filter=simsearch1.fha&filter=simsearch2.ffrft&filter=simsearch3.fft&filter=pubt.review&filter=pubt.systematicreview

[71] Luo Y, Zhang Y, Xiong Z, Chen X, Sha A, Xiao W, et al. Peptides used for heavy metal remediation: A promising approach. International Journal of Molecular Sciences. 2024;**25**(12):6717

[72] Alsafran M, Saleem MH, Al Jabri H, Rizwan M, Usman K. Principles and applicability of integrated remediation strategies for heavy metal removal/recovery from contaminated environments. Journal of Plant Growth Regulation. 2023;**42**(6):3419-3440

[73] Dhaliwal SS, Singh J, Taneja PK, Mandal A. Remediation techniques for removal of heavy metals from the soil contaminated through different sources: A review. Environmental Science and Pollution Research. 2020;**27**(2):1319-1333

[74] Abdullahi A, Lawal MA, Salisu AM. Heavy metals in contaminated soil: Source, accumulation, health risk and remediation process. Bayero Journal of Pure and Applied Sciences. 2021;**14**(1):1-12

[75] Zhou DM, Hao XZ, Xue Y, Cang L, Wang YJ, Chen HM. Advances in remediation technologies of contaminated soils. Journal of Ecology and Environmental Sciences. 2004;**13**:234-242

[76] Sarker A, Kim JE, Islam ARMT, Bilal M, Rakib MRJ, Nandi R, et al. Heavy metals contamination and associated health risks in food webs-a review focuses on food safety and environmental sustainability in Bangladesh. Environmental Science and Pollution Research International. 2022;**29**(3):3230-3245

[77] Kumar S, Prasad S, Yadav KK, Shrivastava M, Gupta N, Nagar S, et al. Hazardous heavy metals contamination of vegetables and food chain: Role of sustainable remediation approaches - a review. Environmental Research. 2019;**179**(Pt A):108792. DOI: 10.1016/j.envres.2019.108792

[78] Bhattacharyya KG, Sen Gupta S. Adsorption of chromium (VI) from water by clays. Industrial and Engineering Chemistry Research. 2006;**45**:7232-7240

[79] Yin K, Wang Q, Lv M, Chen L. Microorganism remediation strategies towards heavy metals. Chemical Engineering Journal. 2019;**360**: 1553-1563

[80] Mathur S, Singh D, Ranjan R. Remediation of heavy metal (loid) contaminated soil through green nanotechnology. Frontiers in Sustainable Food Systems. 2022;**6**:932424

[81] Mansoor S, Kour N, Manhas S, Zahid S, et al. Biochar as a tool for effective management of drought and heavy metal toxicity. Chemosphere. 2020;**271**:129458

[82] Abd Elnabi MK, Elkaliny NE, Elyazied MM, Azab SH, Elkhalifa SA, Elmasry S, et al. Toxicity of heavy metals and recent advances in their removal: A review. Toxics. 2023;**11**(7):580

[83] Khalid S, Shahid M, Niazi NK, Murtaza B, Bibi I, Dumat C. A comparison of technologies for remediation of heavy metal contaminated soils. Journal of Geochemical Exploration. 2017;**182**:247-268

Chapter 10

Perspective Chapter: The Dual Nature and Applications of Uranium

Angelica A. Chacon and Carlos R. Cabrera

Abstract

Uranium is a unique element, characterized by its dual role as a valuable resource and a source of significant hazards. Widely used in industry, research, and medicine, uranium has played a pivotal role in shaping modern science and technology. Since the discovery of its fission properties in the early twentieth century, uranium has gained attention as a powerful and influential element. Historically, uranium's role in the discovery of radioactivity marked a turning point in scientific research, and its ability to release immense energy through nuclear fission has transformed human history. From its destructive potential in warfare to its peaceful applications in electricity generation via nuclear power plants, uranium has had a profound impact. Furthermore, it has advanced medical science through the production of isotopes for diagnostic imaging and therapeutic treatments. However, increased demand for uranium has led to expanded mining and milling processes, which pose public health risks due to environmental contamination. This chapter explores uranium's legacy, highlighting its scientific significance while balancing the potential benefits with the inherent risks associated with its use. The discussion underscores the need for careful consideration and management of uranium's dual nature as both a powerful resource and potential source of harm.

Keywords: uranium, use of uranium, duality, environment, health effects

1. Introduction

Uranium, the heaviest naturally occurring element, stands at the intersection of immense potential and significant risk. With an atomic weight surpassing those of all other naturally occurring elements, uranium contains 92 protons and 92 electrons, including 6 valence electrons [1]. As an alpha particle emitter, uranium undergoes a complex decay process, with its decay chain beginning with Uranium-238 (U-238) culminating in stable Lead-206 (Pb-206) [2]. Discovered in 1789 by the German chemist Martin Heinrich Klaproth in a pitchblende specimen and named after the planet "Uranus," uranium initially gained attention for its ability to create vibrant pigments [3]. However, its radioactive properties later unveiled far deeper significance, leading to transformative events in history, science, and medicine.

Uranium's evolution from a material used in decorative arts to cornerstone of scientific discovery exemplifies the complex interplay of beauty, power, and scientific importance [4]. It is most widely recognized today for its critical role in fueling nuclear reactors, providing a powerful source of energy that drives electricity generation around the world [4]. This potent energy offers an alternative to fossil fuels and contributing to the global push for cleaner energy.

However, as the world transitions toward sustainable energy solutions, the global demand for uranium is expected to rise, driving a surge in mining activities [5]. This, coupled with other anthropogenic factors such as waste and the use of agricultural fertilizers, is contributing to higher uranium levels in the environment, thereby heightening environmental and health concerns [6].

This duality presents a profound challenge: how to maximize the benefits of uranium while minimizing its dangers. As the global demand for energy and medical advancements continues to increase, so too does the need for uranium. This chapter explores the dual nature of uranium, tracing its historical significance, scientific contributions, and environmental implications of its use. By understanding both the opportunities and the risks associated with uranium, we can better navigate the path forward in managing this powerful yet perilous element.

2. Uranium in nature and its evolution

Uranium is a naturally occurring element that is ubiquitously found in soils, minerals, and water bodies across the earth [7]. It ranks as the 49th most abundant element in Earth's crust and is lithophile, radioactive, and one of the heaviest naturally occurring elements [7]. Uranium is considered one of the primordial elements that were present in the Earth's formation [8]. Its significant role in the planet's natural radioactivity is largely influenced by the long half-lives of its isotopes [8]. Uranium's abundance is approximately four parts per million (ppm), making it relatively common and widely distributed in the Earth's crust [9]. Uranium is typically found in minerals, such as carnotite, uraninite, and autunite [9].

The geological evolution of uranium, along with thorium, began approximately 4.5–3.5 billion years ago (Ga), a critical era in Earth's history characterized by significant transformations within the planet's crust [10]. During this period, the Earth transitioned from its initial molten state to a more solidified form as it cooled. This cooling process led to the reduction of magmatic materials in the crust, which in turn facilitated the formation of early tetravalent actinide minerals. Notable among these early minerals are uraninite (UO_2), thorianite (ThO_2), and coffinite ($USiO_4$), which began to precipitate as the Earth's crust began to solidify [10].

These early minerals, in particular uraninite, are indicators of the redox conditions prevalent during the planet's formative years. The oxidation states of uranium found in these minerals—such as IV^{4+}, V^{5+}, and VI^{6+}—reflect the changing environment conditions driven by interactions with microorganisms [10]. These interactions between geological and biological process not only influenced the mineral composition but also played a significant role in shaping oxidative states of uranium emerging from redox conditions of the time.

Containing 92 protons, uranium is the element with the highest atomic number naturally occurring in the periodic table. The primary isotopes of uranium are Uranium-238 (U-238) with a half-life of 4.51×10^9 years and Uranium-235 (U-235), with a half-life of 7.13×10^8 years [11]. These isotopes differ in the number of neutrons, affecting their stability and radioactivity. U-238 is the most abundant isotope, comprising about 99.3% of natural uranium, while U-235 accounts for approximately 0.7% [11]. Although these isotopes are chemically identical, they exhibit different behaviors in nuclear reactions. U-235, with 143 neutrons, is capable of undergoing fission when it absorbs a neutron, releasing energy in the process. In contrast, U-238, with 146 neutrons, absorbs a neutron to transform into Uranium-239. Uranium-239 subsequently undergoes two beta decays to become Plutonium-239, which is also a fissionable material [12]. The different nuclear behaviors of these isotopes play a significant role in their applications and the understanding of uranium's role in natural radioactivity.

Both U-238 and U-235 isotopes are alpha emitters, meaning they emit alpha particles during radioactive decay [12]. The decay series of uranium ultimately result in stable, non-radioactive lead isotopes: Lead-206 (Pb-206) from U-238 and Lead-207 (Pb-207) from U-235 [12]. The alpha decay process involves the emission of helium nuclei (alpha particles), reducing the mass and atomic mass and atomic numbers of the parent isotopes, and is a critical aspect of uranium's radioactivity [12].

Uranium came into the spotlight with the discovery of nuclear fission in 1938 and subsequent demands of the nuclear industry [13]. Uranium has become, and remains, an ideal nuclear fuel due to its high density and excellent conductivity [13]. Controlling these properties was a key focus during the years 1943–1955, leading to worldwide efforts in several laboratories to separate single crystals of uranium to distinguish its intrinsic properties from extrinsic ones [13]. This challenge was ultimately met by the Argonne National Laboratory [ANL), demonstrating that uranium's fundamental characteristics were indeed intrinsic [13].

In conclusion, uranium's role in natural radioactivity underscores its dual nature. The long half-lives of U-238 and U-235 contribute to the planet's sustained natural radioactivity, influencing various geological and environmental processes, such as the generation of heat within the Earth's crust and the radiometric dating of rock and minerals. Uranium's radioactive decay is a key factor in understanding Earth's geological history and the evolution of its crust.

Uranium's duality is a product of its chemical consistency and its varied nuclear properties, alongside its significant roles in both geological and biological contexts. This duality illustrates the complex and multifaceted nature of uranium, encompassing its chemical uniformity, divergent nuclear behaviors, and influential interactions in natural systems.

3. Uranium: From artistic medium to powerhouse of nuclear science

Uranium has been used as a colorant in glass and ceramic glazes since ancient times, long before its radioactive properties were understood. Initially, uranium's primary applications were artistic, particularly in glassmaking, ceramics, and glazes [14]. However, as scientific understanding advanced, uranium's role extended far beyond esthetics. The groundwork for this transformation was laid by the French chemist Eugène-Melchior Péligot in 1841, when he first successfully isolated metallic uranium, marking a significant advancement in uranium chemistry [8]. Péligot's work transitioned uranium from a material prized for its decorative use to a substance with potential scientific significance, a potential that would soon be realized with the discovery of radioactivity.

In 1896, French physicist Henri Becquerel's groundbreaking discovery of uranium's natural radioactivity unveiled uranium's true scientific importance [8]. Becquerel observed that uranium salts emitted rays capable of exposing photographic plates, even in the absence of light [8]. This discovery marked the beginning of a new era for uranium, connecting its earlier artistic applications to its pivotal role in the development of nuclear science. Uranium was no longer just a medium for vibrant glassware; it became a fundamental element in the understanding of atomic structure and the development of nuclear energy.

This discovery marked the birth of our understanding of radioactivity, a term later coined by Marie Curie. Curie, along with her husband Pierre, developed quantitative techniques to measure the radioactivity of uranium, further unraveling its mysterious nature. Through their meticulous research, they identified and isolated other radioactive elements within uranium ores, such as thorium, polonium, and radium, laying the foundation for modern nuclear science [8].

The Curies' recognition about the high radioactivity of uranium ores spurred further discoveries by radiochemists, leading to the identification of new radioactive elements and their chemical and nuclear compositions. The luminescent properties of radium created a demand for uranium ores in the medical field. It 1939, uranium became a central focus in the field of nuclear energy research following the discovery of nuclear fission by Otto Hahn and Fritz Strassman. The naturally occurring fissionable materials, such as Uranium-235, Uranium-238, and Uranium-233 (U-233), have since become of great scientific and technical importance in the pursuit of nuclear energy [8].

The past and present applications of uranium highlight its dual nature: on one hand, a source of vivid color in artistic creations, and on the other, a powerful source of penetrating radiation that would revolutionize science and medicine. This duality, where uranium transitions from an esthetic material to a critical element in the study of atomic structure and energy, underscores its complex and multifaceted nature.

4. Uranium applications

Uranium's diverse properties make it indispensable across a wide range of fields, from energy production to medical advancements and industrial processes. Its ability to undergo nuclear fission has made it a key component in clean energy alternatives, while its radioactive isotopes are crucial in medical diagnostics and treatments. Additionally, uranium plays a significant role in research and agricultural studies, underscoring its far-reaching impact. This section explores the various applications of uranium, highlighting its contributions to energy, medicine, and industry.

4.1 Clean energy using uranium

Nuclear energy is regarded as one of the safest and most sustainable methods for producing electricity, playing a critical role in the global transition to clean energy [15]. At the heart of nuclear power is uranium, the primary fuel of nuclear reactors [15]. Uranium's unique properties, particularly the fissile U-235, make it the only naturally occurring element capable of sustaining a chain reaction necessary for nuclear fission [15].

The potential of nuclear energy as a long-term renewable source is immense. It is estimated that with advancements in nuclear technology and more efficient fuel cycles, nuclear power could provide up to 9000 years of renewable energy [16]. This future includes the potential for reusing nuclear waste as fuel for advanced reactors, significantly reducing the overall waste produced by current nuclear technologies [16].

In addition to reducing carbon emissions, nuclear energy offers a stable and reliable electricity source, as it is not dependent on weather conditions like solar or wind energy [17]. With ongoing innovations in nuclear technology, including next-generation reactors and improved safety protocols, uranium-based nuclear energy remains a key component in achieving global energy sustainability. This clean energy source has the potential to balance the growing demand for electricity [17].

4.2 Uranium in medicine

Uranium and its isotopes play a pivotal role in the field of nuclear medicine. Particularly, Uranium-230 (U-230) that has emerged along with Thorium-226 as critical elements in cancer therapy through their applications in "targeted alpha therapy" [18]. These isotopes decay by emitting alpha particles, which are highly effective in destroying cancer cells due to their high-energy and short-range penetration [19]. This makes them ideal for delivering precise and potent doses of radiation to tumors while minimizing damage to the surrounding healthy tissues [19, 20].

Uranium-230's therapeutic potential extends beyond targeted alpha therapy. It serves as a valuable source to produce several important radionuclides, such Iodine-131, Molybdenum-99, Strontium-89, and Xenon-133 [21]. Iodine-131 is extensively used in the treatment of thyroid diseases, including thyroid cancer [22]. Molybdenum-99 is essential for generating Technetium-99 m (Tc-99 m), a widely employed diagnostic tool in medical imaging [23]. Strontium-89 has been utilized since 1993 to alleviate pain in patients with bone metastasis, [24] while Xenon-133 is used in lung imaging to aid in the diagnosis of respiratory conditions [25].

When combined with targeted delivery mechanisms, such as antibodies, these alpha-emitting isotopes can potentially offer a promising approach for treating challenging and resistant cancer type [19]. In clinical settings, the alpha particles emitted during decay of Uranium-230 and Thorium-226 provide substantial localized radiation that can more effectively destroy cancer cells compared to conventional beta or gamma radiation treatments [21]. These advancements are part of a broader initiative in nuclear medicine to develop more effective and less invasive cancer treatments, building on the unique radioactive properties of isotope like uranium.

4.3 Uranium in industry

Uranium's versatility extends far beyond its well-known applications in medicine and energy; it also plays a significant role in various industrial processes due to its unique properties, which make it valuable in several sectors:

4.3.1 Research and development

Uranium is central in advancing research into nuclear science and materials. In particular, studies involving uranium have significantly contributed to our understanding of organo-actinide chemistry, which explores the bonding, reactivity, and molecular structures of uranium-containing organic compounds [26]. Research into uranium is essential to refining technologies related to nuclear energy and materials [27].

4.3.2 Agriculture

Uranium has applications in agriculture, particularly in soil and water studies. Uranium isotopes, such as Uranium-238, can be used as tracers to study water movement [28, 29]. Additionally, uranium's radioactive properties can be employed to study the effects of uranium on plant growth [30] and fauna [31], examining potential effects on biodiversity and ecosystems.

Overall, uranium's role industry extends well beyond the nuclear sector, underscoring its diverse applications in research, agriculture, and environmental science.

5. Uranium in the environment

While uranium is naturally present as a part of the Earth's crust, its mobility in the environment is influenced by factors, such as pH levels, redox conditions, and the presence of other minerals and organic matter [2]. These conditions can either immobilize uranium in sediments and soils or increase its solubility, allowing it to migrate into water systems [2]. In aquatic environments, uranium can exist in both dissolved and particulate forms [2]. Groundwater in certain regions, particularly those with uranium-rich geological formations, may exhibit elevated levels of uranium due to the natural leaching of the element from rocks [2].

Typically, concentrations of naturally occurring uranium that are found in the environment are low, posing minimal risk to human health under normal circumstances [32]. In the Earth's crust, uranium concentration ranges from 1 to 4 mg/kg. Surface uranium strongly binds to oxygen forming hexavalent uranium (UO_2^{2+}), which is highly stable, soluble, and mobile [32]. Uranium can be detected in various water bodies, with concentrations in rivers, seawater, and groundwater typically being <4, 3.3, and < 5 μg/L, respectively [7]. In soils, uranium concentrations vary from 0.7 to 10.7 mg/kg [33].

However, uranium's radioactive nature raises concerns about environmental contamination and health risks, particularly in areas affected by human activities, such as oil extraction, mining, nuclear energy production, and agriculture [34]. In these sectors, uranium can become more concentrated in waste products, increasing the risks of environmental contamination and public health hazards, if not properly managed [34]. The persistence of uranium in the environment, along with its radioactive decay products, necessitates careful monitoring and regulation to prevent harmful exposure and contamination [34].

Uranium-238, a naturally occurring radioactive material (NORM), is particularly problematic in industries such as oil and gas, where extraction processes generate waste like sludge, scales, and produced water [35]. These byproducts contain radioactive elements, including uranium, which can contaminate soil, surface water, and groundwater [35]. In regions like the Permian Basin in the U.S., studies have shown elevated levels of uranium in groundwater near drilling sites, with uranium concentrations sometimes exceeding safe drinking water limits [36]. These contaminants pose long-term risks, as the radioisotopes from uranium have half-lives extending millions of years, contributing to persistent environmental hazards [36].

Uranium mining, the first step in nuclear power generation, also poses significant environmental and health risks [37]. Mining operations generate dust and use large volumes of water for dust control and uranium extraction, both of which can spread long-lived radioisotopes into the surrounding environment [37]. Dust particles containing uranium and its decay products can be inhaled, ingested, or absorbed through the skin [38]. Gamma-ray emissions from radioactive materials present in mining waste further elevate exposure risks, particularly for workers [38]. Occupational exposure in mining regions, such as in Namibia, has been linked to increased cases of lung cancer and other radiation-induced illnesses [39].

To illustrate the scale of uranium's environmental impact, a case study from the Navajo Nation in the U.S. provides a sobering example [40]. Over 500 abandoned uranium mines have contaminated local water supplies, leading to uranium concentrations in excess of the U.S. Environmental Protection Agency's (EPA's) 30 μg/L drinking water limit [41]. As a result, there has been a documented increase in health complications among residents, including kidney disease and cancer [38]. This highlights the profound and complex health implications of uranium exposure, especially in communities living near mining operations.

Recent research into uranium's environmental effects has revealed long-term contamination pathways, including groundwater transport, soil accumulation, and bioaccumulation in plants and animals [38]. Once mobilized into water systems, uranium contamination can spread across vast distances, potentially impacting downstream ecosystems and human populations [7, 38]. Furthermore, phosphate fertilizers, a significant source of uranium contamination, exacerbate these issues. In countries like Brazil, uranium concentrations in fertilizers have been recorded at levels ranging from 5.17 to 54.3 μg/g, which pose a substantial risk for groundwater contamination [2].

Addressing these challenges requires a multifaceted approach, including advancements in remediation technologies. Emerging techniques, such as bioremediation, where microbes are used to immobilize uranium in contaminated soils and water, have shown promise [42]. Current research is focusing on remediation strategies, which leverage microbial processes to immobilize uranium and use electrochemical methods and adsorbent materials that can capture and remove uranium from contaminated water [43–47]. Additionally, new adsorbent materials, like modified biochars [48] and metal-organic frameworks (MOFs) [49], are being explored to efficiently remove uranium from contaminated water. These innovative strategies offer hope for managing uranium's environmental footprint more effectively.

In conclusion, uranium embodies a duality of immense potential and significant risk. Its role in nuclear energy underscores its values, but the environmental and health risks associated with its radioactive nature and industrial activities demand vigilant management. The balance between harnessing uranium's benefits and mitigating its dangers remains a critical challenge for the future.

6. Uranium demand and economy

Uranium, with its exceptional energy potential, lies at the heart of one of the most powerful energy sources: nuclear power. As countries move toward adopting sustainable energy solutions and committing to carbon-free emissions, the demand for uranium remains fundamental fuel that powers nuclear reactors.

6.1 Uranium demand in nuclear energy

The demand for uranium is mainly determined by its use in nuclear reactors. In 2020, several countries relied heavily on uranium-based nuclear energy, with France generating 69.11%, Sweden 31.24%, and United States 18.72% of their electricity from nuclear power plants [47]. As of 2023, there were approximately 450 nuclear reactors operating globally, supplying approximately 10% of the world's electricity [50]. Major uranium consumers included the United States, France, China, and Russia, as these countries depended significantly on nuclear power for their electricity generation needs [50].

The International Energy Agency (IEA) projects that global uranium consumption will rise by 39% by 2050 as nuclear energy continues to be integrated into low-carbon energy strategies [51]. This surge is driven by the increasing global demand for electricity and the need to meet climate goals while reducing greenhouse gas emissions from fossil fuels [51].

6.2 Geopolitical and economic influences on uranium demand

Geopolitical factors also play a crucial role in shaping uranium demand. In response to the growing urgency of climate change, countries like the United States have committed to reduce reliance on fossil fuels. Legislation efforts are underway to develop the next generation of nuclear power plants, further boosting uranium demand [51]. This shift toward nuclear energy, however, brings with it challenges related to the environmental agreement.

Historically, the uranium industry has contributed to significant environmental pollution, especially in regions where mining has been extensive [52]. New legislation in the United States seeks to balance the need for uranium with environmental stewardship by exploring the opening of new mines and the potential reopening of old ones [52]. Despite these initiatives, ongoing pollution issues persist, particularly in states like Arizona and New Mexico, which have historically supplied much of the country's uranium [52].

The challenge moving forward lies in balancing uranium's critical role in reducing carbon emissions with environmental and ethical responsibilities that come with its extraction and use. As the demand for uranium rises alongside the global transition to sustainable energy, efforts to minimize its environmental impact and ensure safe, regulated use will become more crucial.

7. Conclusion

Uranium remains a critical resource, essential not only for energy generation but also for its applications in technological advancements across industry, research, and medicine. Its role in the global energy landscape is complex and multifaceted. On one hand, uranium fuels nuclear reactors, providing a low-carbon energy source that helps to meet the rising global demand for electricity. On the other hand, its extraction, radioactive properties, and waste management pose significant environmental and health risks, raising important concerns about the sustainability and safety of its continued use.

As the global energy transition unfolds, the duality of uranium—as both a valuable source and a potential hazard—will become even more pronounced. The future of uranium demand will depend on how the world balances the need for clean energy with the imperative of minimizing environmental and public health risks. Sustainable mining practices, innovative nuclear technologies, and international cooperation will be key to navigating the challenges and opportunities associated with uranium in the twenty-first century. By navigating this delicate balance, we can harness uranium's potential while mitigating its dangers, ensuring that its dual nature is managed responsibly for future generations.

Looking ahead, technological advancements in nuclear waste management, such as more efficient containment systems and recycling methods, will be critical to minimizing uranium's environmental footprint. Additionally, research into alternative energy systems may provide new pathways to harnessing nuclear power while reducing dependency on traditional uranium-based technologies.

In conclusion, uranium's dual role as both a key resource and a potential hazard must be managed responsibly. By adopting forward-looking strategies and fostering international collaboration, we can harness uranium's potential while mitigating its dangers, ensuring that future generations can benefit from its energy potential without compromising environmental and public health.

Acknowledgements

The authors would like to acknowledge the Department of Energy-National Nuclear Security Administration-Minority Serving Institution Partnership Program (MSIPP) Scholarly Partnership in Nuclear Security (SPINS) Consortium, Grant number DE-NA0003980, and Graduate Research Training Initiative for Student Enhancement (G-RISE) program at the University of Texas at El Paso, Grant number: 5T32GM144919-03.

Conflict of interest

The authors declare no conflict of interest.

Author details

Angelica A. Chacon* and Carlos R. Cabrera
The University of Texas at El Paso, El Paso, Texas, United States

*Address all correspondence to: aachacon3@miners.utep.edu

References

[1] U.S Department of Energy Office of Nuclear Energy. Nuclear Energy Facts: Uranium. Available from: https://www.energy.gov/ne/nuclear-fuel-facts-uranium

[2] Balaram V, Rani A, Rathore DPS. Uranium in groundwater in parts of India and world: A comprehensive review of sources, impact to the environment and human health, analytical techniques, and mitigation technologies. Geosystems and Geoenvironment. 2022;**1**(2):100043

[3] Morss LR, Edelstein NM, Fuger J, Katz JJ, editors. The Chemistry of the Actinide and Transactinide Elements. 3rd ed. Dordrecht: Springer; 2006. 5 p

[4] Brook BW, Alonso A, Meneley DA, Misak J, Blees T, Van Erp JB. Why nuclear energy is sustainable and has to be part of the energy mix. Sustainable Materials and Technologies. 2014;**1-2**:8-16

[5] Li G, Yao J. A review of In situ leaching (ISL) for uranium mining. Mining. 2024;**4**(1):120-148

[6] Zheng S, Guo D, Yang X, Liu Y, Liu S, Tian W, et al. Full life cycle of uranium toward ore-soil-root-shoot system: A comprehensive review. Critical Reviews in Environmental Science and Technology. 2024:1-23

[7] Smedley PL, Kinniburgh DG. Uranium in natural waters and the environment: Distribution, speciation and impact. Applied Geochemistry. 2023;**148**:105534

[8] Krishnaswami S, Cochran JK. Uranium-thorium radionuclides in ocean profiles. In: Encyclopedia of Ocean Sciences. 3rd ed. Academic Press; 2016. pp. 377-391. Available from: https://linkinghub.elsevier.com/retrieve/pii/B9780124095489097608

[9] Gandhi TP, Sampath PV, Maliyekkal SM. A critical review of uranium contamination in groundwater: Treatment and sludge disposal. Science of the Total Environment. 2022;**825**:153947

[10] Degueldre C, Joyce MJ. Evidence and uncertainty for uranium and thorium abundance: A review. Progress in Nuclear Energy. 2020;**124**:103299

[11] Fiorentini G, Lissia M, Mantovani F. Geo-neutrinos and earth's interior. Physics Reports. 2007;**453**(5-6):117-172

[12] Kragh H. Quantum Generations: A History of Physics in the Twentieth Century. 5th Printing, and 1. Paperback Printing. Princeton, NJ, Chichester: Princeton University Press; 2002. 494 p (Princeton paperbacks)

[13] Gosling FG. The Manhattan Project: Making the Atomic Bomb (DOE/MA-0001). Washington: History Division, Department of Energy; 1999. 66 p. Available from: https://www.osti.gov/opennet/manhattan-project-history/publications/DE99001330.pdf

[14] U.S Environmental Protection Agency. Radioactivity Antiques. 2023. Available from: https://www.epa.gov/radtown/radioactivity-antiques#:~:text=For%20hundreds%20of%20years%2C%20glassmakers,glass%20to%20glow%20bright%20green

[15] World Nuclear Association. Uranium and Depleted Uranium. 2024. Available from: https://world-nuclear.org/information-library/

nuclear-fuel-cycle/uranium-resources/uranium-and-depleted-uranium

[16] Rehm TE. Advanced nuclear energy: The safest and most renewable clean energy. Current Opinion in Chemical Engineering. 2023;**39**:100878

[17] Galindo A. What is Nuclear Energy? The Science of Nuclear Power. International Atomic Energy Agency Home Page; 2022. Available from: https://www.iaea.org/newscenter/news/what-is-nuclear-energy-the-science-of-nuclear-power

[18] Isotope R&D and Production (DO IP). Harnessing the Power of Uranium to Treat Disease. National Isotope Development Center Home Page. U.S Department of Energy Isotope Program; 2021

[19] Pandit-Taskar N. Targeted radioimmunotherapy and theranostics with alpha emitters. Journal of Medical Imaging and Radiation Sciences. 2019;**50**(4):S41-S44

[20] Friend MT, Mastren T, Parker TG, Vermeulen CE, Brugh M, Birnbaum ER, et al. Production of 230Pa by proton irradiation of 232Th at the LANL isotope production facility: Precursor of 230U for targeted alpha therapy. Applied Radiation and Isotopes. 2020;**156**:108973

[21] World Nuclear Association. Radioisotopes in Medicine. 2014. Available from: https://world-nuclear.org/information-library/non-power-nuclear-applications/radioisotopes-research/radioisotopes-in-medicine

[22] Sgouros G, Bodei L, McDevitt MR, Nedrow JR. Radiopharmaceutical therapy in cancer: Clinical advances and challenges. Nature Reviews. Drug Discovery. 2020;**19**(9):589-608

[23] National Academy of Sciences, Opportunities and Approaches for Supplying Molybdenum-99 and Associated Medical Isotopes to Global Markets: Proceedings of a Symposium. Washington, DC: National Academies Press; 2018

[24] McKay RR, Taplin ME, Choueiri TK. Optimizing bone health and minimizing skeletal morbidity in men with prostate cancer. Hematology/Oncology Clinics of North America. 2013;**27**(6):1261-1283

[25] Ziessman HA et al, editors. Nuclear Medicine. 4th ed. W.B. Saunders, Pulmonary System; 2014. pp. 204-226. Available from: https://linkinghub.elsevier.com/retrieve/pii/B9780323082990000109

[26] Hartline DR, Meyer K. From chemical curiosities and trophy molecules to uranium-based catalysis: Developments for uranium catalysis as a new facet in molecular uranium chemistry. JACS Au. 2021;**1**(6):698-709

[27] Kumar R, Qureshi M, Vishwakarma DK, Al-Ansari N, Kuriqi A, Elbeltagi A, et al. A review on emerging water contaminants and the application of sustainable removal technologies. Case Studies in Chemical and Environmental Engineering. 2022;**6**:100219

[28] Rovan L, Zuliani T, Horvat B, Kanduč T, Vreča P, Jamil Q, et al. Uranium isotopes as a possible tracer of terrestrial authigenic carbonate. Science of the Total Environment. 2021;**797**:149103

[29] Hadj Ammar F, Deschamps P, Chkir N, Zouari K, Agoune A, Hamelin B. Uranium isotopes as tracers of groundwater evolution in the complexe terminal aquifer of southern Tunisia. Quaternary International. 2020;**547**:33-49

[30] Fallatah O, Khattab MR. Determination of uranium concentrations and 234U/238U isotopic ratios in plants and the groundwater used in their cultivation in an area with high background radiation. Sustainability. 2023;**15**(2):1590

[31] Cain DJ, Croteau MN, Fuller CC, Beisner KR, Campbell KM, Stoliker DL, et al. Aquatic insect accumulation of uranium at spring outflows in the grand canyon region as influenced by aqueous and sediment geochemistry and biological factors: Implications for monitoring. Environmental Monitoring and Assessment. 2023;**195**(7):841

[32] Gupta DK, Walther C, editors. Uranium in Plants and the Environment. Cham: Springer International Publishing; 2020. (Radionuclides and Heavy Metals in the Environment). Available from: http://link.springer.com/10.1007/978-3-030-14961-1

[33] Vodyanitskii YN. Chemical aspects of uranium behavior in soils: A review. Eurasian Soil Science. 2011;**44**(8):862-873

[34] Randhawa JS, Robin KP, Meehnian H. A comprehensive review on health and environmental hazards of uranium: Analytical techniques, mitigation strategies and its toxicity treatments. Journal of Radioanalytical and Nuclear Chemistry. 2024;**333**:3693-3711. DOI: 10.1007/s10967-0.24-09527-1

[35] MacIntosh A, Oldfield DT, Cendón DI, Langendam AD, Howell N, Howard DL, et al. Naturally occurring radioactive materials in offshore infrastructure: Understanding formation and characteristics of baryte scale during decommissioning planning. Journal of Hazardous Materials. 2024;**465**:133506

[36] Jiang W, Xu X, Hall R, Zhang Y, Carroll KC, Ramos F, et al. Characterization of produced water and surrounding surface water in the Permian Basin, the United States. Journal of Hazardous Materials. 2022;**430**:128409

[37] Dewar D, Harvey L, Vakil C. Uranium mining and health. Canadian Family Physician. 2013;**59**(5):469-471

[38] National Academies Press. Uranium Mining in Virginia: Scientific, Technical, Environmental, Human Health and Safety, and Regulatory Aspects of Uranium Mining and Processing in Virginia. 2011. Available from: https://www.ncbi.nlm.nih.gov/books/NBK201047/

[39] Agius R, Batistatou E, Gittins M, Jones S, McNamee R, Liu H, et al. An epidemiological study of lung cancer and selected other cancers among Namibian uranium workers. Radiation Research. 2023;**200**(4). Available from: https://bioone.org/journals/radiation-research/volume-200/issue-4/RADE-23-00051.1/An-Epidemiological-Study-of-Lung-Cancer-and-Selected-Other-Cancers/10.1667/RADE-23-00051.1.full

[40] Brugge D, Goble R. The history of uranium mining and the Navajo people. American Journal of Public Health. 2002;**92**(9):1410-1419

[41] U.S Environmental Protection Agency. Abandoned Mines Cleanup. 2024. Available from: https://www.epa.gov/navajo-nation-uranium-cleanup/aum-cleanup#:~:text=Today%20the%20mines%20are%20closed%2C%20but%20a,from%20exposure%20to%20radionuclides%20in%20drinking%20water

[42] Bala S, Garg D, Thirumalesh BV, Sharma M, Sridhar K, Inbaraj BS, et al.

Recent strategies for bioremediation of emerging pollutants: A review for a green and sustainable environment. Toxics. 2022;**10**(8):484

[43] Li S, Zhu Q, Luo J, Shu Y, Guo K, Xie J, et al. Application progress of deinococcus radiodurans in biological treatment of radioactive uranium-containing wastewater. Indian Journal of Microbiology. 2021;**61**(4):417-426

[44] Verma S, Kim KH. Graphene-based materials for the adsorptive removal of uranium in aqueous solutions. Environment International. 2022;**158**:106944

[45] Raj SK, Carrier AJ, Youden BC, Servos MR, Oakes KD, Zhang X. Electrochemical techniques for uranium extraction from water. Chemical Engineering Journal. 2024;**492**:152341

[46] Shuibo X, Chun Z, Xinghuo Z, Jing Y, Xiaojian Z, Jingsong W. Removal of uranium (VI) from aqueous solution by adsorption of hematite. Journal of Environmental Radioactivity. 2009;**100**(2):162-166

[47] Shen C, Pan J, Chen M, Su M, Chen D, Song G. Statistically and visually analyzing the latest advancements and future trends of uranium removal. Environmental Research. 2023;**239**:117280

[48] Chen X, Ma X, Jiang J. Phosphate-modified hydrothermal biochar: Green and efficient uranium adsorption. Materials Letters. 2024;**377**:137363

[49] Rani L, Srivastav AL, Kaushal J, Shukla DP, Pham TD, Van Hullebusch ED. Significance of MOF adsorbents in uranium remediation from water. Environmental Research. 2023;**236**:116795

[50] World Nuclear Association. Nuclear Power in the World Today. 2024. Available from: https://world-nuclear.org/information-library/current-and-future-generation/nuclear-power-in-the-world-today

[51] U.S Department of Energy Office of Nuclear Energy. Office of Nuclear Energy. Newly Signed Bill Will Boost Nuclear Reactor Deployment in the United States. 2024. Available from: https://www.energy.gov/ne/articles/newly-signed-bill-will-boost-nuclear-reactor-deployment-united-states

[52] Robins J. A Nuclear Power Revival is Sparking a Surge in Uranium Mining. Yale Environment 360; 2024. Available from: https://e360.yale.edu/features/us-uranium-mining-nuclear-power

Section 3

Remediation of Heavy Metals Pollution

Chapter 11

Perspective Chapter: Utilization of Activated Carbon Derived from Biowaste for Heavy Metal Removal in Wastewater Treatment

Tsenbeni N Lotha, Latonglila Jamir, Ketiyala Ao and Lemzila Rudithongru

Abstract

Industrial human activities have resulted in the release of significant quantities of heavy metals into the environment. The contamination of water by heavy metals such as lead (Pb), cadmium (Cd), copper (Cu), zinc (Zn), chromium (Cr), and nickel (Ni) poses a serious threat due to their toxicity. Increasingly stringent discharge regulations on heavy metals have hastened the search for highly efficient yet economically feasible or alternative methods for their removal. Utilizing low-cost biowaste as biosorbents for the removal of dissolved metal ions has demonstrated potential to offer economical solutions to this environmental issue. The use of activated carbon (AC) for wastewater remediation is gaining significant attention among researchers, as it not only reduces production costs but also improves the efficiency of biowaste disposal. This chapter reviews recent developments and findings on the use of AC derived from biowaste for treating wastewater contaminated with heavy metals. The origins and toxicity of heavy metal ions are discussed, highlighting that utilizing biowaste to remove these metals is environmentally beneficial. This approach addresses the issues of incineration and promotes recycling, aligning with the principles of circular economy.

Keywords: heavy metal, biowaste, toxicity, activated carbon, reusability, wastewater

1. Introduction

Wastewater pollution from industrial activities is a global issue, posing severe risks to aquatic environments. Uncontrolled pollutant release leads to chronic and acute toxicity for aquatic plants and animals, ecology accumulation, loss of biodiversity, and health hazards for humans [1]. Industrial wastewater contains heavy metals like Cu, Fe, Cr, Ni, Cd, Zn and Pb. These non-biodegradable metals accumulate in streams and lakes, leading to bioaccumulation in living organisms causing various health issues that include metabolic acidosis, kidney failure, oral ulcers, stomach damage in rodents, and cancer [2]. It is deemed harmful when their concentrations reach or exceed 5 mg/L as the ions can pass in water, soil, and air. The immediate environmental impact includes the destruction of marine habitats and soil degradation, leading to lower crop yields [3]. Cu^{2+}, Ni^{2+}, Zn^{2+}, Pb^{2+}, and Cd^{2+} are extensively utilized in industries such as chemical production, electroplating, mining, metallurgy, and tanning. Toxic heavy metals such as Cd^{2+}, Pb^{2+}, and Cr^{2+} also persist in the atmosphere, leading to bioaccumulation in the livers and kidneys of both vertebrates and invertebrates, thus causing cancers in the kidneys, colon, brain, rectum, and bladder [4].

In recent decades, various methods have been developed to devise efficient techniques for removing heavy metals, including ion exchange [5], photocatalytic degradation [6], chemical precipitation [1], electrochemistry [7], membrane filtration [8], and adsorption [9]. Among these approaches, carbon-based adsorption is widely employed in environmental protection, where activated carbon (AC) derived from biomass is recognized as highly cost-effective and promising for removing heavy metals. To enhance the adsorption capability of AC for removing toxic contaminants, a range of physical, chemical, and biological treatments are used [10]. The development of surface functional groups is crucial in the adsorption process. Chemical modifications of AC with alkali or acid treatments have been shown to greatly improve the efficiency of heavy metal removal compared to alternative methods. To remove heavy metals from diverse wastewater sources, zeolites, commercial AC, and biomass-derived adsorbents like sawdust, banana peel, coconut shells, animal horn, cow dung, eggshell, fish scales, and animal hooves have been utilized. Therefore, this chapter explores recent progress in the preparation of AC derived from biomass and its use in removing heavy metals from wastewater. It covers the processes of carbonization, activation, and modification of AC, along with pertinent research on its efficacy in heavy metal adsorption. Moreover, it outlines the critical factors influencing the adsorption process and addresses the challenges encountered by biomass-derived AC in this application.

2. Heavy metal sources and its toxicity

The excessive release of these metals into the environment poses significant health hazards to water, soil, and organisms that primarily originate from gases and dust

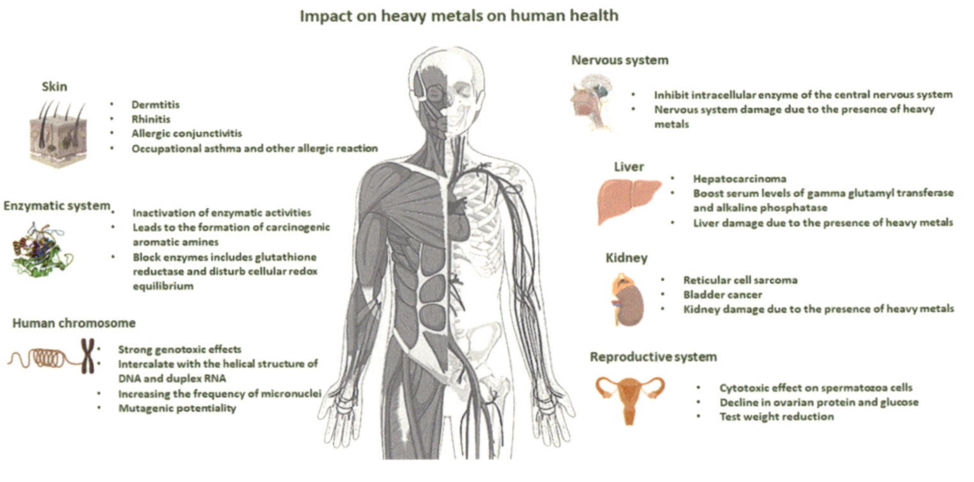

Figure 1.
Impact of heavy metals on human health.

emitted by energy production, metallurgical processes, and the production of construction materials. Industrial and mining effluent is often discharged directly into sewers, where it mixes with domestic wastewater without any prior treatment. This unregulated discharge has led to a gradual increase in heavy metal content in the soil of areas irrigated with this sewage over time. While pesticides, chemical fertilizers, and plastics are essential for agricultural development, their excessive and prolonged use can result in soil pollution from heavy metals. Essential metals can promote growth and improve feed utilization in many species, but surpassing their maximum allowable limits can disrupt normal physiological and ecological balance in aquatic environments [11]. Most heavy metals are highly carcinogenic and can lead to severe health issues such as liver disorders, cardiovascular problems, kidney dysfunction,

Heavy metal	Main source	Health impacts	Target organs	Reference
Arsenic	Industrial activities, particularly mining, smelting, pesticides, and herbicides.	Respiratory and cardiovascular problems, liver and kidney failure	Liver, heart, and kidney	[14]
Iron	Iron and steel production, use of iron-containing products such as infrastructure.	Nausea, vomiting, diarrhea, and abdominal pain	Cardiovascular system, liver, and kidney	[15]
Lead	Steel and battery production and paint manufacturing	Brain damage, muscle and joint pain, and kidney and nervous disorder	Bones, brain, kidney, and digestive system	[16]
Manganese	Industrial activities such as mining, steel production, and battery manufacturing	Headache, anxiety, and memory loss	Brain, kidney, and digestive system	[17]
Mercury	Smelting of precious metals, lighting fixtures, and dental materials	Liver and neural injury, gastrointestinal toxicity	Liver and brain	[18]
Cadmium	Plastic stabilizers, waste batteries, coating corrosion, and pigments	Renal disorder, Itai-Itai disease, cancer, weight loss, and kidney damage	Liver, kidney, lung, and brain	[19]
Zinc	Organic synthesis, galvanizing, and papermaking	Nausea, diarrhea, and bone damage	Respiratory and hematological system	[12]
Chromium	Steel and textile industry	Carcinogenic, nausea, vomiting, and diarrhea	Respiratory and digestive system	[20]
Nickel	Chemical, food, and metallurgical industries	Chronic bronchitis, lung cancer, and dermatitis	Respiratory and immune system	[21]
Copper	Copper cooking pots, copper jewelry, water pipes, and pesticides	Neurotoxicity, dizziness, diarrhea, Wilson disease, and liver damage	Liver and nervous system	[19]

Table 1.
Origins and health impacts of several hazardous heavy metals.

and, in extreme cases, death (**Figure 1**). Heavy metal pollution significantly impacts the physiology of many aquatic organisms, particularly fish [12, 13].

Table 1 displays the origins and health impacts of several hazardous heavy metals. Exposure to heavy metals can result in a range of health issues such as headaches, dizziness, insomnia, memory loss, nervous disorders, joint pain, kidney stones, and cancer [10]. For instance, Pb^{2+} is primarily ingested by humans through meat, fruits, fish, shrimp, and eggs. In the environment, Cu^{2+} is crucial for the advancement of industrial and urban development, required by over 90% of modern industrial enterprises. However, Cu^{2+}-smelting operations have led to severe pollution. For instance, the farmland adjacent to China's largest non-ferrous metal smelters has been contaminated, spanning more than 130 hectares [22]. Cr^{6+} poisoning can result in numbness in limbs and mental abnormalities, whereas Zn^{2+} poisoning is characterized by symptoms such as diarrhea, vomiting, loss of appetite, tiredness, and exhaustion.

3. Process of heavy metal adsorption

Adsorption mechanisms of heavy metals by AC include ion exchange, chemical precipitation, electrochemical interactions, and adsorption (**Figure 2**). **Table 2** outlines frequently used methods for heavy metal removal with their benefits.

Ion exchange between protonated oxygen-containing functional groups (e.g., –OH, –COOH) and divalent metal cations (M^{2+}) depends on the cation's ion exchange capacity. According to Yuan and co-workers [27], it is the primary mechanism for heavy metal adsorption by carbon-based adsorbents. Electrostatic interaction happens when the surface of the adsorbent, which carries a charge, attracts metal ions that have an opposite charge. Studies on heavy metal removal by AC have identified electrostatic interaction as a promising adsorption mechanism. Notably, carbon materials can exhibit surfaces with different electrical charges. Furthermore, electrostatic interaction is influenced by pH levels and the zero-point charge (pH_{ZPC}). For instance, scientists have noted electrostatic interactions amid different protonated or deprotonated functional groups on carbon-based adsorbents and metal ions with opposite charges, such as NH^{3+} with $HCrO^{4-}$. Surface complexation refers to interactions between heavy metal ions (M^{2+}) or complexes (e.g.,

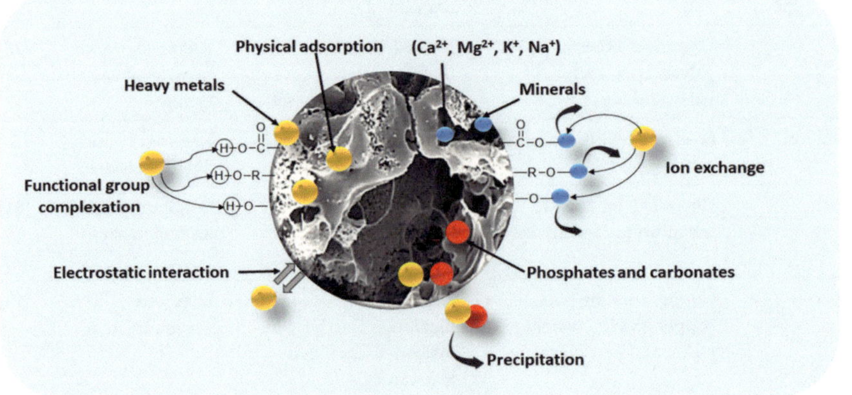

Figure 2.
Adsorption mechanism of heavy metals.

Methods	Benefits	Reference
Ion exchange	High removal rates, upfront equipment, effective control, and easy operation	[23]
Electrochemical	Equipment is compact with a small footprint; require no oxidants, flocculants, or additional chemicals, ensuring no secondary pollution. Flexible and convenient operation, low operating costs, and excellent removal efficiency	[24]
Chemical precipitation	Simple and easy to operate, minimal investment in equipment	[25]
Adsorption	High stability, good selectivity, simple installation, reusability, and minimal material replacement	[26]

Table 2.
Benefits of various techniques for removing metal ions.

$Pb(OH)^+$, $Cd(OH)^+$) and functional groups such as –OH, –CO–NH-, and -COOH. Consequently, these interactions enable the creation of complex structures via surface complexation reactions. Multiple studies have observed possible complexation reactions between metal cations and oxygen-containing functional groups such as –OH and –COOH. Solid precipitates or coprecipitates can be generated through the reaction of heavy metal ions with functional groups present on the surface of adsorbent materials. This phenomenon typically occurs in aqueous solutions containing high concentrations of heavy metal ions. In certain instances, the reduction process can greatly impact the removal of heavy metals. Some heavy metal ions of high valence may undergo initial reduction to lower valence states before being extracted from solutions via complexation or ion exchange mechanisms. The efficiency of adsorption depends largely on the distribution of pore sizes and the surface area of the adsorbent material. A greater number of mesopores and a larger surface area facilitate better dispersion of pollutants, thereby speeding up physical adsorption and its associated kinetics. Physical adsorption entails heavy metal ions being deposited into the pores of adsorbent materials through purely physical mechanisms, devoid of chemical bond binding forces.

4. AC for environmental pollutant remediation

Carbon materials are esteemed for their exceptional properties, such as resistance to acids and bases, corrosion, and stability in varying temperatures, alongside their biocompatibility, making them highly regarded as some of nature's most remarkable materials. Furthermore, carbon can adopt diverse forms and exhibit a wide range of properties through sp. electron orbital hybridization such as being an excellent electrical conductor, semiconductor, or insulator, as well as appearing as opaque black or remarkably transparent material. With the expansion of coal, oil, and modern industry, various types of AC have been developed and are widely utilized in manufacturing raw materials. Due to its favorable physicochemical properties, AC finds extensive use in environmental remediation across various industries such as pharmaceuticals, automobiles, and fabrics. It also finds widespread applications in gas adsorption, solvent recovery, and wastewater treatment, particularly for the removal of organic dyes and as a catalyst in the production of biodiesel. Irwan and co-workers [28] synthesized AC from coconut shells to remove methylene blue with an adsorption capacity of 79.31 mg/g.

Xiaoli and co-workers [29] produced AC from tea residue and effectively applied it to treat wastewater containing 16 organophosphorus pesticides as the main pollutant. Similarly, Joseph and co-workers [30] synthesized AC from rice husks for removing organic pollutants like phenol from wastewater. The literature indicates that AC is a highly cost-effective material that can be easily prepared and has extensive applications in environmental remediation. Biomass stands out among various raw materials for its notable advantages in producing AC, which effectively controls environmental pollution, improves rural environments, and protects the environment through waste reuse.

4.1 Biomass carbonization

Advancements in producing AC from biomass include a carbonization process, where it is heated in an oxygen-free environment typically between 400 and 600°C, thereby removing O_2, H_2, N_2, and other elements, resulting in materials with high carbon content. It undergoes thermochemical transformation to create a carbon-rich solid known as biomass-derived AC, with carbonization occurring in three stages. Initially, at temperatures below 110°C, the raw material undergoes gradual water loss without significant internal molecular changes. Subsequently, heating to 350°C causes the temperature of -CHO and -COOH functional groups to reach 400°C, leading to the breakage of numerous chemical bonds like O-H, C-H, and C-O bonds, as well as benzene rings and ether bonds. During the carbonization process, chemical bonds rearrange, leading to decarboxylation, dehydration, and polycondensation, which generate minor molecular gases and molecular groups, such as -OH, CH_3, CH_2OH, and OCH_3. These small molecular groups eventually combine to form H_20, CH_4, and CH_3OH. Additionally, a significant number of benzene-free radicals lead to the formation of polycyclic aromatic compounds. And between 450 and 500°C, volatile substances are removed to increase the proportion of fixed carbon in the carbonized products.

4.2 Activating biomass-derived AC

Carbon materials are frequently subjected to activation through physical and chemical methods. Chemical activation methods are classified into two types depending on the procedure involved. In one approach, the activator is blended with the biomass in a precise ratio, followed by simultaneous carbonization and activation at high temperatures. Li and co-workers [4] prepared corncob AC by a one-step chemical activation method using H_3PO_4. Under these conditions, Cr^{6+} exceeded 98% removal, with 9.99 mg/g adsorption capacity. In another approach, the carbonized material is activated at a high temperature in a specific ratio to create specific surface porosity, significantly enhancing the adsorptive potential of AC. Thus, the type of biological reagent and the impregnation duration influence the physical and chemical properties of AC. The influence of $ZnCl_2$ and KOH on lignocellulose AC was individually investigated [31] to produce micropores and wide pore size distribution by $ZnCl_2$ and KOH impregnation. Both activation methods yielded a large specific surface area of the AC, primarily composed of mesopores and micropores.

5. AC modification for better adsorption of heavy metals

AC, a highly promising adsorbent material, renowned for its high porosity, specific surface area, significant pore volume, and surface properties, including morphology and functional groups, finds wide application in various fields, particularly in water treatment [32]. The raw material, processing conditions, and manufacturing technique significantly influence the efficiency of AC adsorption. It has uniformly distributed functional groups and pore sizes that are more effective in adsorbing specific types of heavy metals. The choice of precursor material for AC depends on factors such as availability, safety, and cost of production. Numerous studies have noted improved structural and textural qualities when using materials with high fixed carbon and low ash [33, 34]. The surface area of AC generally falls between 150 and 1500 m^2/g, but this can vary depending on their specific properties [35]. Various surface functional groups, including -CHO and -COOH, enhance AC's ability to adsorb pollutants from aqueous solutions [33]. Recent research has demonstrated that the adsorption capacity can be enhanced through many modification approaches. Surface functional group properties of AC are determined by the precursor material, activation process, and post-treatment and are typically conducted after carbon activation, with various methods explored in the literature, including physical, chemical, and biological approaches [10].

5.1 Physical activation methods

Physical activation methods such as microwave heating, ultraviolet treatment, and steam modification have been employed to modify AC as it enhances the physical characteristics by increasing its BET surface area and pore volume [36]. Ultrasound irradiation acts as a driving force for chemical processes by accelerating mass transfer through acoustic cavitation, ultimately resulting in the generation of acoustic streaming. Sonic waves create microscopic turbulence inside the AC, significantly increasing the identified diffusion coefficient value regardless of the carbon's shape [37]. Microwave heating has also garnered significant interest for preparing AC and has been employed to produce mesoporous AC with a higher specific surface area. For instance, Shuheng and co-workers [38] modified AC with HNO3 under microwave heating to remove Pb2+ from water. Li and co-workers [39] prepared AC by using microwave activation for high removal of Hg. Steam activation is also employed to alter AC, boosting its pore surface to improve the adsorption of heavy metals. Similarly, Shim and co-workers [32] revealed that steam activation at 800°C significantly enhanced the Cu^{2+} adsorption capacity of Miscanthus AC compared to the original material produced through slow pyrolysis at 500°C.

5.2 Chemical modification methods

Surface functional groups on AC impart a degree of polarity, alkalinity, and acidity that significantly influence the adsorption capacity and the types of pollutants that can be adsorbed. Chemical modification of AC involves altering its surface functional groups to create adsorbents with tailored adsorption properties. Typical chemical modification methods include plasma treatment and alkali and acid treatment. Acid modification involves oxidizing AC with agents like HNO_3, H_2SO_4, or $C_6H_8O_7$. Surface functional groups on AC like -OH, -COOH, and -C=O are typically introduced through oxidation processes. Li and co-workers [40] used in-situ modification of AC with H_3PO_4 and disodium EDTA salts to enhance Ni^{2+} adsorption. The Ni^{2+} were adsorbed through cation exchange, electrostatic interaction, and surface complexation mechanisms, with the adsorption amount proportional to the surface functional groups under suitable conditions with an increase in adsorption capacity from 15.4 to

27.9 mg/g. In another experiment, spent coffee grounds were utilized as a precursor for producing AC modified with H_3PO_4, which increased the specific surface area to 720 m^2/g and the pore volume to 0.334 cm^3/g. The resulting carbons all featured acidic (primarily -COOH) groups and displayed an amorphous structure [41]. The findings suggest that AC with -COOH groups holds great promise for adsorption applications. Li and co-workers [42] studied the efficacy of carbonized sewage sludge at 500°C modified with HNO_3 to remove metal ions like Cu^{2+} and Zn^{2+} at removal rates of 98.9%, and 42.6%, respectively. Functional group complexation with metal ions not only reduced adsorption time but also improved removal efficiency. Alkali modification of AC involves using alkalis like NaOH, KOH, or NH_3 [43]. For example, ZnO-AC nanocomposites were synthesized by activating sugarcane bagasse-derived AC with NaOH, followed by carbonization using the hydrothermal method, resulting in the removal of 97% of Cr^{6+} [44]. Wei and co-workers [45] also synthesized a biomass-derived AC with a Schiff base structure by NH_3-assisted hydrothermal treatment. This AC was employed for the adsorption of Cr^{6+}, showing a high adsorption capacity of 349.6 mg/g [46]. Jiang and co-workers [47] also studied the removal of Pb^{2+} using polyethyleneimine-loaded AC, which increased the Pb^{2+} removal rate from 32.6 to 214 mg/g. Plasma modification is an effective, user-friendly, and environmentally conscious surface alteration method that has emerged as a key technology for modifying AC by altering its pore structure and creating surface functional groups like -OH, -CHO, and -COOH groups [48]. For instance, peat soil-derived AC particles were modified using a flying jet plasma torch by drying at 70°C for 48 h, followed by carbonization at 800°C for 4 h. This treatment significantly altered surface morphology and chemical composition, increasing the mass oxygen percentage and carbonyl C=O group content with enhanced polarity and solubility in polar liquids [49].

5.3 Microbial modification

In recent years, scientists have discovered that the unique structure of AC surfaces creates a favorable atmosphere for the colonization and proliferation of microorganisms [50, 51]. Microbial modification of AC involves adsorbing microorganisms onto its surface to alter its adsorption characteristics and offers numerous advantages in applications, including high activity, rapid reaction rates, robust toxicity resistance, easy product separation, and the ability for continuous operation. As a result, microbial modification technology has seen rapid development and widespread use in water treatment. Microorganisms adhering to AC surfaces can pre-oxidize certain organic substances, minimizing contact amid organic matter and adsorption sites on the AC. This process helps extend the AC service life [52]. Yet, microbial-modified AC has drawbacks, like microorganism proliferation on its surface and membrane thickening, which can impede the diffusion of adsorbed substances within its pores, lowering adsorption efficiency [10].

6. Adsorption process

Adsorption techniques utilizing solid sorbents are frequently employed to eliminate diverse chemical pollutants from water, especially those that are not effectively treated by biological wastewater methods. Among all the sorbent materials studied, AC is the most highly recommended for removing pollutants from wastewater (**Table 3**). It can be created by the thermochemical conversion of organic molecules at

AC	Heavy metal	Adsorption capacity (mg/g)	Adsorption isotherm	pH	Reference
Olive stones	Cu^{2+}	0.27	Langmuir	5	[53]
	Ni^{2+}	0.40			
Goat teeth	Pb^{2+}	62.32	Langmuir	5	[13]
Coconut shell	Zn^{2+}	—	Langmuir	7	[54]
Plum stones	Pb^{2+}	172.43	Langmuir	6	[55]
	Cd^{2+}	112.74			
	Ni^{2+}	63.74			
Chicken bone	Pb	263.2	Langmuir	5	[16]
Chicken feather	Cd^{2+}	76.3	Langmuir	5.5	[56]
	Cu^{2+}	56.5			
	Pb^{2+}	113.3			
	Ni^{2+}	32.6			
	Zn^{2+}	45.5			
Pig bone	Pb^{2+}	120	Langmuir	6	[57]
	Cu^{2+}	30			
Bovine bone	Cu^{2+}	83.71	Elovich	5.5	[58]
kenaf fiber	Pb^{2+}	12.7	Freundlich	6	[59]
	Cu^{2+}	12.0			
Cow dung	Cd^{2+}	5.8800	—	—	[60]
Eggshell	Pb^{2+}	25.189	Langmuir	5	[15]
Cow manure	Cu^{2+}	78.63	Langmuir	6	[61]
Chicken feather	Cr^{6+}	90.90	Langmuir	2	[62]
Bovine bone	Cd^{2+}	78.969	Langmuir	—	[63]

Table 3.
Adsorption capacity of AC for heavy metal removal.

high temperatures in the absence of oxygen. The raw material, production technique, and conditions all have a considerable impact on the adsorption efficacy [64].

7. Adsorption isotherm, kinetics, and thermodynamic studies

Different isotherm modeling has been proposed for heavy metal sorption studies onto AC. Majority of the studies stated the Langmuir and Freundlich isotherm as the best fit. Freundlich isotherm was reported for Cu^{2+} adsorption with $R^2 = 0.93$ [65], Cr^{6+} with $R^2 = 0.992$ [66], and Cr^{6+} with $R^2 = 0.9886$ using H_3PO_4 as activating agent [67] as the best fit. Several studies have also shown Langmuir isotherm as the best fit in the uptake of Cu^{2+} with R^2 of 0.992 [68], Cd^{2+} with R^2 of 0.9982 [56], Cu^{2+} with R^2 of 0.999, Pb^{2+} with R^2 of 0.993 [69], and Cr^{6+} with R^2 of 0.912 [66]. Uptake of Pb^{2+} using goat hair, human hair, and sheep wool with R^2 of 0.09, 0.87, and 0.58, respectively, was also observed. Ibrahim and co-workers [70] have also synthesized AC from marine macroalga *Ulva lactuca*, showing maximum removal efficiency of 64.5 and 84.7 mg/g for Cu^{2+}, 62.5 and 84.6 mg/g for Cd^{2+}, 60.9 and 82 mg/g for Cr^{3+}, and

68.9 and 83.3 mg/g for Pb^{2+} following Langmuir isotherm. Various kinetic models were used for studying the equilibrium isotherm and kinetic modeling of Cu^{2+}, Cd^{2+}, Ni^{2+}, Pb^{2+}, and Zn^{2+} adsorptions. Among the many models used, PSO revealed better fits of the experimental results based on statistical indices such as χ^2 (Chi square), R^2 values, and the HYBRID values, which evaluate the goodness of fit for adsorption models by combining multiple error metrics into a single value, reflecting how well the model matches experimental data [56]. Maria and co-workers [69] studied AC synthesized from cow bones for Pb^{2+} adsorption. Both the Langmuir and Freundlich models were utilized, showing correlations, but the Langmuir model provided a more accurate description of the equilibrium, signifying a higher R^2 value. Initial Pb^{2+} concentrations were tested using the pseudo-second-order (PSO) kinetic model. Desorption using HCl achieved a maximum recovery rate of about 50%, reaching equilibrium in just 3 h. Efficient cationic exchange was attributed to the presence of acid functional groups on the AC surface [69]. Tonetti and co-workers [68] utilized AC derived from thermally-treated "horn–hoof" powder for the removal of Cu^{2+} ions. The Langmuir model was used to calculate the adsorption isotherm, while thermodynamic and kinetic data suggested a pseudo-second-order (PSO) model, indicating an endothermic process. Thermodynamic studies can be employed to examine various aspects of the adsorption process, such as changes in enthalpy, Gibbs-free energy (ΔG), and entropy (ΔS). The change in enthalpy (ΔH) reveals the thermal nature of the adsorption process, indicating whether it is exothermic or endothermic, while the change in Gibbs-free energy (ΔG) signifies the spontaneity of the process. Additionally, the change in entropy (ΔS) reflects the system's level of disorder or its reversibility [71]. A negative ΔG value indicates that the adsorption process at higher temperatures is spontaneous, feasible, and therefore more favorable [56]. C. Tonetti and co-workers [68] investigated the uptake of Cu^{2+} using horn-hoof powder, whereas J.O. Tijani and co-workers [72] studied on the feasible yet non-spontaneous (positive ΔG) uptake of Ni^{2+}, Mn^{2+}, and Cd^{2+} using AC derived from cow horns. Spontaneity in a system indicates that a process occurs without any external input, which means that the sorption of the pollutant happens naturally upon introducing the biosorbent into the pollutant solution, even without a catalyst or additional chemical, physical, or thermal intervention. Thus, spontaneity was observed for the uptake of Cu^{2+} [57] and for heavy metal ions like Cd^{2+}, Pb^{2+}, Zn^{2+}, and Ni^{2+} [56]. When adsorption occurs more readily at lower temperatures, it indicates an exothermic process or a negative enthalpy change (ΔH). Conversely, if adsorption is more favorable at higher temperatures, it suggests an endothermic sorption with a positive ΔH. Uptake of Mn^{2+} and Cr^{6+} [66] was demonstrated to occur through an exothermic adsorption process. Endothermic sorption was observed for uptake of Mn^{2+} [72], Cr^{6+} [62], and metal ions like Cd^{2+}, Pb^{2+}, Zn^{2+}, Ni^{2+} [56], and Cu^{2+} [68]. Also, a positive entropy change (ΔS) indicates increased disorder at the solid/solution interface, suggesting a stronger affinity between adsorbate and biosorbent. Conversely, a negative entropy change (ΔS) suggests that the sorption process is driven by enthalpy [71].

8. Key factors for regulating the adsorption of heavy metals by biomass-derived AC

8.1 Specific surface area and pore structure

A large specific surface area typically indicates better adsorption performance for any adsorbent, as it provides ample space for adsorption during the process.

Yu and co-workers [73] analyzed N_2 adsorption-desorption curves for four samples with ultra-high specific surface areas: 1348, 2004, 1324, and 1974 m^2/g. The findings suggest that a high specific surface area improves the wettability amid the electrode material and the electrolyte, leading to excellent adsorption capacity for pollutant removal. Zhang and co-workers [74] carbonized rape straw powder using a hydrothermal method to create magnetized AC, which was modified to achieve a high specific surface area of 553 m^2/g demonstrating maximum adsorption capacities of 253 mg/g and 73.3 mg/g for Pb^{2+} and Cd^{2+}, respectively. Several techniques are commonly employed to modify carbon adsorbents and enhance their surface area, including heat, acid alkali, microwave, ozone, and plasma treatment. An example is the successful synthesis of humic acid-coated, N_2-doped magnetic porous carbon using lignin extracted from black liquor as the primary carbon source [75], which exhibited a remarkable adsorption capacity of 131 mg/g for Cr^{6+}, with a BET-specific surface area of 748 m^2/g. This high surface area is expected to provide numerous adsorption sites for metal ions. Additionally, the adsorption material, produced by impregnating zeolite with Fe^{2+} salt, possessed a high specific surface area of 1358 m^2/g and removal rate of 99% for Pb^{2+} within 20 minutes. Its maximum adsorption capacity of 790 mg/g far exceeded that of most previously reported unmodified zeolites [76]. The porous nature of biomass-derived AC makes it an effective adsorbent, with its morphological structure, particularly the distribution of pore sizes, playing a crucial role in determining its heavy metal adsorption capacity [77]. It features a range of pore sizes under an optical microscope, with the smallest comparable to the size of the adsorbed molecules (**Figure 3**). Micropores serve as the main adsorption sites, while mesopores facilitate intraparticle diffusion, reducing adsorption time. The pore size of biomass-derived AC can influence its heavy metal adsorption capacity. The macropores, also known as transport pores, function as transport channels during the adsorption process. Adsorbed molecules initially enter the macropores, then move through the mesopores, and finally reach the micropores, where they are adsorbed within the AC particles [78]. Mesopores, also known as transition pores, are smaller than macropores but significantly larger than the adsorbed molecules. The adsorption process begins with a single molecular layer, followed by the formation of

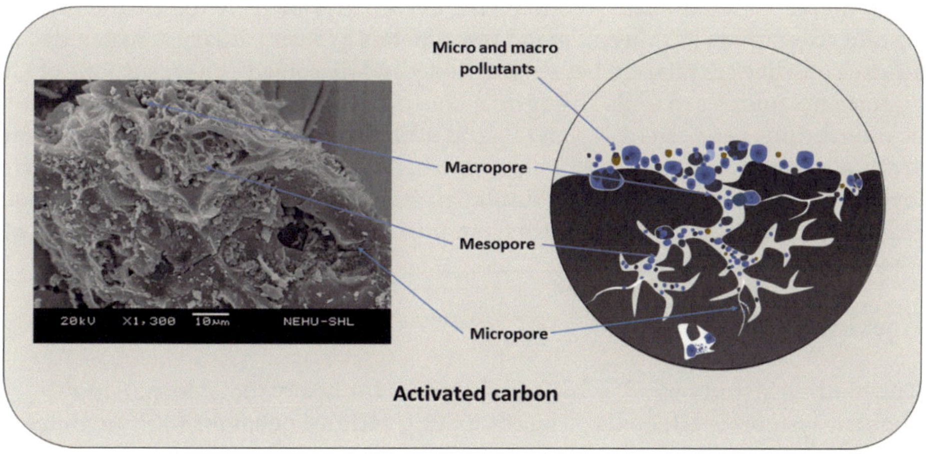

Figure 3.
Pore structure of AC.

a multi-molecular layer, creating a continuous adsorption layer and the smallest pores in carbon, known as micropores that have a very small effective radius [79].

8.2 Functional groups

O_2- and N_2-containing functional groups serve as adsorption sites on AC [33]. Pollutants attach to the adsorbent's surface through ion exchange, electrostatic interactions, hydrogen bonding, acid-base interactions, and complexation. The adsorption of heavy metals by carbon adsorbents encompasses both physical and chemical processes. Consequently, along with the morphological structure of the carbon, the surface chemical functional groups are crucial for heavy metal adsorption. These factors depend not only on the biomass source but also on the activation or modification method employed. Surface functional groups on carbon, including O_2 and N_2, play a critical role in determining surface chemistry and are particularly significant for adsorption on porous carbon.

9. Factors affecting adsorption capacity

The diverse characteristics and conditions of water sources impact the effectiveness of adsorbents in removing heavy metals, which can be equally threatening [80]. Compounds like acids, carboxyl groups, nitrogen, halogens, and bacteria in wastewater can inhibit adsorption. Accurate assessment of adsorption capacity requires considering heavy metal and adsorbent properties in various physiochemical conditions of the water source, including pH, temperature, and ionic strength.

9.1 Ionic strength

The ionic strength of water can significantly impact the behavior of heavy metals and their interaction with adsorbents during treatment. High Cl^- concentrations can lead to the formation of water-soluble heavy metal- Cl^- complexes that are harder to remove [81]. Thuan and co-workers [82] reported that the adsorption effectiveness of Ni^{2+} and Cu^{2+} decreased with an increase in ionic strength due to the formation of heavy metal- Cl^- complexes, reducing the adsorbents' affinity for the metal ions. Wang and co-workers [83] investigated trace metals in river estuaries, which had a significant positive correlation between salinity and dissolved concentrations of Cu^{2+}, Cd^{2+}, and Zn^{2+}. Increased ionic strength in aqueous solutions can enhance electrostatic interactions between heavy metals and adsorbent surfaces, altering heavy metal behavior and adsorption capacity. This can reduce the electric double layer, affecting the overall adsorption process [84]. Similarly, Yang and co-workers [85] noted that increasing the solution's ionic strength from 0.01 to 1.0 M improved the removal of Ni^{2+} and As^{3+} by up to 25%.

9.2 Coexisting ions

The binding affinity of adsorbent surface sites for heavy metal ions in multicomponent systems is affected by the competing cations' behavior [86]. In multicomponent systems, coexisting heavy metal ions compete for a limited number of adsorption sites, reducing the adsorption effectiveness of the same adsorbent compared to a single-ion system. Park and co-workers [87] noted that the adsorption capacity in a mixed solution containing both Pb^{2+} and Cd^{2+} is lower than in a system with only one of these components. Higher concentrations of Na^{+}, K^{+}, Mg^{2+}, and Ca^{2+} in the aqueous solution directly compete with heavy metal ions, leading to a decrease in overall adsorption efficiency. The presence of SO_4^{2-} promotes the precipitation of Cr^{6+}, while HCO^{3-} reduces the efficiency of Cr^{6+} removal and the valence transition of Fe. SO_4^{2-} also reduces the adsorption of Cr^{6+} by AC due to increased competition and lower attraction compared to Cl^{-} and NO^{3-} [88].

9.3 Initial concentration

Higher initial concentrations can enhance the ability of absorbents as they increase the attraction force that pulls metal ions onto the surface of the AC [89]. Once a certain threshold is crossed, higher initial concentrations hinder further adsorption because there are fewer accessible binding sites on the adsorbent surface. This limits the efficacy of the adsorbent as the operation progresses. Additionally, the initial heavy metal concentration affects the fitness of the selected adsorption isotherm model. As the initial concentration increases, the best-fit model might shift from single to multimolecular layer adsorption.

9.4 pH level

The pH, which determines the concentration of H^{+} in the solution, significantly influences ion exchange and electrostatic interactions during adsorption. The pH_{zpc} values, representing the pH at which the zeta potential of adsorbents is zero, can vary depending on the nature of the AC. When the pH is lower than the pH_{ZPC}, the higher H^{+} concentration facilitates stronger ion exchange and promotes the protonation of functional groups ($-NH_2$, -OH) on the adsorbent surface, leading to a more positively charged surface. This enhances the adsorption of anions (e.g., $HCrO_4^{-}$) but repels heavy metal cations due to electrostatic repulsion [80]. In contrast to lower pH values, pH levels over pH_{ZPC} allow the return of negative charges to functional groups, which has an inverse effect on cations and anions [36]. The pH_{ZPC} of an adsorbent should be more acidic or alkaline depending on whether it removes cations or anions from aqueous solutions. Heavy metal ions can interact with H^{+}, forming different ionic forms with varying degrees of attraction for surface functional groups, thereby affecting overall adsorption effectiveness. Studies on higher pH solutions have shown that -COOH plays a role in the adsorption of heavy metal ions onto the surface of AC [36, 90]. -COOH groups on the surface of AC aid in acidic dissociation, leading to the formation of chelating agents in complexes formed by heavy metal ions interacting with surface functional groups.

9.5 Temperature

Temperature significantly affects the adsorption process. Higher temperatures lead to lower solution viscosity, allowing better dispersion of adsorbate molecules on the adsorbent surface, thereby enhancing adsorption capacity. However, this capacity can decrease at elevated temperatures in an exothermic process, with the opposite true for endothermic reactions. Higher temperatures can also destabilize metal ions bound to the adsorbent surface, reducing the amount of metal ions that can be effectively connected to AC. Additionally, in exothermic adsorption processes, electrostatic attraction between the adsorbent and ions decreases at high temperatures 4.

10. Challenges and future prospects

AC stands out among current treatment methods for its exceptional ability to adsorb and its cost-effectiveness in removing significant water contaminants. Numerous potential applications have been explored and validated through laboratory studies; however, there is a pressing need for additional research on scaling up production and exploring commercial applications of AC. One potential drawback of using AC as adsorbents is that some contaminants, such as sodium, nitrate, fluoride, and bacterial pathogens, may not be effectively targeted by the attractive forces of carbon atoms. Future research could explore the modification processes to improve the efficiency of AC against these challenging pollutants. Once AC becomes saturated with contaminants, it reaches the end of its operational lifespan and requires either replacement or regeneration. Furthermore, there is a need for more research into the regeneration to mitigate secondary pollution. Hence, it is crucial to conduct comprehensive evaluations of the long-term applications of AC to assess its efficacy in wastewater treatment strategies. Most of the research conducted thus far, both current and historical, has focused on the adsorption capacity of individual heavy metals by AC. Given the diversity of heavy metal pollutants found in potentially contaminated water sources, future research should explore how AC behaves in systems where multiple metals are present. This is crucial because interactions among different metal ions may occur, affecting their adsorption behavior. Research might investigate the feasibility of employing various types of adsorbents. Moreover, current optimization studies on the activation process of AC have traditionally focused on one parameter individually, which can be difficult and time-intensive given the requirement for multiple iterations. Response surface methodology could potentially address these technical challenges by providing optimal conditions for multiple variables when treating AC with pollutants. Until recently, the majority of studies on heavy metal removal has concentrated exclusively on batch adsorption. Experts predict that as AC gains recognition as a highly efficient adsorbent, future research will increasingly shift toward column studies and industrial-scale bed adsorption. Present circumstances favor the large-scale production and utilization of AC derived from diverse biomass wastes. Despite significant challenges and obstacles in research, development, and application, continued advancement is warranted due to recent breakthroughs in the field. Utilizing biomass waste as feedstock for AC production contributes to waste reduction. Additionally, developing various modified AC-based adsorbents is essential for effectively removing heavy metals from aqueous solutions.

11. Conclusion

Human activities have caused heavy metals to accumulate faster than they can be effectively treated or managed. Over the past five years, challenges in addressing heavy metal pollution have included raising public awareness about environmental protection and identifying appropriate sorbents for removal. While numerous methods for heavy metal removal have been proposed, many are unsuitable for commercial use due to inefficiency, high energy use, or lack of selectivity. Biomass-derived activated carbon stands out for its promising potential because of its extensive specific surface area, abundant pore structure, excellent stability, and relatively low production cost. In this chapter, the sources of heavy metals, their removal approaches, and the production techniques for biomass-based AC are reviewed. The key factors influencing heavy metal adsorption include adsorbent characteristics, heavy metal properties, and adsorption conditions. The type of biomass, carbonization method, and activation process affect the carbon's porosity and surface area. With suitable conditions or modifications, most biomass-derived AC exhibit high heavy metal adsorption capacity, influenced by their physical form and chemical functional groups. Generally, larger specific surface areas and smaller pore sizes enhance adsorption, while conditions like temperature and pH significantly impact heavy metal uptake. With the increasing focus on "green" solutions, biomass-derived AC is showing great promise in wastewater treatment. Despite this, there are knowledge gaps that need to be addressed to enhance the adsorption capacity of biomass-AC for heavy metals in complex samples. Currently, most modifications of AC are limited to laboratory studies, making commercial reproducibility challenging. However, future advancements are expected to simplify and streamline the preparation process.

Acknowledgements

The author Tsenbeni N Lotha (Award No- 202021-NFST NAG-03968) is grateful to the Ministry of tribal Affairs (under NFST Scheme) Gov of India. Ketiyala is grateful to the UGC-NET JRF fellowship (NTA Ref. No.: 190520065608) and Lemzila Rudithongru (NU/RDC-200(1)-11 EvS/NNF/2022) is grateful to the UGC Non-NET fellowship, for financial assistance provided.

Conflict of interest

The authors declare no conflict of interest.

Author details

Tsenbeni N Lotha, Latonglila Jamir*, Ketiyala Ao and Lemzila Rudithongru
Department of Environmental Science, Nagaland University Lumami, India

*Address all correspondence to: latongli.jamir@gmail.com

References

[1] Hoang AT, Kumar S, Lichtfouse E, Cheng CK, Varma RS, Senthilkumar N, et al. Remediation of heavy metal polluted waters using activated carbon from lignocellulosic biomass: An update of recent trends. Chemosphere. 2022;**302**:134825

[2] Van Thuan T, Quynh BTP, Nguyen TD, Ho VTT, Bach LG. Response surface methodology approach for optimization of Cu2+, Ni2+ and Pb2+ adsorption using KOH-activated carbon from banana peel. Surfaces and Interfaces. 2016;**6**:209-217

[3] Priyan V, Kumar N, Narayanasamy S. Toxicological assessment and adsorptive removal of lead (Pb) and Congo red (CR) from water by synthesized iron oxide/activated carbon (Fe_3O_4/AC) nanocomposite. Chemosphere. 2022;**294**:133758

[4] Li H, Gao P, Cui J, Zhang F, Wang F, Cheng J. Preparation and Cr (VI) removal performance of corncob activated carbon. Environmental Science and Pollution Research. 2018;**25**:20743-20755

[5] Sharma V, Upadhyay LSB, Vasanth D. Extracellular thermostable laccase-like enzymes from bacillus licheniformis strains: Production, purification and characterization. Applied Biochemistry and Microbiology. 2020;**56**:420-432

[6] Aravind M, Amalanathan M. Structural, morphological, and optical properties of country egg shell derived activated carbon for dye removal. Materials Today Proceedings. 2020;**43**:1491-1495

[7] Arla SK, Godlaveeti SK, Sana SS, Jwuiyad A, Konidena NSS, Boya VKN. Synthesis of diverse titanium dioxide (TiO_2) nanostructures for enhanced photocatalytic degradation and supercapacitor performance. Journal of Solid State Electrochemistry. 2023;**11**:1-11

[8] Behera M, Nayak J, Banerjee S, Chakrabortty S, Tripathy SK. A review on the treatment of textile industry waste effluents towards the development of efficient mitigation strategy: An integrated system design approach. Journal of Environmental Chemical Engineering. 2021;**9**:105277

[9] Grich A, Bouzid T, Naboulsi A, Regti A, Tahiri AA, El Himri M, et al. Preparation of low-cost activated carbon from Doum fiber (Chamaerops humilis) for the removal of methylene blue: Optimization process by DOE/FFD design, characterization, and mechanism. Journal of Molecular Structure. 2024;**1295**:136534

[10] Lotha TN, Sorhie V, Bharali P, Jamir L. Advancement in sustainable wastewater treatment: A multifaceted approach to textile dye removal through physical, biological and chemical techniques. ChemistrySelect. 2024;**11**:e202304093

[11] Khan S, Malik A. Toxicity evaluation of textile effluents and role of native soil bacterium in biodegradation of a textile dye. Environmental Science and Pollution Research. 2018;**25**:4446-4458

[12] Stevens MGF, Batlokwa BS. Environmentally friendly and cheap removal of Lead (II) and zinc (II) from wastewater with fish scales waste remains. International Journal of Chemistry. 2017;**9**:22

[13] Anuradha J, Andal NM, Gayathri NS. Utilization of a material derived from

animal waste in the biosorption of Pb(II) ions – Isothermal/thermodynamic, kinetics and statistical studies. Current World Environment: An International Research Journal of Environmental Sciences. 2017;**12**:445-455

[14] Babaei K, Salimi S, Mahmou-dabadi ZS, Tavasoli A. Synthesis and application of chicken manure biochar as an effective nanoporous adsorbent for removal of arsenic from wastewater: A cost-effective approach. Journal of the Iranian Chemical Society. 2023;**20**:541-550

[15] Praipipat P, Ngamsurach P, Pratumkaew K. The synthesis, characterizations, and lead adsorption studies of chicken eggshell powder and chicken eggshell powder-doped iron (III) oxide-hydroxide. Arabian Journal of Chemistry. 2023;**16**:104640

[16] Jim JP, Kim JWS, Ronald JC, Yong DD, Ok S. Lead sorption characteristics of various chicken bone part- derived chars. Environmental Geochemistry and Health. 2018;**8**:1675-1685

[17] Thakor R, Mistry H, Tapodhan K, Bariya H. Efficient biodegradation of Congo red dye using fungal consortium incorporated with Penicillium oxalicum and aspergillus tubingensis. Folia Microbiologia (Praha). 2022;**67**:33-43

[18] Li Y, Liu Y, Yang W, Liu L, Pan J. Adsorption of elemental mercury in flue gas using biomass porous carbons modified by microwave/hydrogen peroxide. Fuel. 2021;**291**:120152

[19] Manzoor J, Sharma M. Impact of textile dyes on human health and environment. Impact of Textile Dyes on Public Health and the Environment. 2019;**11**:162-169

[20] Zeng H, Zeng H, Zhang H, Shahab A, Zhang K, Lu Y, et al. Efficient adsorption of Cr (VI) from aqueous environments by phosphoric acid activated eucalyptus biochar. Journal of Cleaner Production. 2021;**286**:124964

[21] Natsir M, Ika Y, Dwiprayogo P, Maulidiyah W, La M, Agus O, et al. Effects of Ni–TiO_2 pillared clay–montmorillonite composites for photocatalytic enhancement against reactive Orange under visible light. Journal of Inorganic and Organometallic Polymers and Materials. 2021;**31**:3378-3388

[22] Zhang J, Sun X, Deng J, Li G, Li Z, Jiang J, et al. Emission characteristics of heavy metals from a typical copper smelting plant. Journal of Hazardous Materials. 2022;**424**:127311

[23] Elgarahy AM, Elwakeel KZ, Mohammad SH, Elshoubaky GA. A critical review of biosorption of dyes, heavy metals and metalloids from wastewater as an efficient and green process. Cleaner Engineering and Technology. 2021;**4**:100209

[24] Kristanti RA, Hadibarata T, Syafrudin M, Yılmaz M, Abdullah S. Microbiological contaminants in drinking water: Current status and challenges. Water, Air, and Soil Pollution. 2022;**233**:1-17

[25] Adegoke KA, Bello OS. Dye sequestration using agricultural wastes as adsorbents. Water Resources and Industry. 2015;**12**:8-24

[26] Khan TA, Fuzail Siddiqui M, Abbasi N, Alharthi SS. Adsorptive decolouration of anionic dye from water by goat dropping activated carbon prepared via microwave-assisted H3PO4 activation: Process optimization using response surface methodology,

isotherm and kinetics modelling. Biomass Conversion and Biorefinery. 2022;**12**:5409-5425

[27] Yuan Y, An Z, Zhang R, Wei X, Lai B. Efficiencies and mechanisms of heavy metals adsorption on waste leather-derived high-nitrogen activated carbon. Journal of Cleaner Production. 2021;**293**:126215

[28] Kurnia I, Karnjanakom S, Irkham I, Haryono H, Situmorang YA, Indarto A. Enhanced adsorption capacity of activated carbon over thermal oxidation treatment for methylene blue removal: Kinetics, equilibrium, thermodynamic, and reusability studies. The Royal Society of Chemistry. 2023;**13**:220-227

[29] Bai X, Quan B, Kang C, Zhang X, Zheng Y, Song J, et al. Activated carbon from tea residue as efficient absorbents for environmental pollutant removal from wastewater. Biomass Conversion and Biorefinery. 2022;**15**:1-10

[30] Nwabanne JT, Iheanacho OC, Obi CC, Onu CE. Linear and nonlinear kinetics analysis and adsorption characteristics of packed bed column for phenol removal using rice husk-activated carbon. Applied Water Science. 2022;**12**:1-16

[31] Nayak A, Bhushan B, Gupta V, Sharma P. Chemically activated carbon from lignocellulosic wastes for heavy metal wastewater remediation: Effect of activation conditions. Journal of Colloid and Interface Science. 2017;**493**:228-240

[32] Shim T, Yoo J, Ryu C, Park Y-K, Jung J. Effect of steam activation of biochar produced from a giant Miscanthus on copper sorption and toxicity. Bioresource Technology. 2015;**197**:85-90

[33] Supong A, Sinha UB, Sinha D. Density functional theory calculations of the effect of oxygenated Functionals on activated carbon towards cresol adsorption. Surfaces. 2022;**5**:280-289

[34] Supong A, Bhomick PC, Karmaker R, Ezung SL, Jamir L, Sinha UB, et al. Experimental and theoretical insight into the adsorption of phenol and 2,4-dinitrophenol onto Tithonia diversifolia activated carbon. Applied Surface Science. 2020;**529**:147046

[35] Saraswat SK, Demir M, Gosu V. Adsorptive removal of heavy metals from industrial effluents using cow dung as the biosorbent: Kinetic and isotherm modeling. Environmental Quality Management. 2020;**30**:51-60

[36] Hoang AT, Nižetić S, Cheng CK, Luque R, Thomas S, Banh TL, et al. Heavy metal removal by biomass-derived carbon nanotubes as a greener environmental remediation: A comprehensive review. Chemosphere. 2022;**1**:131959

[37] Ondarts M, Reinert L, Guittonneau S, Baup S, Delpeux S, Lévêque J, et al. Improving the adsorption kinetics of ibuprofen on an activated carbon fabric through ultrasound irradiation: Simulation and experimental studies. Chemical Engineering Journal. 2018;**343**:163-172

[38] Yao S, Zhang J, Shen D, Xiao R, Gu S, Zhao M, et al. Removal of Pb(II) from water by the activated carbon modified by nitric acid under microwave heating. Journal of Colloid and Interface Science. 2016;**463**:118-127

[39] Li G, Wang S, Wang F, Wu Q, Tang Y, Shen B. Role of inherent active constituents on mercury adsorption capacity of chars from four solid

wastes. Chemical Engineering Journal. 2017;**307**:544-552

[40] Li Y, Zhang J, Liu H. In-situ modi fi cation of activated carbon with ethylenediaminetetraacetic acid disodium salt during phosphoric acid activation for enhancement of nickel removal. Powder Technology. 2018;**325**:113-120

[41] Zięzio M, Charmas B, Jedynak K, Hawryluk M, Kucio K. Preparation and characterization of activated carbons obtained from the waste materials impregnated with phosphoric acid (V). Applied Nanoscience. 2020;**10**:4703-4716

[42] Li LY, Gong X, Abida O. Waste-to-resources: Exploratory surface modification of sludge-based activated carbon by nitric acid for heavy metal adsorption. Waste Management. 2019;**87**:375-386

[43] Jawad AH, Kumar U, Mohd S, Mastuli S, Alothman ZA, Wilson LD. Multivariableoptimizationwithdesirability function for carbon porosity and methylene blue adsorption by watermelon rind activated carbon prepared by microwave assisted H3 PO4. Biomass Conversion and Biorefinery. 2024;**14**:577-591

[44] Sharma M, Joshi M, Nigam S, Shree S, Avasthi DK, Adelung R, et al. ZnO tetrapods and activated carbon based hybrid composite: Adsorbents for enhanced decontamination of hexavalent chromium from aqueous solution. Chemical Engineering Journal. 2018;**358**:540-551

[45] Wei Y, Wang H, Zhang X, Liu C. Ammonia-assisted hydrothermal carbon material with schiff base structures synthesized from factory waste hemicelluloses for Cr(VI) adsorption. Journal of Environmental Chemical Engineering. 2021;**9**:106187

[46] Li X, Wang X, Wang L, Ning P, Ma Y, Zhong L, et al. Efficient removal of carbonyl sulfur and hydrogen sulfide from blast furnace gas by one-step catalytic process with modified activated carbon. Applied Surface Science. 2022;**579**:152189

[47] Jiang Q, Xie W, Han S, Wang Y, Zhang Y. Enhanced adsorption of Pb(II) onto modified hydrochar by polyethyleneimine or H3PO4: An analysis of surface property and interface mechanism. Colloids and Surfaces A: Physicochemical and Engineering Aspects. 2019;**583**:123962

[48] Haye E, Job N, Wang Y, Pennickx S, Stergiopoulos V, Tumanov N, et al. ZnO/carbon xerogel photocatalysts by low-pressure plasma treatment, the role of the carbon substrate and its plasma functionalization. Journal of Colloid and Interface Science. 2020;**570**:312-321

[49] Abdul WS, Khamis M, Al Riyami O. Activation of peat soil carbon and production of carbon nanostructures using a flying jet cold plasma torch. Environmental Chemistry Letters. 2019;**17**:1383-1390

[50] Custodio M, Peñaloza R, Espinoza C, Espinoza W, Mezarina J. Treatment of dairy industry wastewater using bacterial biomass isolated from eutrophic lake sediments for the production of agricultural water. Bioresource Technology Reports. 2022;**17**:1-9

[51] Sharma M, Agarwal S, Agarwal Malik R, Kumar G, Pal DB, Mandal M, et al. Recent advances in microbial engineering approaches for wastewater treatment: A review. Bioengineered. 2023;**14**:2184518

[52] Elitas M, Dhar N, McKinney JD. Revealing antibiotic tolerance of the mycobacterium smegmatis xanthine/uracil permease mutant using microfluidics and single-cell analysis. Antibiotics. 2021;**10**:794

[53] Bohli T, Ouederni A, Villaescusa I. Simultaneous adsorption behavior of heavy metals onto microporous olive stones activated carbon: Analysis of metal interactions. Euro-Mediterranean Journal for Environmental Integration. 2017;**2**:1-15

[54] Adeleke ARO, Latiff AAA, Daud Z, Mat Daud NF, Aliyu MK. Heavy metal removal from wastewater of palm oil mill using developed activated carbon from coconut shell and cow bones. Key Engineering Materials. 2017;**737**:428-432

[55] Pap S, Šolević Knudsen T, Radonić J, Maletić S, Igić SM, Turk Sekulić M. Utilization of fruit processing industry waste as green activated carbon for the treatment of heavy metals and chlorophenols contaminated water. Journal of Cleaner Production. 2017;**162**:958-972

[56] Rahmani-sani A, Singh P, Raizada P, Claudio E, Anastopoulos I, Giannakoudakis DA, et al. Use of chicken feather and eggshell to synthesize a novel magnetized activated carbon for sorption of heavy metal ions. Bioresource Technology. 2020;**297**:122452

[57] Liu Y, Xu J, Cao Z, Fu R, Zhou C, Wang Z, et al. Adsorption behavior and mechanism of Pb(II) and complex Cu(II) species by biowaste-derived char with amino functionalization. Journal of Colloid and Interface Science. 2020;**559**:215-225

[58] Wang M, Liu Y, Yao Y, Han L, Liu X. Comparative evaluation of bone chars derived from bovine parts: Physicochemical properties and copper sorption behavior. Science of the Total Environment. 2019;**1**:134470

[59] Mandal S, Calderon J, Marpu SB, Omary MA, Shi SQ. Mesoporous activated carbon as a green adsorbent for the removal of heavy metals and Congo red: Characterization, adsorption kinetics, and isotherm studies. Journal of Contaminant Hydrology. 2021;**243**:103869

[60] Van Phuong N, Hoang NK. Assessment of cadmium ion adsorption capacity in water by biochar produced from pyrolysis of cow dung. International Journal of Emerging Trends in Engineering Research. 2021;**9**:203-210

[61] Xu Q, Xiong C, Fan J, Zhang F, Wang J, Xu Q, et al. Assessing the efficiency and mechanism of copper adsorption onto biochars derived from corn straw and cow manure. Water, Air, and Soil Pollution. 2023;**234**:1-13

[62] Chakraborty R, Asthana A, Singh AK, Yadav S. Comparison of biosorption efficiency for hexavalent chromium remediation in synthetic wastewater using unmodified and chemically modified chicken feathers. Journal of Dispersion Science and Technology. 2023;**5**:1-15

[63] Wang M, Ye H, Zheng X, Chen S, Xing H, Tao X, et al. Adsorption behaviors and mechanisms of simultaneous cadmium and fluoride removal on waste bovine bone from aqueous solution. Journal of Environmental Chemical Engineering. 2023;**11**:109035

[64] Khan AM, Usmani MA, Yasmeen K, Ahmed MN, Obaid M, Afshan Naz S, et al. Conversion of waste animal bones to biofertilizer and adsorbent for wastewater treatment: An innovative

approach to develop zero-waste technology. Research Square. 2023;**8**:1-28

[65] Idrees M, Batool S, Kalsoom T, Yasmeen S, Kalsoom A, Raina S, et al. Animal manure-derived biochars produced via fast pyrolysis for the removal of divalent copper from aqueous media. Journal of Environmental Management. 2018;**213**:109-118

[66] Mondal NK, Basu S. Potentiality of waste human hair towards removal of chromium(VI) from solution: Kinetic and equilibrium studies. Applied Water Science. 2019;**9**:1-8

[67] Prabu D, Kumar PS, Rathi BS, Sathish S, Anand KV, Kumar JA, et al. Feasibility of magnetic nano adsorbent impregnated with activated carbon from animal bone waste: Application for the chromium (VI) removal. Environmental Research. 2022;**203**:111813

[68] Tonetti C, Aluigi A, Selmin F, Cilurzo F, Mazzuchetti G. Removal of Cu(II) ions from water using thermally-treated horn–hoof powder as biosorbent. Desalination and Water Treatment. 2015;**55**:1105-1115

[69] Cechinel MAP, Ulson De Souza SMAG, Ulson De Souza AA. Study of lead (II) adsorption onto activated carbon originating from cow bone. Journal of Cleaner Production. 2014;**65**:342-349

[70] Ibrahim WM, Hassan AF, Azab YA. Biosorption of toxic heavy metals from aqueous solution by Ulva lactuca activated activated carbon. Egyptian Journal of Basic and Applied Sciences. 2016;**3**:241-249

[71] Eletta OAA, Tijani IO, Ighalo JO. Adsorption of Pb(II) and phenol from wastewater using silver nitrate modified activated carbon from groundnut (Arachis hypogaea L.) shells. West Indian Journal of Engineering. 2020;**43**:26-35

[72] Tijani J, Temitope BM. Development of low cost adsorbent from cow horn for the biosorption of Mn (II), Ni (II) and Cd (II) ion from aqueous solution. Research Journal of Applied Sciences, Engineering and Technology. 2014;**1**:9-17

[73] Yu B, Jiang G, Cao C, Lei N, Li C, Evariste U, et al. Preparation and electrochemical properties of ultra-high specific surface area N-doped biomass-porous carbon. Journal of Energy Storage. 2020;**30**:101537

[74] Zhang Z, Wang T, Zhang H, Liu Y, Xing B. Adsorption of Pb(II) and Cd(II) by magnetic activated carbon and its mechanism. Science of the Total Environment. 2021;**757**:143910

[75] Zhang T, Wei S, Waterhouse GIN, Fu L, Liu L, Shi W, et al. Chromium (VI) adsorption and reduction by humic acid coated nitrogen-doped magnetic porous carbon. Powder Technology. 2020;**360**:55-64

[76] Jafari Z, Avargani VM, Rahimi MR, Mosleh S. Magnetic nanoparticles-embedded nitrogen-doped carbon nanotube/porous carbon hybrid derived from a metal-organic framework as a highly efficient adsorbent for selective removal of Pb(II) ions from aqueous solution. Journal of Molecular Liquids. 2020;**318**:113987

[77] Gayathiri M, Pulingam T, Lee KT, Sudesh K. Activated carbon from biomass waste precursors: Factors affecting production and adsorption mechanism. Chemosphere. 2022;**294**:133764

[78] Mopoung S, Moonsri P, Palas W, Khumpai S. Characterization and properties of activated carbon prepared from tamarind seeds by KOH activation

for Fe(III) adsorption from aqueous solution. Scientific World Journal. 2015;**2015**:1-9

[79] Phyo S, Kim S, Lee H, Han SS, Oh Y, Lee J. Adsorption enhancement of hazardous odor gas using controlled thermal oxidation of activated carbon. Journal of Cleaner Production. 2023;**393**:136261

[80] Wang H, Wang S, Chen Z, Zhou X, Wang J, Chen Z. Engineered biochar with anisotropic layered double hydroxide nanosheets to simultaneously and efficiently capture Pb2+ and CrO42− from electroplating wastewater. Bioresource Technology. 2020;**306**:123118

[81] El Batouti M, Al-Harby NF, Elewa MM. A review on promising membrane technology approaches for heavy metal removal from water and wastewater to solve water crisis. Water (Switzerland). 2021;**13**:1-42

[82] Van Tran T, Bui QTP, Nguyen TD, Le NTH, Bach LG. A comparative study on the removal efficiency of metal ions (Cu2+, Ni2+, and Pb2+) using sugarcane bagasse-derived ZnCl2-activated carbon by the response surface methodology. Adsorption Science and Technology. 2017;**35**:72-85

[83] Wang Y, Ling M, Liu R, Yu P, Tang A, Luo X, et al. Distribution and source identification of trace metals in the sediment of Yellow River estuary and the adjacent Laizhou Bay. Physics and Chemistry of the Earth, Parts A/B/C. 2017;**97**:62-70

[84] Jurado LA, Espinosa-Marzal RM. Insight into the electrical double layer of an ionic liquid on graphene. Scientific Reports. 2017;**7**:1-12

[85] Yang S, Wu Y, Aierken A, Zhang M, Fang P, Fan Y, et al. Mono/competitive adsorption of arsenic(III) and nickel(II) using modified green tea waste. Journal of the Taiwan Institute of Chemical Engineers. 2016;**60**:213-221

[86] Park J-H, Wang JJ, Kim S-H, Kang S-W, Jeong CY, Jeon J-R, et al. Cadmium adsorption characteristics of biochars derived using various pine tree residues and pyrolysis temperatures. Journal of Colloid and Interface Science. 2019;**553**:298-307

[87] Park JH, Ok YS, Kim SH, Cho JS, Heo JS, Delaune RD, et al. Competitive adsorption of heavy metals onto sesame straw biochar in aqueous solutions. Chemosphere. 2016;**142**:77-83

[88] Islam MM, Mohana AA, Rahman MA, Rahman M, Naidu R, Rahman MM. A comprehensive review of the current Progress of chromium removal methods from aqueous solution. Toxics. 2023;**11**:1-43

[89] Yang X, Wan Y, Zheng Y, He F, Yu Z, Huang J, et al. Surface functional groups of carbon-based adsorbents and their roles in the removal of heavy metals from aqueous solutions: A critical review. Chemical Engineering Journal. 2019;**366**:608-621

[90] Wang B, Lan J, Bo C, Gong B, Ou J. Adsorption of heavy metal onto biomass-derived activated carbon: Review. RSC Advances. 2023;**13**:4275-4302

Chapter 12

Sustainable Remediation of Heavy Metals through Agricultural Waste-Derived Adsorbents

Robert Birundu Onyancha

Abstract

Heavy metals (HMs) are known to be non-biodegradable and non-metabolic agents, hence detrimental to human health. Then, innovative and novel techniques must be formulated to mitigate and remove these contaminants. Several HMs removal techniques are expensive, have low removal efficiency, produce secondary pollutants, and require high operational energy. The adsorption technique has solved these drawbacks, especially when low-cost and naturally available adsorbents derived from agricultural wastes are used. Reviewed papers show that the adsorption of HMs by these adsorbents depends on pH, contact time, temperature, adsorbent dosage, adsorbent particle size, HM ion concentration, agitation speed, and interfering ions. Removal efficiency ranged between 61 and 100% with contact time between 20 and 240 min and pH of 3–10. The adsorption process was well described by pseudo-second-order and Langmuir models. Also, thermodynamic studies revealed endothermic or exothermic processes depending on the type of agricultural waste material, adsorption factors, and adsorption conditions.

Keywords: adsorption, heavy metals, remediation, isotherms, kinetics, regeneration, wastewater, agricultural waste adsorbent (AWA)

1. Introduction

Generally, heavy metals (HMs) can be classified using atomic weight, density, chemical properties, and toxicity [1]. According to [2], HMs represent elements whose atomic weights range from 33.5 to 222.5 g mol^{-1} with a specific weight greater than 5.2 g cm^{-3}. Thus, on the period table, HM can be grouped into transition metals, rare earth metals, and some metals from group P. They can be toxic, radioactive, pseudo-metals, and essential metals for metabolism, as shown in **Figure 1**. Of particular interest here are the toxic HMs like lead, cadmium, chromium, mercury, arsenic, nickel, copper, and zinc. Despite being categorized as toxic, copper, Magnesium, Nickel, selenium, Zinc, molybdenum, manganese, iron, chromium, and cobalt are vital nutrients when consumed within allowable levels since they are necessary for diverse physiological and biochemical roles in the human body [3]. The World Health Organization and the Occupational Safety and Health Administration Agency report

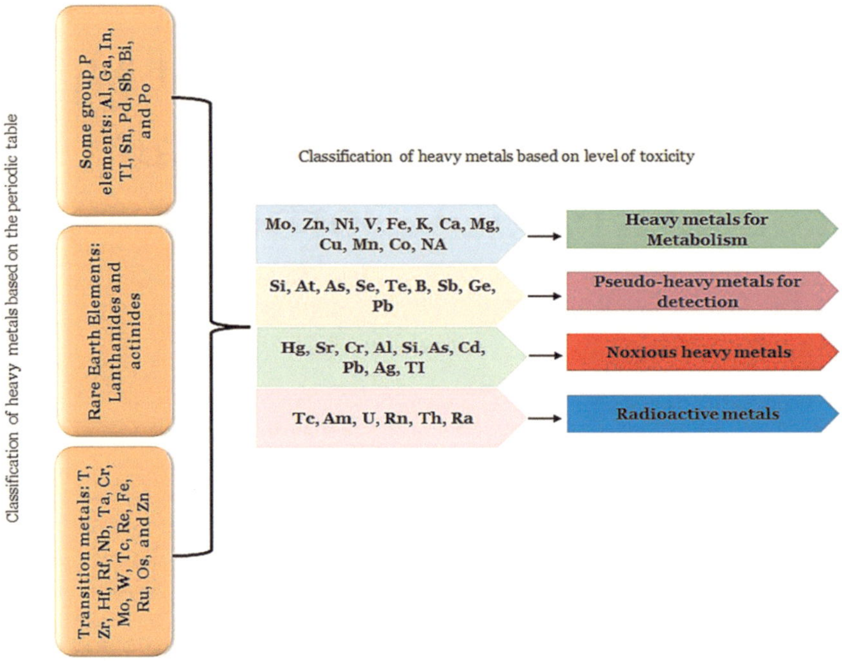

Figure 1.
Classification of HM based on atomic weights, toxicity, and chemical properties.

that Cd, As, Hg, Cr, Be, and Pb are highly toxic even at low concentrations and cautiously indicating that other HMs cannot be overlooked [3].

Primarily, HMs can enter the human body by inhaling polluted air and consuming contaminated food and water. Once inside the body, they form ions that react with bioparticles to generate harmful substances whose properties rely on concentration and biomagnification [4]. Chronic exposure and consumption of above permissible limits of these metals can lead to respiratory issues, neurological disorders, kidney damage, cancer, reproductive system disorders, brain damage, skin irritation, and immune dysfunction, among many more (**Figure 2**) [3]. In aquatic systems, HM can interrupt marine life, collect in the food chain, lead to oxygen deficiency, and interfere with biodiversity by modifying water's essential biological and chemical properties. They can cause plant impairment, lead to a reduction in agricultural production, and contaminate crops while in the soil [5].

It is imperative to note that HMs are introduced into the environment via diverse natural sources like geological breaking of rocks, volcanoes, soil erosion, forest fires, and biological activities coupled with industrial activities such as mining, chemical processing, construction, transport, pesticides, and fertilizer applications, households discharge, and animal and medical wastes (**Figure 2**) [3]. These activities discharge raw HMs-ridden materials and effluents to water systems like oceans, lakes, rivers, and streams, leading to wastewater that is unfit for human and aquatic life. Notably, freshwater usage in various essential sectors such as agriculture, industries, and households has increased by 70%, 22%, and 8%, respectively, leading to massive wastewater discharge, hence aggravating the situation even more [6].

As a scarce resource, the continuous obliteration of water, which has been available freely over time, signals the danger to the human community [7]. According to

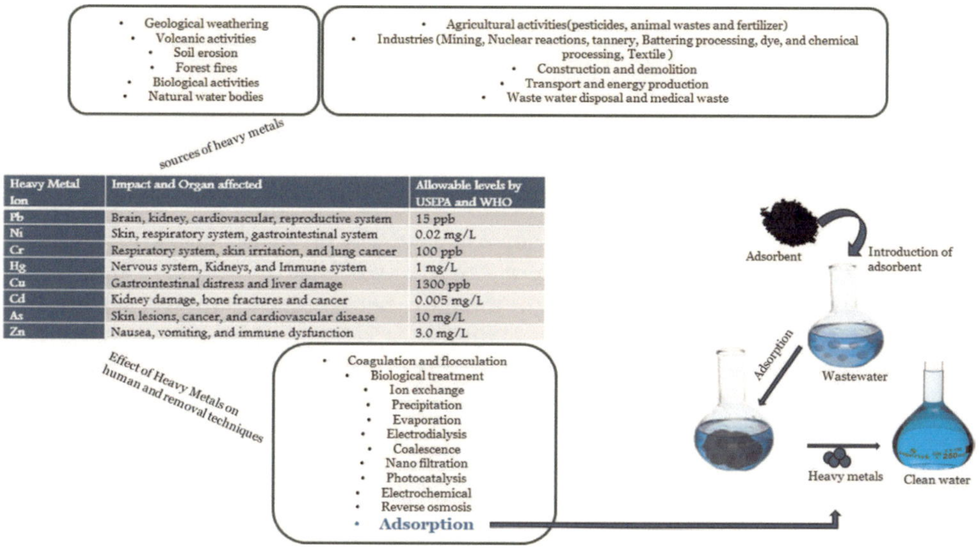

Figure 2.
Sources, effects on humans, and removal techniques of HMs.

the statistical calculations on global water scarcity, it is estimated that in one year, more than two-thirds of the worldwide population does not have access to clean water for more than 30 days, with China and India representing 50% of this population. Overall, the report indicates that a total of 0.5 billion human beings are adversely affected annually by this problem [8].

Therefore, with a rapidly increasing global population and industrialization, an increase in demand for freshwater consumption and converse contamination from HMs is a reality. As a result of this, there is an urgent need for more diverse and effective remedial measures to decontaminate wastewater. This can be attested by the ever-increasing yearly research publications on HMs emphasizing their dangers, mitigation, and removal (**Figure 3**). Notably, conventional methods such as ion

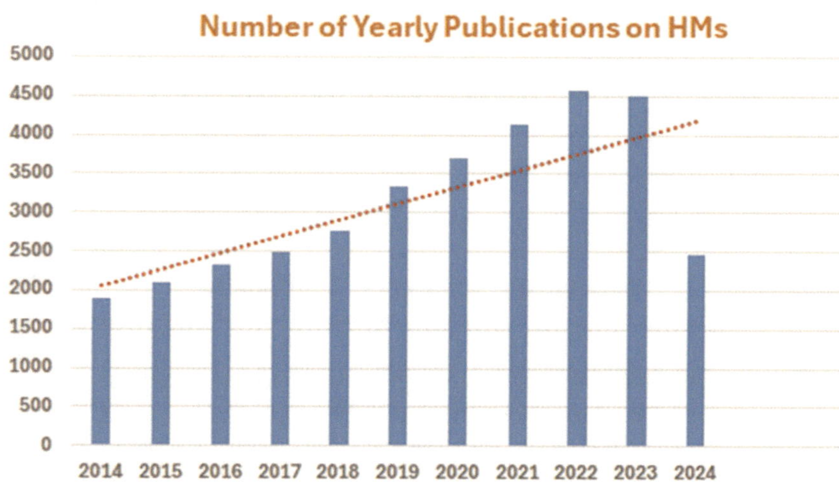

Figure 3.
Representation of yearly published articles with the word 'heavy metals' published between 2014 and June 2024 obtained from "Scopus".

exchange, precipitation, coalescence, filtration, evaporation, biological treatment, coagulation, and flocculation have been employed for HM removal [5].

Nevertheless, these strategies are inherently expensive to assemble, have high operational costs, produce secondary pollutants and sludge, require expertise to operate, and often do not attain complete removal, specifically at low concentrations of contaminants. Imperatively, the adsorption method leverages the natural capabilities of certain materials to capture and bind HMs, thereby naturally purifying contaminated water. The technique is popular since it is simple, efficient, cost-effective, and eco-friendly and offers the potential to regenerate and reuse adsorbents [9].

The ideal adsorbent should be environmentally benign, have superior selectivity, be easily recyclable, have outstanding adsorption capacity, and be extremely efficient even at low concentrations of HMs contaminants. Adsorbents like metal oxides, carbon-derived materials, dendrimers, and nanocomposites with the particle size range of micro-nano range, have been found to be effective [10–12]. However, they are inherently expensive and produce secondary chemicals which pollute the environment. Alternatively, adsorbents derived from agricultural waste materials, biological materials, and natural clay have been proven to address these drawbacks.

Research studies show that clove leaves, sunflower leaves, guava leaves, mandarin garden grass, orange peels, sugar cane waster, maize residues, onion skin, sesame hull, wood biomass, groundnut shell, cotton shells, lobeira fruit, neem bark, mustard waste, eggshell [13–20], among others have successfully been synthesized and utilized in the adsorption process to remove contaminants from water [9]. They are abundantly available, low-cost alternatives to commercial adsorbents, biodegradable, environmentally friendly, and exhibit high adsorption capacity. Furthermore, they can undergo surface modification and functionalization to enhance selectivity and affinity, hence increasing removal capacity [3].

Consequently, confiscating HMs from aqueous systems via the adsorption technique remains the ideal solution that harnesses natural processes and materials to tackle major environmental contamination from HMs. Therefore, this book chapter endeavors to review the available studies on the adsorption of HMs. The chapter provides concise reviews of recent developments in the use of adsorbents produced from agricultural waste materials, which are naturally biodegradable, effective, biocompatible, non-sludge generators, and inexpensive. Discussion on the adsorption process, suitability of these materials, mechanisms of adsorption through isotherm and kinetics models, factors affecting adsorption, thermodynamics, and regeneration are well articulated.

2. Adsorption mechanisms for sustainable HMs remediation

Adsorption involves molecule transfer from the bulk aqueous system to the solid surface due to a difference in concentration gradient. The transfer process depends on the molecular weight, the shape, and the adsorbent polarity, which is responsible for binding adsorbate and necessitating easier separation with the binding rate proportional to the square root of the contact time [21]. Notably, the outer surface molecules do not fully occupy the adsorbent surface hence it is capable of binding adsorbate molecules. The adsorbate molecules in the bulk solution have lower energies compared to adsorbent surface molecules, resulting in residual attractive forces. Consequently, it is this surface energy (attractive force per unit surface area) that is responsible for adsorbent-adsorbate adsorption.

The adsorption experiments can be conducted as batch, continuous, or semi-batch. In batch, a fixed amount of adsorbent is mixed with a known adsorbate concentration, and an HM decrease in concentration is measured over time. On the other hand, continuous involves feeding a system with the adsorbent and adsorbent mixture coupled with continuous removal of the treated solution. A semi-batch is simply a hybrid of batch and continuous processes where one component is fixed, and the other is continuously fed [22]. How an adsorbate is attracted to the adsorbent surface is determined by an attraction force, which can be either physisorption or chemisorption.

Physisorption adsorption (**Figure 4a**) is realized when the adsorbate materials attach to the surface due to feeble Van-der-Waals interactions. This process is characterized by dependence on temperatures and adsorbent surface area. Its low heat of adsorption implies that activation energy is not required, the multilayer adsorption is reversible, and the adsorption enthalpy is significantly low (20–40 kJ/mol). The implication on the surface area is that the rate of adsorption increases with the increase of the surface [23].

Chemisorption adsorption (**Figure 4a**) occurs when the attractive force between the pollutant and the adsorbent surface is due to covalent or ionic bonds. It occurs at high temperatures and hence demands activation energy. Also, new compounds are formed with the process accompanied by elevated enthalpy, rendering the process irreversible. This process is equally dependent on temperature and surface area [23].

Imperatively, the mechanisms of adsorption of HMs on the surface of the adsorbent can be through ion exchange, chelation, precipitation, complexation, reduction, or a combination of different mechanisms operating concomitantly [3]. The mechanisms are governed by parameters such as pH, temperature, adsorbent mass, volume, and type of adsorption [24].

- The study of adsorption mechanisms using banana peels as an adsorbent in the uptake of Cu(II) and Pb(II) is through ion exchange where acid groups are

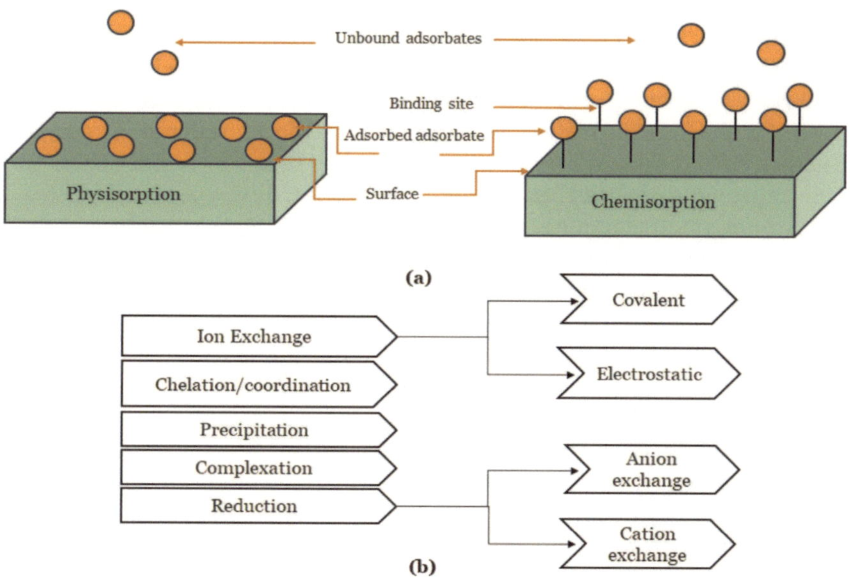

Figure 4.
(a) Type of adsorption (Physisorption and Chemisorption) (b) Adsorption mechanisms.

deprotonated leading to a negatively charged adsorbent surface, which necessitates HMs uptake via electrostatic repulsive interactions [25].

- A similar reversible ion exchange mechanism was noted in the adsorption of 33.04% of Cd(II) compounded with a non-reversible precipitation mechanism in the removal of 66.97% of Ca(II) ions using an eggshell [26].
- A combination of precipitation and complexation mechanisms has been ascribed to the adsorption of Cd(II) on Turkish Coffee Grounds [27].
- Also, complexation and ion exchange mechanisms have been involved in the sequestration of Pb(II) by cotton shell biomass with several surface moieties of functional groups [28].
- The elimination of Cr(II) and Cu(II) by soybean meal waste material was through ion exchange, chelation, and precipitation by hydroxyl and carboxyl groups [29].
- The ion exchange process in the acid pH and physisorption mechanism in the alkaline pH have also been identified in Pb(II) removal by rice husk [30].

It is so evident that these mechanisms work independently or synergically, in that way, they enhance adsorption efficiency.

3. Low-Cost agricultural waste adsorbents (AWA) for HMs removal

The need to develop an environmentally friendly, inexpensive, effective, and readily available adsorbent to decontaminate HMs from wastewater systems is on the rise due to the commercially expensive adsorbents. The AWA are produced from freely available agricultural waste materials through modification, treatment, and transformation to activated carbon, which has high affinity and selectivity to HMs [31]. Also, they are characterized by high porosity, small pore diameter, enhanced adsorption rate, large surface area, and ease of regeneration and retrieval hence, guaranteeing high HMs adsorption capacity [32]. **Table 1** compares AWA with commercial activated carbon and synthetic adsorbents in terms of surface area, cost, adsorption capacity, efficiency, environmental impact, and regeneration.

Waste from agricultural materials contains cellulose, lignin, water, starch, lipids, proteins, hemicellulose, hydrocarbons, and simple sugar [50]. They also have surface functional groups such as amide, alcohols, carboxyl, acetamido, esters, sulfhydryl, and phenolics with high attractions and selectivity for HMs species, thus capable of binding and safely sequestrating them from aqueous solution [51].

The synthesis or conversion of agricultural wastes into an adsorbent undergoes many steps which include selection of raw material, pre-treatment, carbonization, activation, characterization, and application, as explained in **Figure 5**.

Reports on the removal of HMs from contaminated water and effluents by employing AWA is well documented in **Table 2**. For example, biochar produced from wood biomass was used to remediate lead-polluted water. With a dose of 5 g/L and contact time of 4-h, the Pb(II) percentage removal observed was 100%, and the adsorption process was well fitted by the pseudo-second-order (PSO) and Langmuir (LMR) models [22].

Parameter	AWA	Activated carbon	Synthetic adsorbents
Surface area	500–1200 [33, 34]	900–1500 [35, 36]	400–1000 [37]
Cost	Low (\$0.1–\$1)/kg [38]	Medium-High (\$2–\$5)/kg [39]	High (\$10–\$30)/kg [37, 40]
HMs adsorption capacity (mg/g)	100–250 [38, 41]	200–450 [39]	100–300 [42]
Dye adsorption capacity (mg/g)	150–476.19 [43]	250–600 [44]	100–1000 [45]
Efficiency (%)	60–100 [46]	80–95 [9]	70–90 [37, 40]
Environmental Impact	Low [46]	Medium [47]	High [37]
Method of Regeneration	Thermal, Washing [48]	Thermal, chemical [49]	Thermal, pressure change [40]
Regeneration cost	Low [46, 48]	Medium [47]	High [40]

Table 1.
Various Parameters for AWA, Activated Carbon, and Synthetic Adsorbents.

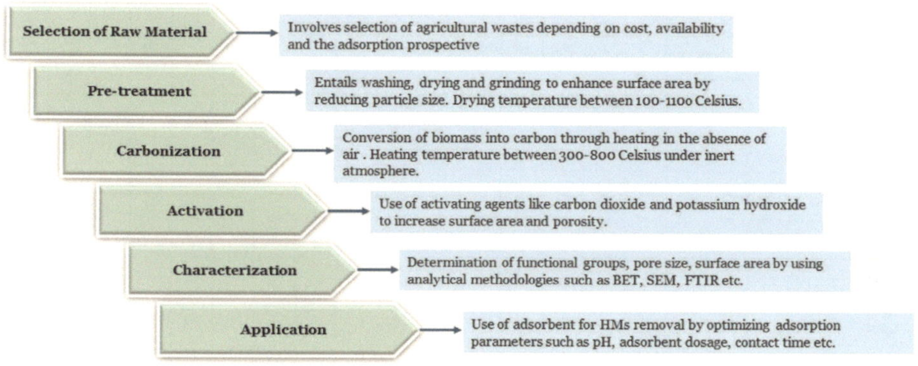

Figure 5.
Steps of converting agricultural wastes into adsorbent [52, 53].

Similar studies on the elimination of Pb(II) from an aqueous system of 20 mg/L of concentration, with a pH of 3, contact time of 210 min, and temperature of 50°C using adsorbent from pomelo peels demonstrated 99.9% removal of Pb(II). Equilibrium adsorption studies on kinetics were well represented by PFO with monolayer adsorption mechanisms attributed to chemical bonding and physical adsorption [69].

Through batch conditions, an adsorbent produced by a modified sesame hull was employed to eradicate Cr(VI) from aqueous solutions. Tartaric acid was used to alter the sesame hull to enhance the exposure of cellulose constituents responsible for improved adsorption. With set parameters of pH, contact time, initial concentration, and adsorbent dosage, 3, 120 min, 23.23 mg/L, and 4.94 g/L respectively, the modeled data exhibited high correlation with experimental data as attested by high R^2 (0.9763). Adsorption mechanisms were well described by LMR isotherm and kinetics fitted with the PSO [50].

HMs	AWA	pH	Contact Time	Removal %	Isotherm	Kinematics	Thermodynamics	Ref.
Hg(II), As(III)	Maize residues	4,8	90,120	85.68–98.13, 93.14				[54]
Cd(II)	Shea fruit biomass	5	150	76.86	FNH	PSO	ENDO	[55]
Cd(II)	*Helianthus annuus* (Sunflower) waste	6		99.80		PSO		[56]
Cr(VI)	Sesame hull	3	120	100.00	LMR	PSO		[50]
Hg(II)	Sugarcane bagasse	8	160	61				[57]
Pb(II)	Bean husk	7	60	90	LMR	PSO	ENDO	[58]
Pb(II)	Wood biomass	—	240	100	LMR	PSO	—	[22]
Cd(II)	Eggshell	7	120	98.76	FNH	PSO		[26]
Pb(II)	Groundnut shell	6	120	96.61	Tempkin			[59]
Pb(II)	Cotton shells	6	90	90.00	FNH			[28]
Cd(II)	Turkish coffee	8.87	60	96.00	LMR	PSO		[27]
Cr(VI)	Spent tea leaves	10	60 and 150	95.42	LMR, FNH			[60]
As(III)	water chestnut shell	7	120	75	LMR	PSO		[61]
As(III)	corn cob	7	120	67	LMR	PSO		[61]
As(III)	Tea waste	7	120	74	LMR	PSO		[61]
As(III)	pomegranate peel	7	120	65	LMR	PSO		[61]
As(III)	Mung bean husk	3		98.75	LMR			[62]
Pb(II)	Lobeira fruit	2	60	93.00	LMR			[63]
As(III)	Neem bark	6	20	89.96	LMR	PSO	ENDO	[64]
Pb(II)	Rice hunk ash	3	—	75.00	LMR	PFO	EXO	[30]
Pb(II), Zn(II), Cd(II)	Mustard waste	5.5	120	94.56. 96.15, 76.48	LMR	PSO	EXO	[65]
Pb(II)	Banana peels	5.5	60	80.65	LMR	PSO	EXO	[66]
Pb(II)	Banana Peels	6	60	>90%	LMR, FNH			[67]

HMs	AWA	pH	Contact Time	Removal %	Isotherm	Kinematics	Thermodynamics	Ref.
Cd(II), Cu(II)	Sorghum bicolor	5–6	—	15.151 $mg.g^{-1}$	LMR	PSO	—	[68]
Pb(II)	Pomelo peels	3	210	99.90	LMR	PFO		[69]
Cu(II)	Watermelon shell, lemon Peel		120	90,78	LMR			[70]
As(III)	Eggshell	7	120	78–87	LMR	PSO		[61]
As(III)	java plum seed	7	120	78–87	LMR	PSO		[61]
NH_3	Sugarcane bagasse ash	8	180	60	LMR	PSO	ENDO	[71]
Cu(II), Pb(II)	Banana peels	5	120	99.76, 88.94	LMR	PSO		[25]

Table 2.
A summary of HM removal using AWA.

A set of batch and column biosorption experiments employing neem bark powder to decontaminate As(III) was investigated. A desirable performance (R^2 = 0.961 and 0.954) modeled from an artificial neural network was realized. Biosorption removal of 89.96% was obtained with optimized individual parameters such as flow rate of 3.0 mL/min, initial concentration of 2000.0 μg/L, and adsorbent dosage of 2.0 g. The thermodynamic analysis revealed an endothermic reaction process. The adsorption rate was determined to rely on the square of unoccupied adsorbent sites [64].

Preparation of rice husk through incineration at 800°C for 6 h followed by 0.5 M HCl activation to produce biosorbent for Pb(II) removal has been reported. Batch experiments were performed at a pH of 3 with LMR and the PFO describing the experimental data. The removal efficiency was temperature-dependent, and the adsorption mechanism was attributed to ion exchange. Thermodynamic study analysis disclosed that the process was both exothermic and spontaneous [30].

Activated TETA-PP biosorbent produced from pea peels activated with triethylenetetramine was used in Cr(VI) adsorption. The absorbent showed a pH dependence with a maximum removal of 99% experience with a pH of 1.6. LMR and PSO determined and discussed the 312.50 mg/g biosorption capacity. The electrostatic attraction mechanism resulting from surface protonation of TETA-PP sites was identified to be responsible for the biosorption of Cr(VI) [8].

Biochar of a specific surface area of 12.26 m^2/g and total pore volume of 0.012 cm^3/g was formed from watermelon peels and employed in removing Cu(II) ions. The adsorption process was identified as chemisorption and dependent on pH, with optimal removal of Cu(II) obtained at 5.6. LMR and PSO models were used to describe the adsorption process with correlation coefficient values of 0.99 and < 0.99. The adsorbent's adsorption capacity was 140.85 mg/g and six regeneration cycles, which clearly shows that this adsorbent is effective, cost-effective, and eco-friendly for Cu(II) [16].

An investigation was conducted on the efficacy of adsorbent derived from modified rice straw for eliminating Cr(VI) and Pb(II) from wastewater. SEM, FTIR, XRD, and EDX realized the determination of the morphology and functional group of the biochar produced. The ideal pH, contact time, initial concentration, and biosorbent dose of 2, 50 min, 20 mg/L, and 2 g/L, respectively, were determined. The efficiency removal of Cr(VI) and Pb(II) was 95.57% and 85.68%, respectively. A multilayer adsorption process on heterogeneous surfaces was witnessed with thermodynamic studies revealing a spontaneous and endothermic system [72].

Similarly, magnetically activated orange peel composite with surface area, volume of pores, and mean pore diameter of 155.09 m^2/g, 0.1768 cm^3/g, and 4.5604 nm, respectively, was fabricated and used to confiscate Cr(VI) from wastewater. Contact time, pH, the Cr(VI) starting concentration ions, and adsorbent dose were found to influence the uptake of Cr(VI). The process was identified as a monolayer with LMR isotherm, determining an adsorption capacity of 277.8 mg/g and a correlation coefficient range between 0.992 and 1 achieved through PSO [73].

Sorghum bagasse activated by NaOH (SB-NaOH) was produced and employed to absorb Cd(II) from an aqueous system. An impressive adsorption percentage removal of 96.48 was realized with optimal conditions of pH of 5, contact time of 70 min, initial concentration of 100 mg/L, and temperature of 30°C. LMR and PSO models described the adsorption mechanism. Thermodynamic studies obtained a positive Gibbs free energy and a negative enthalpy, thus indicating an exothermic process [74].

From **Table 2**, it is evident that these adsorbents exhibit remarkable removal efficiencies ranging between 61 and 100% and the capability to target a wide range of

HMs owing to their distinctive properties, including a high surface area, environmental friendliness, large surface area, high affinity, and selectivity. In addition, they are inexpensive and readily available, rendering them extremely appealing for large-scale HM sequestration. The promising adsorption results coupled with consistent HM removal underscore the possibility of utilizing these adsorbents as efficient remedies for mitigating HMs pollution.

4. Adsorption kinetics and equilibrium studies

4.1 Adsorption isotherm

Isotherm models are designed to descriptively examine the adsorbate accumulation on the adsorbent at equilibrium adsorbate concentration and constant temperature. At the equilibrium of the sorption system, the isotherm models explain the distribution of adsorbate molecules in bulk aqueous solution and solid phases [46].

Establishing dynamic equilibrium is realized once the desorption and adsorption rates are similar. Likewise, when the adsorbate molecules bind on the adsorbent interface and are dynamically at equilibrium with the adsorbate concentration, sorption equilibrium is established. This results in the determination of absorbent-adsorbate interaction, thus guiding optimum utilization of the adsorbent material. Many isotherm models are employed to explain the adsorption mechanisms of HMs, including LMR, Dubinin-Radushkevich, Freundlich, Sips, Redlich-Peterson, Henry, and Temkin [5].

The models can predictably give information on uptake capacity, adsorption mechanisms, sorbent surface properties, selectivity, and affinity of adsorbate molecules. **Table 3** provides some of the predominantly used isotherm models (**Figure 6**).

A comprehensive summary of the reviewed studies shown in **Table 2** regularly demonstrated the following:

- that LMR isotherm provides the best description of the adsorption mechanisms of the AWA. It indicates that the adsorption in these AWA systems is predominantly monolayer on a homogeneous surface with a restricted number of sites. The adsorption occurs evenly on the active adsorbent sites, and the adsorption process is only complete when adsorbate molecules fill all active sites.

- a few studies exhibited some degree of heterogeneous and multilayer types of adsorptions which is well-explained by the Freundlich model.

4.2 Adsorption kinetic studies

The adsorption kinetic studies examine the rate at which pollutants are adsorbed onto an adsorbent surface, giving important information on the efficiency and mechanisms of the adsorption. Generally, the studies entail examining the concentration of pollutants over time and providing an analytical assessment of data to establish adsorption rates and the parameters/factors affecting them [75]. Therefore, the significance of these studies rests on the capability to advise on the design and implementation of efficient approaches to control pollution. Efficient removal of pollutants can be achieved through the development of accurate kinetic models that inform the ideal adsorbent and set the operational parameters. Equally, the studies contribute to

Isotherm Model	Equation	Definition of Terms	Model assumptions	Ref.
Langmuir	$\frac{1}{Q_e} = \frac{1}{Q_{max}K_L}\frac{1}{C_e} + \frac{1}{Q_{max}}$ $R_L = \frac{1}{1+K_L C_e}$	K_L is Langmuir adsorption constant. Q_{max} is monolayer adsorption capacity R_L parameter used to identify the adsorption process $R_L > 1 \rightarrow$unfavorable adsorption $R_L = 1 \rightarrow$ linear adsorption $R_L = 0 \rightarrow$ irreversible adsorption $0 < R_L > 1 \rightarrow$ favorable adsorption	→It indicates monolayer adsorption on homogenous surfaces. → This leads to the creation of a monolayer coating of adsorbate on the adsorbent surface thus obstructing further adsorption	[5, 75]
Freundlich	$log\, Q_e = log\, k_f + \frac{1}{n} log\, C_e$	k_f is constant(Freundlich) C_e is the adsorbate concentration at the equilibrium n is a parameter describing the adsorption process $n < 1 \rightarrow$ chemical process $n = 1 \rightarrow$ linear adsorption $n > 1 \rightarrow$physical process	→ Points to adsorption on the heterogenous surface having its adsorbed molecules interacting resulting in multilayer adsorption. → It shows a decrease in adsorption energy as the adsorption process proceeds.	[5, 75]
Temkin	$q_e = \beta_T(InC_e) + (\beta_{T\beta}InA_T)$	A_T− constant (equilibrium binding) β_T -Temkim isotherm constant β_T < 8kj shows physisorption β_T > 8k indicates chemisorption	The model offers a factor that distinguishes various absorbent-adsorbate interactions. This is achieved through providing the heat of adsorption as a function of temperature.	[5, 75]
Dubinin-Radushkevich	$Inq_e = Inq_s - (\beta\varepsilon)^2$ $\varepsilon = RTIn\left[1 + \frac{1}{C_e}\right]$ $E = \frac{1}{\sqrt{2\beta}}$	q_s -theoretical isotherm saturation capacity β -Dubinin-Radushkevich theoretical isotherm saturation capacity ε Polanyi potential	The model employs Gaussian energy distribution to express the adsorption mechanism on heterogeneous surfaces.	[5, 75]

Table 3.
Isotherm models.

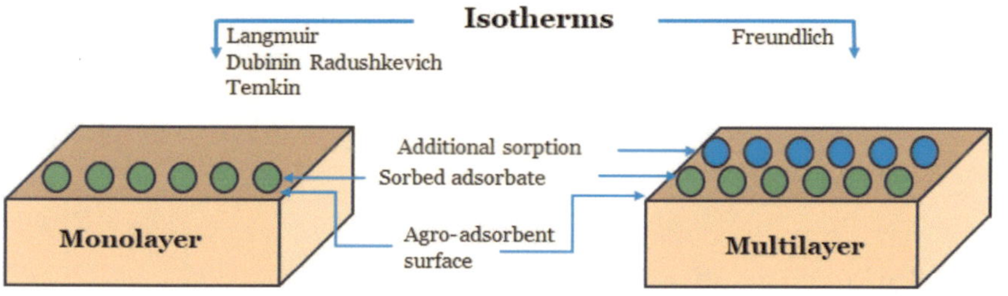

Figure 6.
Monolayer and Multilayer adsorption.

Kinetic Model	Equation	Definition of Terms	Model assumptions	Ref.
Pseudo-first-order	$\frac{dq_t}{dt} = kp_1(q_e - q_t)$	kp_1: PFO rate constant t: reaction time q_t: Adsorption capacity at time t q_e: Adsorption capacity at equilibrium	The derivation of this model presumes that the rate of reaction is directly proportional to the number of freely active binding sites on the adsorbent, and the reaction is reversible at equilibrium. The model describes physisorption adsorption.	[5, 75]
Pseudo-second-order	$\frac{dq_t}{dt} = kp_2(q_e - q_t)^2$	kp_2: Rate constant for PSO	This mode assumes that the rate of adsorption is proportional to the square of many binding sites on the adsorbent surface. The model explains chemisorption	[5, 75]
Elovich	$\frac{dq_t}{dt} = \propto \exp(-\beta q_t)$	∝: Initial adsorption rate β: Desorption constant	The model assumes that the active adsorption sites increase exponentially with adsorption. Used to explain systems that do not follow PFO or PSO models.	[5, 75]
Intraparticle diffusion	$q_t = k_{id}t^{1/2} + C$	k_{id}: Intraparticle diffusion rate constant C: Intercept	The model is used to describe the diffusion in the adsorption process.	[5, 75]

Table 4.
Kinetic study models.

the discovery and establishment of new adsorbent materials, which will be essential for the advancement of environmentally sustainable practices [5].

The adsorption of HMs by porous materials like AWA can be characterized by the three most widely adopted models, namely PFO, PSO, and Elovich, as represented in **Table 4**. However, the intraparticle diffusion model is always invoked to interpret the diffusion of contaminants into the adsorbent surface before concluding the nature of the adsorption mechanisms [5].

Based on **Table 2**, the adsorption kinetic studies show that;

- the PSO model precisely fits and describes the adsorption mechanism of a large number of these adsorbents. This implies that the adsorption process and mechanisms are largely managed by chemisorption involving the sharing or exchange of electrons between the AWA and HMs ions.

- equally, some studies have shown an involvement of the physisorption process which is well described by the PFO model.

5. Factors influencing adsorption efficiency

HMs removal from wastewater via the adsorption method is influenced by several important factors, including pH, adsorbent dosage, contact time, ion concentration, temperature, agitation speed, adsorbent particle size, and interfering ions. Studying and understanding these factors is important for optimizing the adsorption process which leads to optimum removal efficiency and enhanced cost-effectiveness. Also, this helps in continuous modification, reengineering, and development of novel adsorbents with superior characteristics.

5.1 Contact time

The duration of contact between the adsorbent and adsorbate species influences the adsorption process. It has been reported that the maximum uptake of adsorbate and reaching of equilibrium is proportional to shaking time [76]. The implication is that inadequate contact time leads to incomplete adsorption, while prolonged time shows a saturation in the adsorption process owing to exhaustion of the active adsorbent binding sites [77].

For instance, the adsorption kinetics data for Pb(II) by pomelo peels at a pH of 3 and temperature of 45°C established three time-dependent distinct stages. A quick increase in adsorption at the initial stage between 0 and 30 min, a gradual increase between 30 and 120 min, and an equilibrium adsorption value at 210 min. The equilibrium value of 210 min was then employed during the set of experiments as the optimal contact time [69]. Also, the use of cotton biomass powder for the removal of Pb(II) showed a steady increase of adsorption for 60 min, then slowed and eventually attained equilibrium after 90 min because of the saturation of active binding adsorption sites.

Equally, the influence of contact time on the sequestration As(III) was studied by varying time between 10 and 30 min. The study showed a steady increase in the removal of As(III) as temperature increased up to 20 min, and there was no substantial change with a further increase in time, thereby setting 20 min as the equilibrium time. The faster adsorption rate was attributed to abundance in the actively available binding sites [64].

5.2 The pH

The acidity (H+) or alkalinity (OH-) in aqueous systems influences the adsorbent uptake of HMs. It characteristically impacts the speciation of HMs, surface charge on the adsorbent, and degree of ionization [78]. It also leads to the modification of HMs distribution, hence influencing electrostatic interaction between HMs and the adsorbent surface [75].

The removal of Hg(II) using an adsorbent derived from maize plant wastes was conducted on varying pH (2–14). The sequestration of Hg(II) ions and uptake capacity increased with a decrease in pH for monocomponent adsorbent due to the absence of precipitation of hydroxyl ions hence minimizing competition for binding sites. Conversely, the removal increased at higher pH values for the bicomponent adsorbent since the pH solution assisted the Hg(II) ions in being attracted to the existing functional groups on the HGAC adsorbent surface. The HGAC active sites deprotonated as the pH increased, thereby creating negative charges in the adsorbent with a high affinity for the contaminants [54].

Similarly, Cd(II) sequestration using shea fruit shell biomass as an adsorbent showed similar trends of increased removal with an increase in pH values. It has been reported that at higher pH, fewer H^+ ions were bordered by the adsorbent, necessitating Cd(II) to be attracted to available sites, thus attaining removal of 74.06% at a pH of 5. Conversely, at lower pH values, the H^+ ions compete actively with Cd(II) ions for the active binding sites, hence resulting in low removal efficiency [55].

Studies on the dependence of pH in removing Cr(VI) using an adsorbent derived from *Pisum sativum* peels showed a decrease in removal efficiency (from 99–34%) with an increase in pH levels. The optimum pH of 1.6 resulted in the efficiency removal of 99%. Notably, Cr(VI) exists in different permutations at varied pH values of an anionic nature, so at a pH of 1.6, the sorption was attributed to electrostatic force existing between positively charged adsorbent binding sites and negatively charged on $Cr_2O_7^{2-}$ (monovalent chromate). With an increase in pH > 6.5, CrO_4^{2-} becomes predominant, increasing the competition between OH- and CrO_4^{2-} which are negatively charged hence leading to repulsion which signals reduced removal efficiency [8].

5.3 Temperature

Thermodynamically, the adsorption process can be categorized as exothermic or endothermic and it has been demonstrated to depend on the solution temperature [8]. The change in solution temperature affects the solubility of HMs in solution systems, HMs diffusion rate, and impacts the sorbent functional groups [79]. Studies have shown elevated temperatures enhance adsorption kinetic energy and increase HMs removal efficiency [23].

Studies on the influence of temperature on removing Cd(II) from aqueous solution using shea fruit shell biomass were conducted at different temperatures with other parameters held constants. An improvement in removal efficiency from 25% to 62.93% was recorded with an increase in temperature up until 40 0C, which registered an optimum efficiency removal of 66.93%. This has been attributed to the enhanced diffusion rate of the adsorbate [55].

Likewise, the effect of temperature on the removal of Cu(II) by sawdust was carried out by adjusting the temperature from 15 to 60°C. An increase in Cu(II) ion uptake with an increase in temperature until a maximum was attained at 30°C. Any further increase in temperature showed a gradual reduction in efficiency rate and eventually attained the equilibrium state. A decrease in ion uptake was also recorded due to a rise in the desorption rate [80].

5.4 HM ion concentration

Initial HM ion concentration is a powerful parameter that determines the uptake capacity and decontamination of HMs by AWA. It has been established that adsorbent capacity quickly reduces when the initial HM concentration is increased [81]. This has been attributed to the limited adsorption sites, and any increase in concentration renders the available sites fully saturated. Conversely, at lower concentrations, high uptake capacity and removal of HMs are experienced due to the high ratio of binding sites to the available adsorbate molecules [82].

Also, studies have shown that concentration gradient facilitates HMs transportation from the aqueous medium to the adsorbent surface with high uptake capacity witnessed when initial concentration is increased [83]. This has been experienced in the study on the removal of Cd(II) on the dependence on ion concentration where removal efficiency and capacity increased with an increase in concentration between 20 and 100 mg/L and decreased thereafter due to the availability of adsorption sites and the presence of surface area which decreases at high concentrations [27]. Similar ion concentration dependence results have been reported in the depolluting Cd(II) using shea fruit shell biomass [55].

5.5 Interfering ions/ionic strength

Anions such as nitrates(NO_3^-), chlorides (Cl^-), phosphates (PO_4^{3-}), bicarbonates (HCO_3^-), and sulfates (SO_4^{2-}) influence the removal of HMs from aqueous solutions. The progression of inhibition in the removal of HMs is shown as $HCO_3^- \rightarrow SO_4^{2-} \rightarrow Cl^-$. Whereas NO_3^-, Cl^-, and SO_4^{2-} have little effect on the removal of HMs, HCO_3^- has been found to greatly lower the uptake capacity of HMs due to competition for the adsorbent's active binding sites, thereby lowering the overall removal efficiency [84].

Similarly, reports indicate that the removal of HMs is inversely proportional to ionic strength; therefore, any decrease in removal is experienced with an upsurge of the ionic strength because of the increase of competitive cationic amount available in aqueous systems, which affects the binding force between the adsorbent and adsorbate [75, 85].

5.6 Adsorbent particle size

The adsorbent particle size plays a fundamental role in the uptake of the HMs during the adsorption process since it significantly alters the uptake capacity and removal rate. Studies show an enhancement in the uptake capacity of HMs with a reduction in particle size due to a greater surface area [86]. Also, large particle size implies a reduced intraparticle diffusion due to an increased resistance to mass transfer of adsorbate molecules, leading to incomplete utilization of adsorbent surface area and reduced uptake [87].

5.7 Adsorbent dosage

It has been determined that a rise in adsorbent concentration results in the creation of more active binding sites, causing improved removal of HMs from aqueous solutions. This has been attributed to improved surface area and enhanced pore volume availability due to higher dosages [88]. A detailed analysis of the dependence of adsorbent dose on Cr(VI) reveals an increase in the percentage removal of Cr(VI) with a rise in dose up to 14 g/L. The enhanced removal with dosage is due to the availability of exchangeable sites [60].

5.8 Agitation speed

Agitation speed has been established to accelerate or retard the adsorption of HMs from aqueous systems [89]. The determination of the maximum agitation speed is realized through the repetition of a set of experiments.

6. Thermodynamics studies

Thermodynamic studies are conducted to analyze various parameters and draw inferences on spontaneity, feasibility, and the characteristics of the adsorption process, thus underscoring fundamental areas optimizing to ensure the sustainability and effectiveness of HM removal [75]. The studies analyze factors such as temperature, Gibbs free energy change (ΔG), enthalpy change (ΔH), and entropy (ΔS).

The Gibbs energy change provides a measure of the natural occurrence of the adsorption process. A negative value indicates that the process is spontaneous:

$$\Delta G = \Delta H - T\Delta S \tag{1}$$

Enthalpy change provides information on whether the adsorption is endothermic or exothermic. The negative and positive values of ΔH indicate exothermic and endothermic adsorption processes, respectively.

$$\ln K = -\frac{\Delta H}{RT} + \frac{\Delta S}{R} \tag{2}$$

Entropy change provides insights into the randomness at the solid–liquid interface during adsorption. When the ΔS is positive, it designates increased randomness often connected to endothermic processes.

By plotting *lnK* against *1/T*, the slope ($\Delta H/R$) and intercept ($\Delta S/R$) are used to extract(ΔH and (ΔS respectively.

where *K, R, and T* are constants known as the thermodynamic equilibrium, universal gas, and absolute temperature.

Based on the articles reviewed, the thermodynamic process is both exothermic and endothermic depending on the type of AWA, type of HMs, and specific adsorption conditions [30, 71]. An endothermic process implies an increased efficiency at higher temperatures, whereas an exothermic process reveals that the adsorption process is more efficient at lower temperatures [64, 66]. Knowing thermodynamic characteristics is important during material optimization, selection of appropriate AWA, and optimization of adsorption parameters to enhance effectiveness and efficiency.

7. Regeneration

Regeneration involves retrieving and restoring the adsorbent's adsorptive properties after use, ensuring continued application for several cycles. The regeneration of AWA not only reduces the cost associated with sourcing or producing new adsorbents but mitigates waste management through adsorbent lifecycle extension.

Therefore, a good choice of AWA should have both economic and environmental benefits. For example, biochar-NH 2, with a maximum removal of 99% of Cu(II), underwent a regeneration procedure by employing 0.10 of HCl and NaOH coupled with rinsing in hot and distilled water to recharge the already depleted adsorbent. The regeneration abilities of biochar-NH 2 were demonstrated after six regeneration cycles with a maximum removal of 90.23%, which compares well with the before regeneration value and fulfills the economic and environmental benefits [17].

Also, desorption studies on adsorbent derived from exhausted rice straw were performed using 0.3 HCl, and an analysis was conducted after numerous cycles. A high recovery, which decreased with a recovery time resulted in a recovery of 74.17% and 63.29% for Cr(VI) and Pb(II), respectively. These values compare well with the optimal removal efficiency of 95.57% for Cr(VI) and 85.68% for Pb(II), this adsorbent material [72].

A desorption attributed to a reversible mechanism of Cd(II) on the eggshell adsorbent using $CaCl_2$ has been reported. Studies reveal a desorption variation of 45.6% to 33.04% when 1 mol/L of $CaCl_2$ is employed compared to 66.97% adsorbed through non-reversible tendencies like precipitation and chemisorption [26]. With these examples among many, it is evident that this approach not only minimizes the cost related to purchasing new commercially available adsorbent but also reduces the environmental impact through the reduction of wastes, thus ensuring natural resource conservation.

8. Conclusion and future perspectives

This chapter demonstrates that the adsorption technique and use of AWA have numerous benefits in decontaminating HMs from wastewater. It has been established that employing these adsorbents derived from agricultural wastes in the adsorption of HMs is reliable, inexpensive, highly efficient, and environmentally safe; hence, it should often be used to reduce decontamination costs while maximizing HM removal. Reviewed studies show that HM removal from wastewater relies heavily on the pH, temperature, contact time, adsorbent dosage, adsorbent particle size, HM ion concentration, agitation speed, and interfering ions. Based on the reviewed data, the removal efficiency of HMs of the low-cost AWA ranged from 61 to 100%, with a contact time range of 20–240 and an optimum pH of 3–10.

The kinetic and isotherm studies were best described by the PSO model and LMR isotherm. Furthermore, thermodynamic studies demonstrated that the adsorption process was endothermic and exothermic depending on the type of AWA materials and adsorption factors and conditions. Also, the high removal capacity and regeneration imply that these natural and engineered adsorbents offer a promising solution for effectively sequestrating HMs and protecting the environment, thereby ensuring natural resource sustainability for future generations. Therefore, leveraging simplicity, cost-effectiveness, eco-friendliness, and effectiveness, the adsorption technique using AWA materials can be the solution to HMs mitigation and removal from aqueous solutions.

Future research using AWA should be centered on enhancing efficiency, selectivity, and reusability through functionalization or chemical modification thus making them more competitive with commercial adsorbents. Also, scaling up through optimization of AWA toward industrial applications should be encouraged by consistently producing quality adsorbents at large volumes. This will culminate in their integration into existing treatment and purification in HMs-ridden water systems, hence reducing operational costs while enhancing environmental safety. Furthermore, for widespread acceptance in commercial markets, a clear framework for proper standardization, examination, and certification must be developed.

Author details

Robert Birundu Onyancha
Department of Technical and Applied Physics, School of Physics and Earth Sciences Technology, Technical University of Kenya, Nairobi, Kenya

*Address all correspondence to: 08muma@gmail.com

References

[1] Tchounwou P, Yedjou C, Patlolla A, Sutton D. Heavy metal toxicity and the environment in molecular, clinical and environmental toxicology. Clinical and Environmental Toxicology. 2012;**101**: 133-164

[2] Tamjidi S, Esmaeili H, Moghadas B. Application of magnetic adsorbents for removal of heavy metals from wastewater: A review study. Materials Research Express. 2019;**6**(10):102004

[3] Onyancha R, Aigbe U, Ukhurebor K, Kusuma H, Darmokoesoemo H, Osibote O, et al. Influence of magnetism-mediated potentialities of recyclable adsorbents for heavy metal ions removal from aqueous solutions–an organized review. Results in Chemistry. 2022;**4**: 100452

[4] Carolin C, Kumar P, Saravanan A, Joshiba G, Naushad M. Efficient techniques for the removal of toxic heavy metals from aquatic environment: A review. Journal of Environmental Chemical Engineering. 2017;**5**(3): 2782-2799

[5] Ukhurebor K, Aigbe U, Onyancha R, Nwankwo W, Osibote O, Paumo H, et al. Effect of hexavalent chromium on the environment and removal techniques: A review. Journal of Environmental Management. 2021;**280**:111809

[6] Aigbe U, Ukhurebor K, Onyancha R, Okundaye B, Pal K, Osibote O, et al. A facile review on the sorption of heavy metals and dyes using bionanocomposites. Adsorption Science & Technology. 2022;**2022**:8030175

[7] Afshin S, Mokhtari S, Vosoughi M, Sadeghi H, Rashtbari Y. Data of adsorption of basic blue 41 dye from aqueous solutions by activated carbon prepared from filamentous algae. Data in Brief. 2018;**21**:1008-1013

[8] El-Nemr M, Aigbe U, Ukhurebor K, Onyancha R, El Nemr A, Ragab S, et al. Adsorption of Cr6+ ion using activated Pisum sativum peels-triethylenetetramine. Environmental Science and Pollution Research. 2022; **29**(60):91036-91060

[9] Neolaka Y, Riwu A, Aigbe U, Ukhurebor K, Onyancha R, Darmokoesoemo H, et al. Potential of activated carbon from various sources as a low-cost adsorbent to remove heavy metals and synthetic dyes. Results in Chemistry. 2022;**5**:100711

[10] Aigbe U, Osibote O. Carbon derived nanomaterials for the sorption of heavy metals from aqueous solution: A review. Environmental Nanotechnology, Monitoring & Management. 2021;**16**: 100578

[11] Aigbe U, Osibote O. Fluoride ions sorption using functionalized magnetic metal oxides nanocomposites: A review. Environmental Science and Pollution Research. 2022;**29**:1-45

[12] Onyancha R, Aigbe U, Ukhurebor K, Muchiri P. Facile synthesis and applications of carbon nanotubes in heavy-metal remediation and biomedical fields: A comprehensive review. Journal of Molecular Structure. 2021;**1238**: 130462

[13] Sudarni D, Aigbe U, Ukhurebor K, Onyancha R, Kusuma H, Darmokoesoemo H, et al. Malachite green removal by activated potassium hydroxide clove leaf agrowaste biosorbent: Characterization, kinetic, isotherm, and thermodynamic studies.

Adsorption Science & Technology. 2021; **5**:1-15

[14] Benaissa H, Elouchdi M. Removal of copper ions from aqueous solutions by dried sunflower leaves. Chemical Engineering and Processing: Process Intensification. 2007;**46**(7):614-622

[15] Eldeeb T, Aigbe U, Ukhurebor K, Onyancha R, El-Nemr M, Hassaan M, et al. Adsorption of methylene blue (MB) dye on ozone, purified and sonicated sawdust biochars. Biomass Conversion and Biorefinery. 2022;**14**:1-23

[16] Eleryan A, Aigbe U, Ukhurebor K, Onyancha R, Eldeeb T, El-Nemr M, et al. Copper (II) ion removal by chemically and physically modified sawdust biochar. Biomass Conversion and Biorefinery. 2024;**14**(8):9283-9320

[17] El-Nemr M, Aigbe U, Hassaan M, Ukhurebor K, Ragab S, Onyancha R, et al. The use of biochar-NH 2 produced from watermelon peels as a natural adsorbent for the removal of Cu (II) ion from water. Biomass Conversion and Biorefinery. 2022;**14**:1-17

[18] Hossain M, Ngo H, Guo W, Setiadi T. Adsorption and desorption of copper (II) ions onto garden grass. Bioresource Technology. 2012;**121**: 386-395

[19] Khoramzadeh E, Nasernejad B, Halladj R. Mercury biosorption from aqueous solutions by sugarcane bagasse. Journal of the Taiwan Institute of Chemical Engineers. 2013;**44**(2): 266-269

[20] Eldeeb T, Aigbe U, Ukhurebor K, Onyancha R, El-Nemr M, Hassaan M, et al. Biosorption of acid brown 14 dye to mandarin-CO-TETA derived from mandarin peels. Biomass Conversion and Biorefinery. 2022;**14**:1-21

[21] Mathew B, Jaishankar M, Biju V, Beeregowda K. Role of bioadsorbents in reducing toxic metals. Journal of Toxicology. 2016;**1**:4369604

[22] Boni M, Chiavola A, Marzeddu S. Remediation of lead-contaminated water by virgin coniferous wood biochar adsorbent: Batch and column application. Water, Air, & Soil Pollution. 2020;**231**:1-16

[23] Soliman N, Moustafa A. Industrial solid waste for heavy metals adsorption features and challenges; a review. Journal of Materials Research and Technology. 2020;**9**(5):10235-10253

[24] Virgen M, Vázquez O, Montoya V, Gómez R. Removal of heavy metals using adsorption processes subject to an external magnetic field. In: Hosam RA, Saleh El-M, editors. Heavy Metals. London: IntechOpen; 2018. pp. 253-279

[25] Afolabi F, Musonge P, Bakare B. Bio-sorption of a bi-solute system of copper and lead ions onto banana peels: Characterization and optimization. Journal of Environmental Health Science and Engineering. 2021;**19**:613-624

[26] Sabah H, Thouraya T, Melek H, Nadia M. Application of response surface methodology for optimization of cadmium ion removal from an aqueous solution by eggshell powder. Chemical Research in Chinese Universities. 2018; **34**:302-310

[27] Demir Delil A, Gülçiçek O, Gören N. Optimization of adsorption for the removal of cadmium from aqueous solution using Turkish coffee grounds. International Journal of Environmental Research. 2019;**13**:861-878

[28] Akram M, Khan B, Imran M, Ahmad I, Ajaz H, Tahir M, et al. Biosorption of lead by cotton shells

powder: Characterization and equilibrium modeling study. International Journal of Phytoremediation. 2019;**21**(2):138-144

[29] Witek-Krowiak A, Reddy D. Removal of microelemental Cr (III) and Cu (II) by using soybean meal waste–unusual isotherms and insights of binding mechanism. Bioresource Technology. 2013;**127**:350-357

[30] Nnaji C, Ebeagwu C, Ugwu E. Physicochemical conditions for adsorption of lead from water by rice husk ash. BioResources. 2017;**12**(1):779-818

[31] Liu X, Xu X, Dong X, Park J. Competitive adsorption of heavy metal ions from aqueous solutions onto activated carbon and agricultural waste materials. Polish Journal of Environmental Studies. 2020;**29**(1): 749-761

[32] Zhang Y, Song X, Zhang P, Gao H, Ou C, Kong X. Production of activated carbons from four wastes via one-step activation and their applications in Pb2+ adsorption: Insight of ash content. Chemosphere. 2020;**245**:125587

[33] Li Y, Ding J, Luan Z, Di Z, Zhu Y, Xu C, et al. Competitive adsorption of Pb2+, Cu2+ and Cd2+ ions from aqueous solutions by multiwalled carbon nanotubes. Carbon. 2003;**41**(14): 2787-2792

[34] Malkoc E, Nuhoglu Y. Potential of tea factory waste for chromium (VI) removal from aqueous solutions: Thermodynamic and kinetic studies. Separation and Purification Technology. 2007;**54**(3):291-298

[35] Moreno-Castilla C. Adsorption of organic molecules from aqueous solutions on carbon materials. Carbon. 2004;**42**(1):83-94

[36] Khalili N, Campbell M, Sandi G, Golaś J. Production of micro-and mesoporous activated carbon from paper mill sludge: I. Effect of zinc chloride activation. Carbon. 2000;**38**(14): 1905-1915

[37] Dutta L, Sethi G, Dey S. A comprehensive and critical assessment on the efficiency of natural and synthetic adsorbents for the removal of recalcitrant malachite green from water: Present level and future perspectives. Korean Journal of Chemical Engineering. 2024;**41**(3):589-607

[38] Gupta V, Ali I, Saini V. Removal of chlorophenols from wastewater using red mud: An aluminum industry waste. Environmental Science & Technology. 2004;**38**(14):4012-4018

[39] Crini G. Non-conventional low-cost adsorbents for dye removal: A review. Bioresource Technology. 2006;**97**(9): 1061-1085

[40] Lata S, Singh P, Samadder S. Regeneration of adsorbents and recovery of heavy metals: A review. International Journal of Environmental Science and Technology. 2015;**12**:1461-1478

[41] Hameed B, El-Khaiary M. Removal of basic dye from aqueous medium using a novel agricultural waste material: Pumpkin seed hull. Journal of Hazardous Materials. 2008;**155**(3):601-609

[42] Demirbas A. Agricultural based activated carbons for the removal of dyes from aqueous solutions: A review. Journal of Hazardous Materials. 2009; **167**(1–3):1-9

[43] Badawi A, Abd Elkodous M, Ali G. Recent advances in dye and metal ion removal using efficient adsorbents and novel nano-based materials: An

overview. RSC Advances. 2021;**11**(58): 36528-36553

[44] Yagub M, Sen T, Afroze S, Ang H. Dye and its removal from aqueous solution by adsorption: A review. Advances in Colloid and Interface Science. 2014;**209**:172-184

[45] Karunakaran K, Usman M, Sillanpää M. A review on superadsorbents with adsorption capacity≥ 1000 mg g− 1 and perspectives on their upscaling for water/wastewater treatment. Sustainability. 2022;**14**(24):16927

[46] Bayuo J, Rwiza M, Mtei K, Choi J. Adsorptive removal of heavy metals from wastewater using low-cost adsorbents derived from agro-based materials. In: In Heavy Metal Remediation: Sustainable Nexus Approach. Switzerland: Springer Nature; 2024. pp. 237-271

[47] Aigbe U, Ukhurebor K, Onyancha R, Okundaye B, Aigbe E, Obodo K, et al. Activated biosorbents for the removal of metals from aqueous solutions. In: Adsorption Applications for Environmental Sustainability. Bristol, UK: IOP Publishing; 2023. pp. 3-1

[48] Lim SF, Karim SKA, Chua SND, Lim BH. Agricultural waste-derived adsorbents for decontamination of heavy metals. In: Wang LK, Wang MHS, Hung YT, Shammas NK, editors. Integrated Natural Resources Management. Handbook of Environmental Engineering. Vol. 20. Springer; 2021

[49] Sabzehmeidani M, Mahnaee S, Ghaedi M, Heidari H, Roy V. Carbon based materials: A review of adsorbents for inorganic and organic compounds. Materials Advances. 2021;**2**(2): 598-627

[50] Ghasemi S, Ghaderpoori M, Moradi B, Taghavi M, Karimyan K, Mehdipour F. Optimization of Cr (VI) adsorption by modified sesame hull from aqueous solutions using response surface methodology. International Journal of Environmental Analytical Chemistry. 2022;**102**(13):3094-3108

[51] Baby R, Hussein M, Zainal Z, Abdullah A. Preparation of functionalized palm kernel shell bio-adsorbent for the treatment of heavy metal-contaminated water. Journal of Hazardous Materials Advances. 2023;**10**: 100253

[52] Benstoem F, Becker G, Firk J, Kaless M, Wuest D, Pinnekamp J, et al. Elimination of micropollutants by activated carbon produced from fibers taken from wastewater screenings using hydrothermal carbonization. Journal of Environmental Management. 2018;**211**: 278-286

[53] Sen T. Agricultural solid wastes based adsorbent materials in the remediation of heavy metal ions from water and wastewater by adsorption: A review. Molecules. 2023;**28**(14):5575

[54] Bayuo J, Rwiza M, Mtei K. Adsorption and desorption ability of divalent mercury from an interactive bicomponent sorption system using hybrid granular activated carbon. Environmental Monitoring and Assessment. 2023;**195**(8):935

[55] Bayuo J. Decontamination of cadmium (II) from synthetic wastewater onto shea fruit shell biomass. Applied Water Science. 2021;**11**(5):84

[56] Jain M, Garg V, Paliwal R, Kadirvelu K, Chaudhry S. Optimization of cadmium (II) removal from water using sunflower waste carbon–a

statistical approach. Toxin Reviews. 2021;**40**(4):1373-1382

[57] Javidi Alsadi K, Esfandiari N. Synthesis of activated carbon from sugarcane bagasse and application for mercury adsorption. Pollution. 2019; **5**(3):585-596

[58] Onwordi CUCAAPL. Comparative study of the adsorption capacity of lead (II) ions onto bean husk and fish scale from aqueous solution. Journal of Water Reuse and Desalination. 2019;**9**(3): 249-262

[59] Bayuo J, Pelig-Ba K, Abukari M. Isotherm modeling of lead (II) adsorption from aqueous solution using groundnut shell as a low-cost adsorbent. IOSR Journal of Applied Chemistry (IOSR-JAC). 2018;**11**(11):18-23

[60] Nur-E-Alam M, Mia MAS, Ahmad F, Rahman MM. Adsorption of chromium (Cr) from tannery wastewater using low-cost spent tea leaves adsorbent. Applied Water Science. 2018;**8**:1-7

[61] Shakoor M, Niazi N, Bibi I, Shahid M, Saqib Z, Nawaz M, et al. Exploring the arsenic removal potential of various biosorbents from water. Environment International. 2019;**123**: 567-579

[62] Mondal S, Aikat K, Halder G. Biosorptive uptake of arsenic (V) by steam activated carbon from mung bean husk: Equilibrium, kinetics, thermodynamics and modeling. Applied Water Science. 2017;**7**:4479-4495

[63] Araújo C, Almeida I, Rezende H, Marcionilio S, Léon J, de Matos T. Elucidation of mechanism involved in adsorption of Pb (II) onto Lobeira fruit (Solanum lycocarpum) using Langmuir, Freundlich and Temkin isotherms. Microchemical Journal. 2018;**137**: 348-354

[64] Roy P, Dey U, Chattoraj S, Mukhopadhyay D, Mondal N. Modeling of the adsorptive removal of arsenic (III) using plant biomass: A bioremedial approach. Applied Water Science. 2017; **7**:1307-1321

[65] Nemeş L, Bulgariu L. Optimization of process parameters for heavy metals biosorption onto mustard waste biomass. Open Chemistry. 2016;**14**(1):175-187

[66] Parlayıcı Ş, Pehlivan E. Removal of metals by Fe3O4 loaded activated carbon prepared from plum stone (Prunus nigra): Kinetics and modelling study. Powder Technology. 2017;**317**:23-30

[67] Ibisi N, Asoluka C. Use of agro-waste (Musa paradisiaca peels) as a sustainable biosorbent for toxic metal ions removal from contaminated water. Chemistry International. 2018;**4**(1):52

[68] Salman M, Rehman R, Farooq U, Tahir A, Mitu L. Biosorptive removal of cadmium (II) and copper (II) using microwave-assisted Thiourea-modified Sorghum bicolor Agrowaste. Journal of Chemistry. 2020;**2020**(1):8269643

[69] Yu X, He Y. Optimal ranges of variables for an effective adsorption of lead (II) by the agricultural waste pomelo (Citrus grandis) peels using Doehlert designs. Scientific Reports. 2018;**8**(1):729

[70] Mohammed T, Ibrahim R. Remediation of Cu (II) from well water of Iraq by using cortex of fruits and agricultural waste. Arabian Journal for Science and Engineering. 2016;**41**:345-355

[71] Mor S, Negi P, Ravindra K. Potential of agro-waste sugarcane bagasse ash for the removal of ammoniacal nitrogen

from landfill leachate. Environmental Science and Pollution Research. 2019;**26**: 24516-24531

[72] Venkatraman Y, Arunkumar P, Kumar N, Osman A, Muthiah M, Al-Fatesh A, et al. Exploring modified rice straw biochar as a sustainable solution for simultaneous Cr (VI) and Pb (II) removal from wastewater: Characterization, mechanism insights, and application feasibility. ACS Omega. 2023;**8**(41):38130-38147

[73] Khalil A, Salem M, Ragab S, Sillanpää M, El Nemr A. Orange peels magnetic activate carbon (MG-OPAC) composite formation for toxic chromium absorption from wastewater. Scientific Reports. 2023;**13**(1):3402

[74] Putri DIM, Darmokoesoemo H, Supriyanto G, Widyaningrum BA, Kusuma HS. Removal of Cd(II) from aqueous solution using biosorbent based on agricultural waste sorgum bagasse (*Sorghum bicolor* (L.) Moench) activated NaOH. Chemistry and Ecology. 2024:1-27. DOI: 10.1080/02757540.2024.2376044

[75] Aigbe U, Ukhurebor K, Onyancha R, Osibote O, Darmokoesoemo H, Kusuma H. Fly ash-based adsorbent for adsorption of heavy metals and dyes from aqueous solution: A review. Journal of Materials Research and Technology. 2021;**14**:27

[76] Azadegan F, Bidhendi M, Badiei A. Removal of Hg (II) ions from aqueous environment with the use of modified LUS-1 as new nanostructured adsorbent. International Journal of Environmental Research. 2019;**13**:557-569

[77] Rahdar S, Taghavi M, Khaksefidi R, Ahmadi S. Adsorption of arsenic (V) from aqueous solution using modified saxaul ash: Isotherm and thermodynamic study. Applied Water Science. 2019;**9**: 1-9

[78] Lenka S, Shaikh W, Owens G, Padhye L, Chakraborty S, Bhattacharya T. Removal of copper from water and wastewater using dolochar. Water, Air, & Soil Pollution. 2021;**232**: 1-15

[79] Mohubedu R, Diagboya P, Abasi C, Dikio E, Mtunzi F. Magnetic valorization of biomass and biochar of a typical plant nuisance for toxic metals contaminated water treatment. Journal of Cleaner Production. 2019;**209**:1016-1024

[80] Ilavarasan N, Rao YS, Gokulan R, Aravindan A. Investigation of copper ion adsorption using activated sawdust powder: Isotherm, kinetic and thermodynamic studies. Global NEST Journal. 2022;**25**(1):47-56

[81] Zhao J, Yu L, Ma H, Zhou F, Yang K, Wu G. Corn stalk-based activated carbon synthesized by a novel activation method for high-performance adsorption of hexavalent chromium in aqueous solutions. Journal of Colloid and Interface Science. 2020;**578**:650-659

[82] Afroze S, Sen T. A review on heavy metal ions and dye adsorption from water by agricultural solid waste adsorbents. Water, Air, & Soil Pollution. 2018;**229**:1-50

[83] Xia S, Song Z, Jeyakumar P, Bolan N, Wang H. Characteristics and applications of biochar for remediating Cr (VI)-contaminated soils and wastewater. Environmental Geochemistry and Health. 2020;**42**: 1543-1567

[84] Kanaujia S, Singh B, Singh S. Removal of fluoride from groundwater by carbonised Punica granatum carbon ("CPGC") bio-adsorbent. Journal of

Geoscience and Environment Protection. 2015;**3**(4):1-9

[85] He S, Li Y, Weng L, Wang J, He J, Liu Y, et al. Competitive adsorption of Cd2+, Pb2+ and Ni2+ onto Fe3 +−modified argillaceous limestone: Influence of pH, ionic strength and natural organic matters. Science of the Total Environment. 2018;**637**:69-78

[86] Memon A, Ahmed SBZ, Maitlo G, Shah A, Mazari S, Muhammad A, et al. Experimental investigations of arsenic adsorption from contaminated water using chemically activated hematite (Fe 2 O 3) iron ore. Environmental Science and Pollution Research. 2021;**28**: 12898-12908

[87] Ahmaruzzaman M. Industrial wastes as low-cost potential adsorbents for the treatment of wastewater laden with heavy metals. Advances in Colloid and Interface Science. 2011;**166**(1–2):36-59

[88] Pyrzynska K. Removal of cadmium from wastewaters with low-cost adsorbents. Journal of Environmental Chemical Engineering. 2019;**7**(1):102795

[89] Dhoble R, Maddigapu P, Bhole A, Rayalu S. Development of bark-based magnetic iron oxide particle (BMIOP), a bio-adsorbent for removal of arsenic (III) from water. Environmental Science and Pollution Research. 2018;**25**: 19657-19674

Chapter 13

Perspective Chapter: Advanced Nanotechnology Approach for Heavy Metal Toxicity – Analysis, Treatment, and Removal

Yimer Seid Ali, Ibrahim Shaw, Yang Liu and Chuanpin Chen

Abstract

Heavy metal pollution is the contamination of the environment by toxic metals, which can lead to life-threatening diseases such as lung cancer, infertility, cardiovascular diseases, and nervous system disruptions. Monitoring and eliminating heavy metal residues from food, water, and environmental samples is crucial for public health safety. Nanomaterials have emerged as promising tools in dealing with heavy toxicity in the environment, offering numerous advantages over traditional methods. This chapter provides a comprehensive overview of nanotechnology-based approaches to heavy metal toxicity and assesses the environmental impacts of toxic HMs on health. It explores the use of various nanomaterials to detect and safely eliminate heavy metal pollution and discusses different nanotechnology-based methods for monitoring HMs in environmental sources. The chapter is designed to be engaging and informative for a broad audience while providing technical depth for expert readers.

Keywords: heavy metal, nanotechnology, heavy metal analysis, heavy metal remediation, heavy metal toxicity

1. Introduction

Naturally, heavy metals (HMs) are a class of transitional metal and metalloid group with a density of 4 g/mL or higher, relative to water [1, 2]. Heavy metal pollution caused by human activities, such as mining, agriculture, and industrial processes, poses significant environmental threats to humans, ecosystems, and the global environment. Accordingly, the Agency for Toxic Substances and Disease Registry (ATSDR) ranks HMs from 1 to 10th as As>Pb > Hg > Cd > Cr (VI) > Co > Ni > Cu > Zn,> Ag, respectively. This ranking is based on potential toxicity, frequency of occurrence, and human exposure [2–5].

Toxic HMs are greatly solubilized in aqueous environments, contaminate the soil and atmosphere, and are harmful, even at low concentrations [6]. Ingestion of

HMs, dust particulate exposure, and contamination near industrial facilities increase the risk of exposure to trace concentrations, thereby posing various health risks to humans. As shown in **Figure 1**, heavy metal toxicity includes kidney damage [7], liver toxicity [8, 9], brain damage [10], cancer [11], respiratory problems [12, 13], infertility [14], cardiovascular diseases [7, 15], and nervous system interruptions [16]. Therefore, toxic metals above permissible levels usually lead to potential toxic outcomes in humans, other living organisms, and the global environment [6]. Low health hazards are linked to permitted acceptable levels of HMs in food and drink samples.

Conventional methods for detecting and removing heavy metals from the environment are often inadequate due to their low sensitivity, cost, not mobile, and inefficiency. Nanotechnology, with its unique properties, offers a transformative approach to overcome these limitations. It is believed to be more sensitive, selective, and efficient in monitoring and eliminating heavy metals from various environmental matrices, protecting public health and the environment. Therefore, this chapter is intended to explore and present a comprehensive overview of the available literature on monitoring toxic HMs from main scientific databases including Google Scholar, PubMed, Web of Science, and Scopus. The literature was searched for nano-based detection, analysis, treatment, and removal of toxic heavy metals and any related nanomaterial for monitoring of heavy metals. The search terms included "Heavy metal," "Heavy metal treatment," "Heavy metal

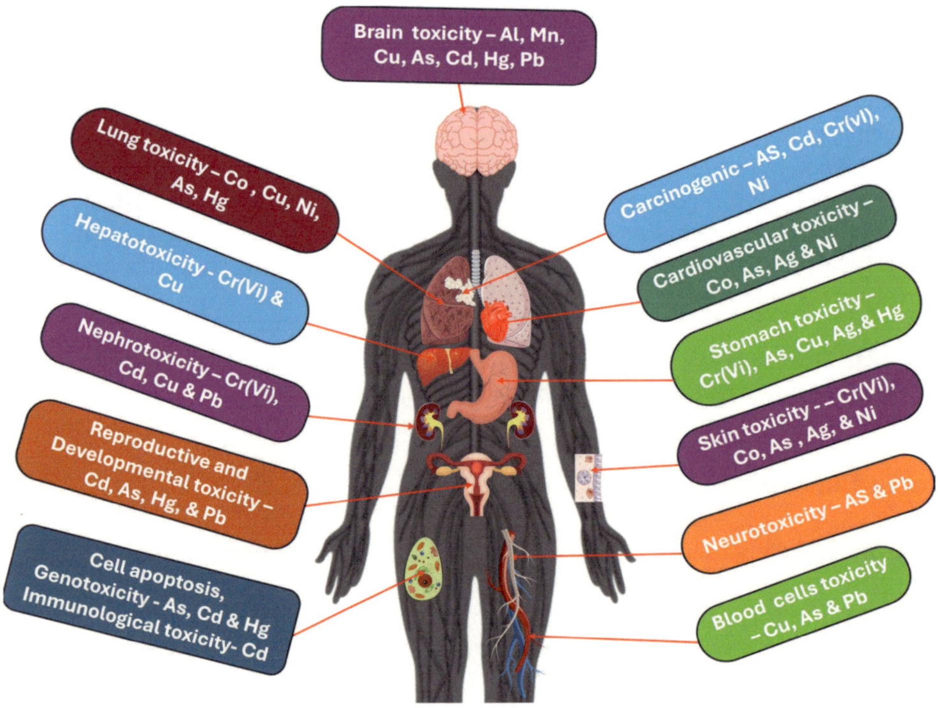

Figure 1.
Health implications of HMs on vital body organs. This figure highlights the diverse and harmful effects of various heavy metals on different human organ or system. The diagram maps out specific toxicities associated with individual metals, linking them to the affected organs or systems within the human body. Heavy metal pollution is known to cause damage to several vital organs, and tissues. Commonly affected organs include those of the nervous, reproductive, respiratory, digestive, and immune systems. It can also cause toxicity to the skin, cells, and genes.

removal" "Carbon nanomaterial," "Metal nanomaterial," "colorimetric nano analysis," "fluorescence nano analysis," "electrochemical nano analysis," "nano heavy metal remediation," "nano heavy metal treatment," and "nano heavy metal removal." Unpublished data, publications unrelated to the topic, and non-English language publications were excluded. Nano-based toxic heavy metal analysis was extracted from the literature, and their similarities and differences regarding sensitivity and selectivity were documented and compared.

This chapter review has six parts. The first explores important details on the causes of harmful heavy metal pollution, as well as its effects on the environment and health risks. The second section focused on principles and nanomaterials used in nanotechnology for controlling heavy metal toxicities. Moreover, it described nanomaterial classification, synthesis, and its application for heavy metal monitoring. The third section aimed at nano-based analytical methods used for heavy metal analysis and pinpoint the sensitivity and selectivity of such methods in practical applications. The fourth section focused on nano-based methods for the remediation of toxic heavy metals from complex environmental samples. Finally, the fifth and sixth section provides future direction and conclusion, respectively.

Heavy metal pollution is a pressing issue requiring innovative solutions. Nanotechnology offers precision, reusability, and effectiveness in detecting and removing toxic pollutants. It can address the immediate crisis and redefine environmental protection in complex global challenges. Future researchers should focus on these promising methods to achieve a cleaner environment and healthier future for all.

1.1 Impacts of HMs on environment

HMs can enter the human body by ingestion, skin contact, or inhalation. Once within the biological system, they accumulate and can be fatal if exposed to for an extended period of time. The quantity, rate of emissions, and duration of exposure to these metals can greatly influence human health. As a result, such toxicity has become a significant ecological concern, posing significant risks to both humans and the environment [17–19]. HMs are naturally less accessible to humans and plants because of their existence in geochemical state [20]. However, as shown in **Figure 2**, heavy metal pollution in the environment is rapidly increasing due to the following human activities:

- The rapid growth of industries and urbanization has resulted in the increased release of chemicals into the atmosphere, lithosphere, hydrosphere, and biosphere, causes high risky for humans and other living organisms [21].

- Anthropogenic activities such as mining, smelting, and other metal-based industries have led to rise in the release of HMs particles, thereby increasing environmental pollution [22];

- Metals leach from various sources, such as HMs in fertilizers, herbicides, and other agricultural products.

- Natural processes, including volcanic eruptions, metal evaporation from soil and water, and rock weathering, can led to a significant escalation in HMs pollution [2, 23, 24]. Consequently, heavy metal toxicity has become an environmental issue that poses a major risk to both humans and the environment.

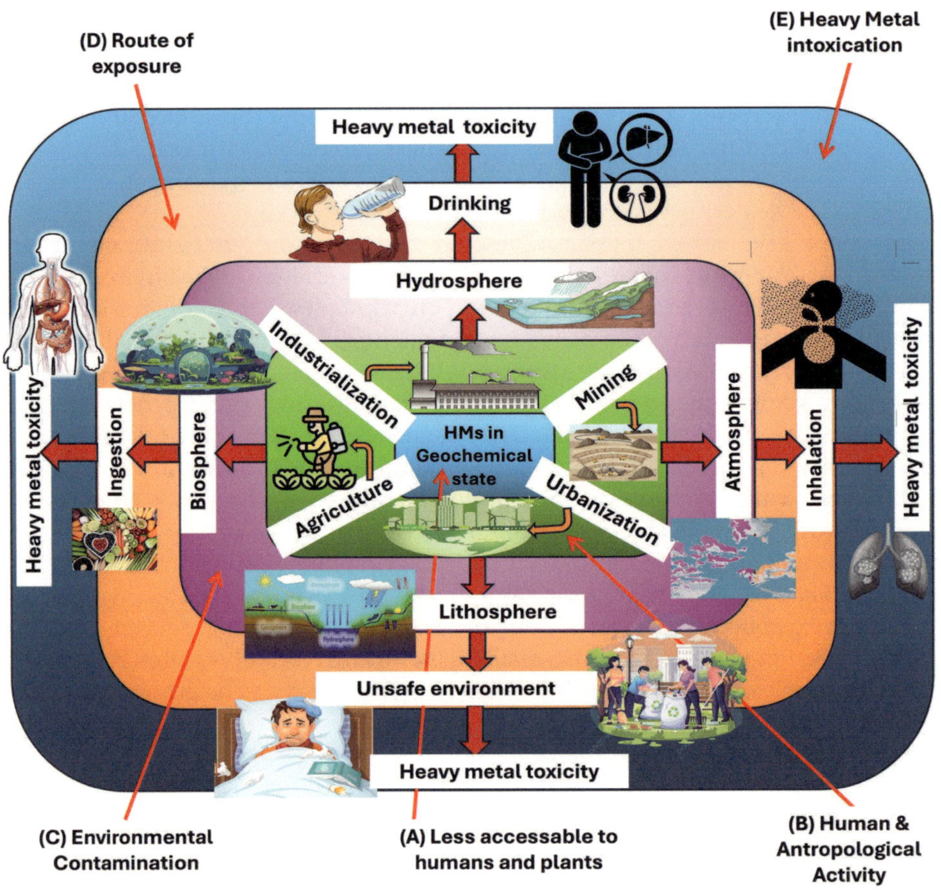

Figure 2.
Origins of heavy metal ions and their pathway from pollutions to human and environmental toxicity. This diagram illustrates the complex interactions between heavy metals (HMs) in the environment and their impacts on ecosystem and human health. (A) The core of the diagram represents heavy metals in their geochemical state, existing naturally in the Earth's lithosphere. (B) Through various natural and anthropogenic activities, these metals are mobilized into other spheres. Activities such as industrialization, agriculture, urbanization, and mining play a significant role in the redistribution of heavy metals. These activities often lead to the release of heavy metals into the environment, shifting them from being less accessible to more bioavailable forms, thereby increasing the potential for contamination. (C) This section shows how these heavy metals spread into various components of the environment, including the lithosphere, hydrosphere, biosphere, and atmosphere, leading to environmental contamination. This contamination can result in unsafe conditions for humans, animals, and plants. (D) Humans are exposed to heavy metals through various routes, prominently including Ingestion and Inhalation (E) Once inside the body, these heavy metals can lead to toxicity, causing adverse health effects. Chronic exposure can affect multiple organs and systems, with symptoms ranging from gastrointestinal distress to more severe effects such as neurological damage, depending on the specific metal and level of exposure.

1.2 Heavy metal limit of standards

International regulatory organizations such as the WHO, the US EPA, and CDC are investigating the harmful effects of HMs on public health. The US EPA describe HMs are potentially harmful owing to their lack of biodegradable properties and persistence in the environment [25]. Even essential elements, such as sodium, iron, zinc, copper, cobalt, and chromium, are toxic at higher concentrations. Accurate and precise analytical techniques are required for detecting and removing these HMs [26, 27]. To control such unintended rise of heavy metal concentration in the environment,

Organization		Different organization recommended limits for toxic HMs (ppm)									
		As (III)	**Pb (II)**	**Hg (II)**	**Cd (II)**	**Cr (VI)**	**Co (II)**	**Ni (II)**	**Zn (II)**	**Cu (II)**	**Ag (II)**
US EPA[a]	Drinking water	0.01	0.015	0.002	0.005	0.1	0.1	0.02	5.0	1.3	0.05
WHO[b]	Drinking water	0.01	0.003	0.006	0.003	0.05	NA	0.07	3.0	2.0	NA
EU[c]	Drinking water	0.01	0.01	0.001	0.005	0.05	NA	0.02	NA	2.0	1.142
ChineseMoH[d]	Drinking water	0.01	0.01	0.002	0.005	0.05	NA	0.02	5.0	1.0	0.05
OSHA[e] [30]	mg/m^3 [f]	0.01	0.2	0.01	0.005	0.005–1	NA	1	NA	0.1–1	0.01

[a]*US EPA—United State environmental protection agency.*
[b]*WHO—World Health Organization.*
[c]*EU—European Union.*
[d]*Chinese Ministry of Health.*
[e]*OSHA—Occupational Safety and Health Administration.*
[f]*Amount of HMs (mg) per cubic meter of workplace air for 8-hour work shifts and 40-hour per weeks.*

Table 1.
Toxic HMs standard limits in drinking water according to US EPA, WHO, OSHA, and Chinese Ministry of Health [28–32].

regulatory bodies have established acceptable limits of standard for several heavy metal ions that vary from ppb to ppm, as shown in **Table 1** [6, 28, 29].

1.3 Health impact of heavy metals on human

The presence and accumulation of HMs in the ecosystem have detrimental effects on humans, animals, and plants. Humans are exposed to these metals through bodily contact, oral consumption, and atmospheric inhalation. As shown in **Figure 1**, large concentrations of these metals have hazardous effects on health and environment [27]. Toxic heavy metal complications such as diabetics, cardiovascular disorder, neurological problems, developmental defects, immunologic diseases, hematologic problems, and various malignancies have negative impacts on human beings. For example, inorganic arsenic compound is well known carcinogenic element. Lead metal ion can increase reactive oxygen species (ROS) in cell and lead to disrupt the function of nucleic acids, cells, proteins resulting in a stressed conditions at cellular level. Beside to organ toxicity, mercury metal ion mainly methyl mercury is neurotoxic, which impairs nerve and muscle functions. In addition to cellular and genotoxicity, cadmium also reported to have organ toxicities such as kidney, liver, and reproductive system toxicities. Chromium exists in two forms: Cr (III) and Cr (VI). Hexavalent chromium, Cr (VI) is water soluble and found to be toxic to humans. Cr (VI) can easily cross the cell membrane and react with biological reductants like thiols lead to production of ROS that disrupts the cellular function. Due to its mutagenic property, it is reported to cause cancer indifferent part of the body. The severity of such adverse effects depends on specific metal exposure, property of chemical, period of exposure, and the amounts. Some research reports have shown that the extents of HMs exceed the upper limits set by international regulations, leading to unwanted complications in vital organs [3, 21, 33–35]. Effective detection and controlling of the levels of such heavy metals are necessary to protect the human health. Failure to monitoring the level HMs exposure will result in severe health complications due to the adverse effects imposed by heavy metals.

1.4 Health risk index for heavy metal pollution

The process of determining the possibility that a person may be negatively impacted by chemicals in a contaminated environment is known as human health risk assessment. Carcinogenic risk (CR) and noncarcinogenic risk (NCR) are two types of hazards to human health that are assessed using the US EPA health risk assessment (HRA) model [17, 36]. CR assesses the probability that an individual will develop cancer because of extended exposure to a certain pollutant or combination of pollutants. NCR is associated with individual chronic exposure, including genetic and teratogenic effects [24, 37].

1.4.1 Non-carcinogenic risk assessment

NCRs were examined by means of the hazard quotient (HQ), which is the proportion of chronic daily intake or dosage to the reference dose (RfD), the greatest amount at which no detrimental health consequences are anticipated [23, 24, 38]. The HQ of a single element is calculated using the following the equation:

$$HQ = \frac{ADI}{RfD} \tag{1}$$

where ADI is the average daily intake per day (mg/kg-day); RfD is the chronic reference dose for the metal element (mg/kg-day).

The potential NCR consequences of numerous substances can be assessed using the Hazard Index (HI) technique [17, 24, 37]. For a multiple of pollutants, HI is calculated using the following equation:

$$HI = \sum HQi = \sum \left(\frac{\text{ADIi}}{\text{RfDi}} \right) \tag{2}$$

1.4.2 Carcinogenic risk evaluation

CR is used to evaluate the possible risk connected with carcinogenic agents' exposure during the lifetime period. The slope factor (SF) measures the incremental risk of an individual developing cancer based on the estimated daily intake of a toxin over a lifetime [24, 37].

$$CR = ADI \times SF \tag{3}$$

where CR refers to the likelihood of an individual developing cancer over their lifetime, and SF is the carcinogenicity slope factor (per mg/kg-day). NCR and CR value were summarized as shown in **Figure 3** [17, 21].

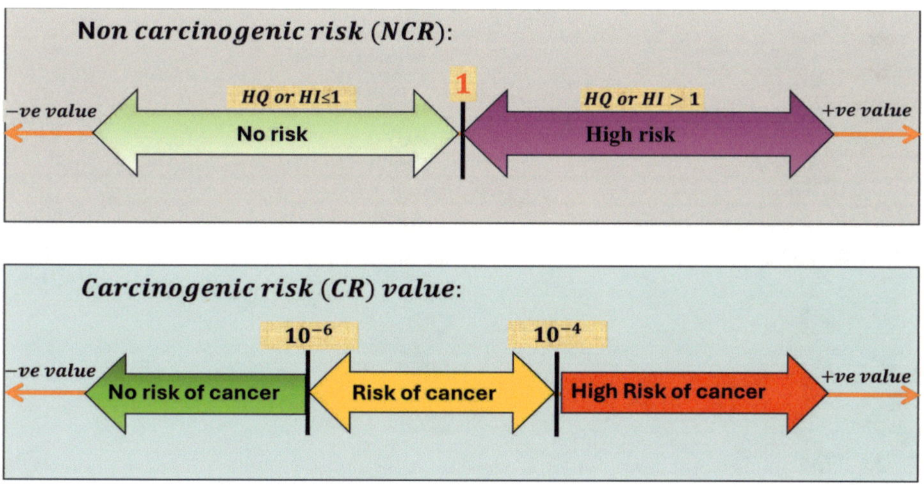

Figure 3.
Description of carcinogenic and noncarcinogenic risk value of heavy metal and other carcinogenic agents. This figure presents the scales used to assess the risk associated with exposure to hazardous substances: For non-carcinogenic risk (NCR), The Hazard Quotient (HQ) or Hazard Index (HI) is used to evaluate non-carcinogenic risks. An HQ or HI value of 1 or less indicates "No Risk," while values greater than 1 indicate a "High Risk" of non-carcinogenic effects. For carcinogenic risk (CR), the scale assesses the likelihood of cancer development. A CR lower value indicates "No Risk of Cancer," while high values suggest a "Risk of Cancer."

2. Nanotechnology-based detection and removal of heavy metal contamination

2.1 Introduction to nanomaterial

In 1959, Richard Feynman predicted the rapid development of an objects and molecules on a small scale [39]. Then after, in 1974, Norio Taniguchi coined the term nanotechnology, which refers to the processing of materials by one atom/molecule [40]. Nanotechnology, according to National Nanotechnology Initiative, is defined as the study and application of science, engineering, and technology at the nanoscale, which allows novel applications in various fields [39, 41]. Hence, a class of materials having dimensions that range, at least in one dimension, from 1 to 100 nm is known as nanomaterials (NMs). The design of sensors for the detection and removal of heavy metals uses a range of metal nanomaterials such as gold, copper, silver, iron, and biological including aptamer, antibodies, as well as carbon-based materials such as graphite, carbon nanotubes, and graphene. The nanomaterials utilized in heavy metal immobilization from complex sample matrices. In such advanced nanotechnology system, there is an enhancement in the sensitivity and selectivity of HM ion detection and removal due to a high surface-to-volume ratio and greater accessibility to the active surface. In accordance with the kind of nanosensor, it significantly raises sensitivity and specificity [42].

2.1.1 Classification of nanomaterials

As shown in **Figure 4**, NMs are categorized by structural makeup: (organic, inorganic, carbon-based, and composite); dimension: (0D,1D,2D, 3D); pore size: (micro, meso, macro); and origin: natural and artificial [43, 44].

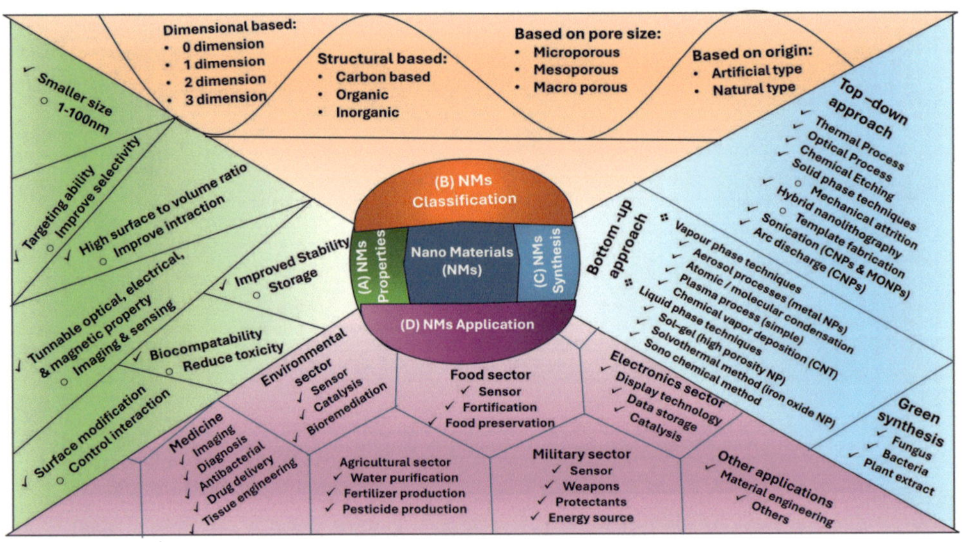

Figure 4.
General properties, classification, synthesis, and application of nanomaterial. This figure provides a comprehensive summary of NMs, focusing on four key areas. (A) The advantages of NMs include their unique size, tunable optical, electrical, and magnetic characteristics, biocompatibility, and increased stability. (B) Classification based on dimensionality (0D, 1D, 2D, 3D), structure (carbon-based, organic, inorganic), pore size (microporous, mesoporous, macroporous), and origin (artificial or natural). (C) Synthesis methods (top-down and bottom-up approaches). (D) Finally, their applications.

Carbon nanotubes, graphene, fullerenes, and carbon nanofibers are the primary components of carbon-based NMs. Graphene and carbon nanotubes are highly desirable for heavy metal detection owing to their remarkable thermal, electrical, optical, and mechanical properties as well as their high surface-area, ease of functionalization, and strong adsorption capabilities for effective detection and removal of HMs. Graphene is a two-dimensional (2D) carbon atom sheet with a large surface area that rapidly absorbs a wide range of chemicals. Owing to its strong electrical conductivity, graphene has been modified to create graphene oxide (GO) or reduced graphene oxide (rGO), which are commonly employed in electrochemical and optical sensors for analysis of toxic heavy metal ions [45–47]. Carbon nanotubes (CNTs) are cylindrical in shape, and the layer of geometrical structure determines the classification as single- and multi-walled carbon nanotubes, SWCNTs and MWCNTs, respectively [45, 48].

Metal nanoparticles (NPs) are composed of metal and metal oxide precursors and exhibit distinctive environmental pollution- and health-monitoring properties. Metal-based nanoparticles, such as AgNPs, ZnONPs, AuNPs, FeNPs, and CuNPs, have excellent nano-dimension, large surface area, optical, electrical, catalytic, photochemical, molecular recognition, and antibacterial characteristics. Metal oxide nanoparticles, composed of positive metallic ions and negative oxygen ions, have remarkable properties such as fluorescence and optical sensors [49, 50]. On top of that metal and metal oxide nanomaterials are developed for the elimination of contaminants from wastewater owing to their well-defined shapes, large specific surface areas, and effective adsorbents for removing HMs and other pollutants [41, 51].

Organic substances such as polymers, proteins, and lipids are used for vesicular (liposomes) and lipid nanoparticles production, which are sensitive to light and heat. However, they serve as carriers in drug delivery as well as flavors in pharmaceutical nanotechnology systems. For example, dendrimers are employed in drug delivery, molecular recognition, and nanosensors in electrochemical detection systems due to their multiple chain ends [52, 53].

2.1.2 Synthesis of nanomaterials

Nanomaterials synthesis involves chemical as well as physical methods to detect HM ions. Feynman's discovery of nanostructure synthesis sparked interest in two manufacturing approaches: top-down and bottom-up approaches [39]. The top-down technique of nanomaterials production involves mechanical ball milling, sonication, mechanical attrition, etc., whereas the bottom-up protocol uses sol-gel method, colloidal dispersion, precipitation, chemical vapor deposition, plasma process, and flame process. Bottom-up methods involve constructing materials by atomic particle-by-atomic particle, particle cluster-by-particle cluster, and molecular cluster-by-molecular cluster. The top-down approach has limitations such as damage to the crystalline structure, nanoparticles (NPs) defects, and internal stress during etching, making the bottom-up method the most used approach (**Figure 4**) [39, 41, 43, 54, 55]. Green synthesis is a sustainable and environmentally friendly method that uses biological materials, such as plant extracts, bacteria, or fungi, to create nanomaterials, reduce the use of hazardous chemicals, and promote sustainability, although it may result in less control over the nanoparticle size and morphology [55]. The advantage of the both approaches is producing smaller-sized nanomaterials with uniform size and increasing the surface-to-volume ratio to achieve better performance for detecting heavy metals.

2.1.3 Application of nanomaterial

Nowadays, the concept of nanotechnology has brought about a dynamic shift in the modern world by connecting various sectors and industries. In all sectors (**Figure 4**), nanotechnology utilizes nanoscale sensors with excellent electrical, mechanical, thermal, and optical properties. It is beneficial in various fields, including pharmaceutical, agricultural, medical, biotechnology, food, military, and electrical fields, owing to its unique physicochemical properties. With rapid industrialization and urbanization, increasing environmental pollution will increase the demand for nanomaterials to monitor released chemicals and heavy metal pollution. Nanosensors offer smart chemical, physical, and mechanical performance that improves selectivity and reactivity compared to their bulk materials owing to their high surface-to-volume ratio, catalytic activity, and affinity [49, 50, 56].

2.2 Conventional method for heavy metal detection and removal

Globally human, natural process and anthropological activities cause the release of toxic substances into the environment, polluting water, soil, and the atmosphere. These pollutants, including release of noxious gases and suspended particles such as HMs and organic materials, pose significant ecological and health-related issues. As shown in **Figure 5**, conventional analytical methods used for HM ions detection include atomic absorption/emission spectroscopy [57], neutron activation analysis [58], X-ray fluorescence spectrometry [59], inductively coupled plasma mass spectroscopy (ICPMS) [60], ICP atomic/optical emission spectroscopy [61], and flame atomic absorption spectrometry [62]. Although such methods have acceptable detection sensitivity, they have unavoidable challenges (expensive, heavy equipment, fixed at lab, and require skilled personnel for operation) [63, 64]. Physicochemical methods such as chemical based precipitation, ion exchange, membrane-based filtration, and adsorption are used for environmental remediations. However, these conventional

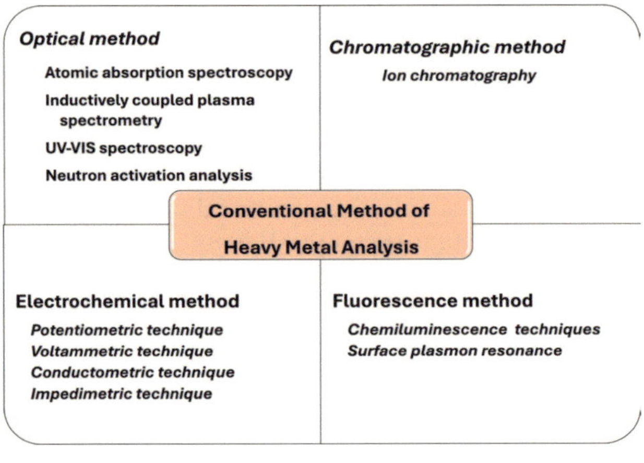

Figure 5.
Existing analytical method for detection of Heavy metal in environmental samples (food, water, air). This figure categorizes the traditional techniques used for analyzing heavy metals into four main methods: Optical, Electrochemical, Chromatography, and Fluorescence Methods. These methods represent the established approaches for detecting and quantifying heavy metals in various samples, each offering unique advantages depending on the analytical needs.

methods have limitations, such as high operation costs, inefficiency, high energy demands, pH sensitivity, and expensive analytical instruments [65, 66].

To date, owing to advancements in science and invention, nanomaterials have been carefully designed and fabricated for heavy-metal detection and removal, and they provide numerous benefits compared to previous analytical methods. The potential uses of nanomaterials have been investigated in several domains, including the detection and monitoring of environmental contamination using various nanotechnology systems. This is due to the progress of nanotechnology, and diverse types of nanomaterials for adsorption and detection [67, 68]. Heavy metal detection and removal are critical biological and environmental issues, with research focusing on user-friendly, sensitive, selective, feasible, and efficient methods. Recent studies have also focused on NMs development, synthesis, and fabrication as reliable tools for analysis, treatment, and removing HMs from environmental samples [18, 66, 69].

3. Nanotechnology-based method for heavy metal analysis

Nanomaterial sensors are small matter with scales ranging from 1 to 100 nm and dimension as zero D, one D, two D, and three D nanostructures [70]. Commonly used nanosensors for heavy metal analysis are organic (carbon nanotube, graphene), inorganic (metal and metal oxide), organic-inorganic (metal organic framework), and silica nanoparticles (**Figure 6**) [70, 71]. Therefore, nanoparticles with a high surface-to-volume ratio, functionalization, and detection characteristics are used in sensor design for heavy metal analysis owing to their high sensitivity and selectivity.

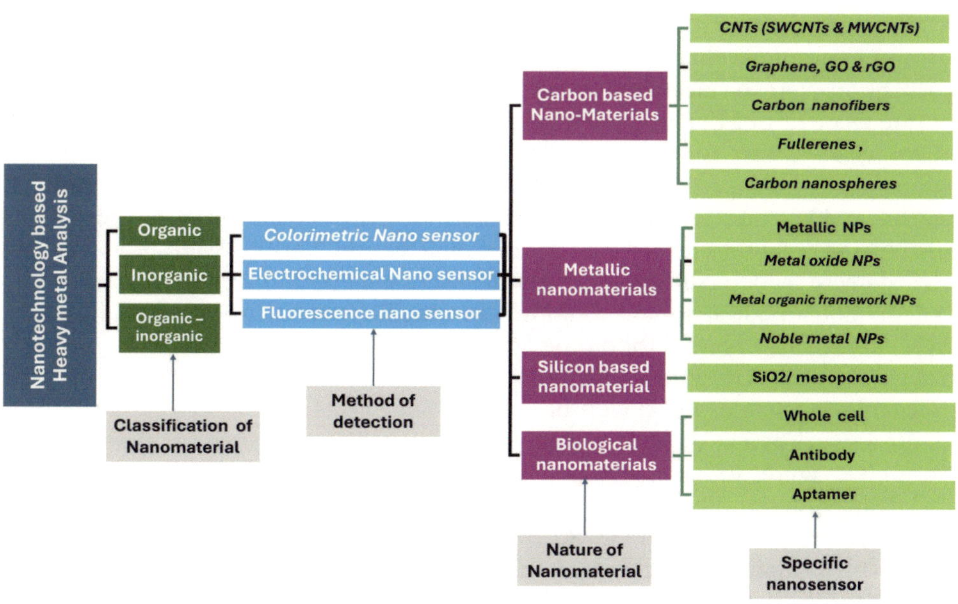

Figure 6.
Nanotechnology-based HM analysis in food, Water and biological samples. This figure outlines the different categories and methods used in nanotechnology for heavy metal analysis, organized into four main sections including Classification of Nanomaterial, Method of Detection, Nature of Nanomaterial, and Specific Nanosensor. This diagram further highlights the integration of nanotechnology in the detection and analysis of heavy metals.

A sensor/probe is a tool that reacts to matter and produces a signal that can be measured. Hence, the main purpose of heavy metal ion detection with nanosensors is to create metal ion identification probes, enhance sensor detection performance, and develop signal transduction pathways [72]. Generally, abiotic chemosensors exhibit changes in physical characteristics, such as color (colorimetric sensors), fluorescence (fluorescent sensors), or redox reaction (electrode-based sensors), when they interact with other species. Therefore, as shown in **Figure 6**, nanomaterials used to detect HMs in the environment include fluorescence, colorimetric, and electrochemical nanosensors [42, 73, 74].

3.1 Electrochemical nanosensors for heavy metal analysis

Water and soil pollution by HMs is a global problem [26, 27]. Tracking of HMs in food samples, drinking water, and other environmental samples by electrochemical (EC) method is preferred due to good sensitivity and selectivity [62]. In addition, EC approach is accurate, direct, inexpensive, and it is a great instrument for detecting HM ions. Nanomaterials such as carbon-based, metallic, and silicon-based materials are utilized to functionalize the electrode as sensors to improve catalysis reaction, conductivity, and fast electrode kinetics for HMs detection [25, 62].

The integration of electrochemical sensors with NMs (**Figure 7**) has become a powerful analytical method for detecting heavy metal ions. Such electrochemical cells use ultra-sensitive sensors to detect redox reactions; conducting wires to collect the signal; signal output transducers to measure electrical properties; and amplifiers to multiply signals, and computer systems for recording. EC cell with three electrode system, Working electrode (WE), Reference electrode (RE), and Auxiliary electrode (AE), is used to detect HMIs. In this system, nanomaterials can be used to modify the WE where the electrochemical reaction occurs on working electrode interface, causing a measurable change in impedance, current, charge and/or resistance. The AE serves to complete the electrical circuit, whereas the RE provide constant potential for accurate recording of the WE potential. Voltammetry, potentiometry, impedemetry, and amperometry are typical examples of analytical detector used for EC detection systems [75–77]. Amperometric detector is commonly used due to high sensitivity, linearity, and low interference, whereas potentiometric detector provides better sensitivity and stability for low concentration detection. Various kinds of voltammetric detector such as differential pulse voltammetry (DPV), cyclic voltammetry (CV), anode/cathode stripping voltammetry (A/CSV), or square wave voltammetry (SWV) are frequently used for online detection of heavy metal contamination due to extraordinary sensitivity and selectivity [77].

Organic, inorganic, and silicon-based nanosensor integrated electrodes offer advantages such as a large surface area, fast electron and mass transfer rates, and reduced NPs aggregation *via* functionalization. NPs functionalization improves colloidal dispersion stability making it a smart alternative method for HM electrochemical detection platform design [78–80]. The distinct electrical, chemical, and physical features of NP-modified electrodes in electrochemical cells have gained important attention for the detection of trace HMs [81, 82]. Electrochemical systems rapidly collect real-time data on heavy metal concentrations in real-world samples, providing valuable dynamic data for biogeochemical research efforts, enabling online monitoring in just a few minutes. Selecting the most appropriate electrochemical detector platform for the target sensor nanomaterials is a crucial step in creating the ideal heavy metal detection system.

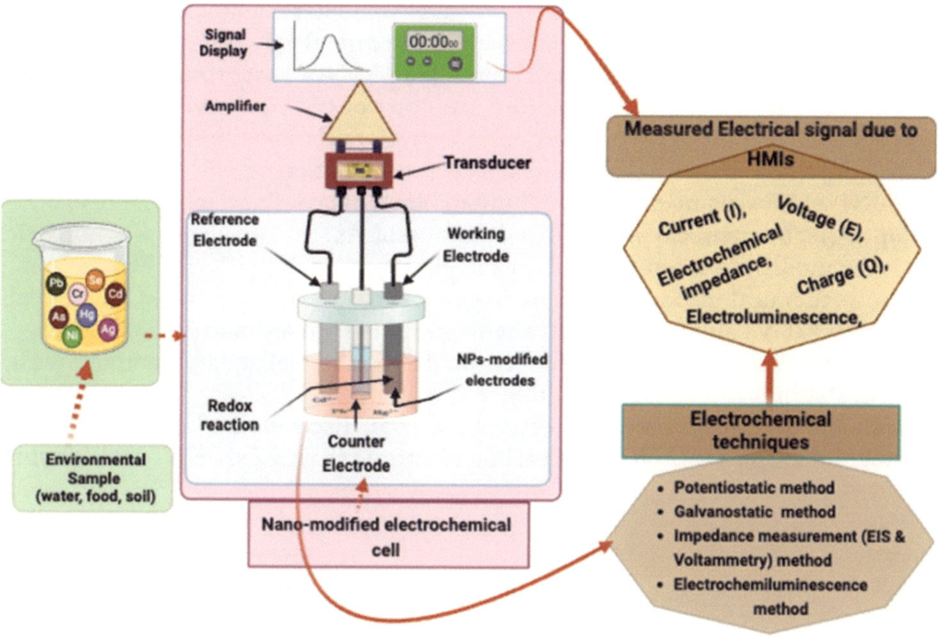

Figure 7.
Schematic representation of nanomodified electrochemical cell for detection of heavy metal ions (HMIs). This diagram illustrates the process of analyzing environmental samples (such as water, food, and soil) using a nano-modified electrochemical cell. The sample containing various heavy metal ions (Pb, As, Cr, Hg, Cd, Co, Ni, Ag) is introduced into the cell where redox reactions occur on nanoelectrode surface. The cell is equipped with nano-particle modified electrodes, including a working electrode, a reference electrode, and a counter electrode. These electrodes facilitate the redox reactions of the heavy metal ions. The electrochemical signals generated are captured by a transducer and amplified. These signals are displayed and measured in terms of current (I), voltage (E), electrochemical impedance, and charge (Q). Various electrochemical techniques are employed to analyze heavy metal include potentiostatic method, galvanostatic method, impedance measurement (using Electrochemical Impedance Spectroscopy (EIS) and voltammetry), and electrochemiluminescence method. These techniques allow for the precise measurement and identification of heavy metal ions in the environmental samples, providing valuable data for monitoring and managing environmental pollution.

3.1.1 Organic/carbon-based nanosensors

Owing to their superior electrical and optical qualities, carbon-based sensors such as carbon nanotubes (CNTs), carbon nanofibers, fullerenes, graphene, and graphitic substances represent potential options for the detection of HMs [25]. Food and aquatic samples are usually treated using organic-based nanoelectrodes to detect metal ions, due to their surface functional groups (**Table 1**).

Inorganic As is carcinogenic, whereas arsenite (As^{+3}) is the most toxic, causing irreversible health problems [83]. Gautam et al. developed a screen-printed carbon electrode (SPE) functionalized with a reduced graphene oxide gold nanoparticle (rGO/AuNP) sensor to detect low concentrations of arsenite (As^{+3}) ion in human serum, with a low detection limit using cyclic and differential pulse voltammetry [84]. In this detection protocol, GO produced through the oxidation of GR, oxygen-containing hydroxyl, carboxyl, and epoxy functional groups. Then, GO is reduced to rGO using a modified Hummers method [85] and AuNPs were synthesized using the Turkevich method [86]. rGO/AuNPs functionalized screen printed carbon electrode (SPCE) was synthesized by sonicating rGO added to citrate-capped AuNPs solution, and finally, rGO/AuNP composite was casted on SPE *via* deposition and ultrasonic irradiation (**Figure 8**) [84]. The rGO/AuNP @SPCE electrochemical sensor exhibited a significant response to As^{+3}, with a detection limit of 10 g/L in standard serum samples and 100 g/L in spiked human serum samples. As^{+3} adsorption is improved by citrate-capped AuNPs, and the porous structure of rGO improves the electrode adsorption and conductivity [84]. Forensic analysis and other applications benefit greatly from the low-cost, sensitive, and selective detection of As^{+3} offered by the rGO/AuNP@SPE sensing platform in human blood assays. Similarly, Ismail et al. developed electrochemical method for detection of As^{3+} ion using silica nanoparticles chemically modified screen-printed carbon electrodes (SiNP/SPCE) (**Figure 8**). The electrochemical responses toward arsenic detection were measured using Cyclic Voltammetry (CV) and linear sweep anodic stripping voltammetry/LSASV/techniques. This method was effective for testing As^{3+} ion in real aquatic samples, with good repeatability and consistency [87].

For highly toxic mercury ion detection, Xia et al. discovered a Ti3C2Tx MXene nanoribbon/carbon nanotube/glass carbon electrode (Ti3C2TxR/CNT/GCE) hybrid

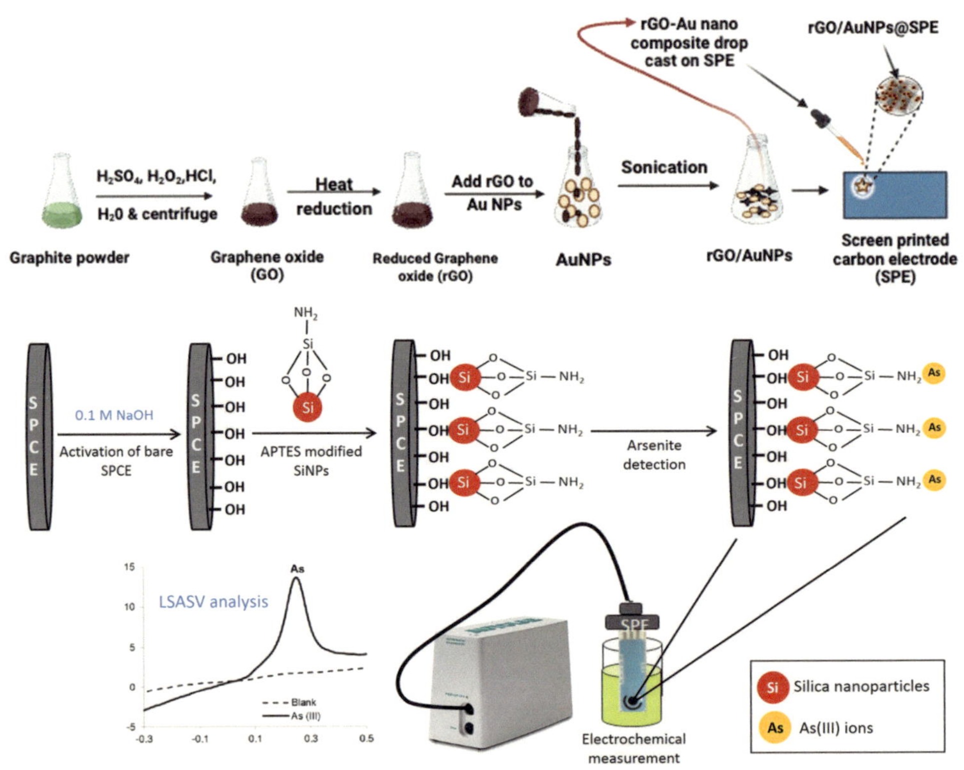

Figure 8.

Schematic representation of rGO/AuNPs modified and APTES modified SiNps coated on SPCE for Arsenite Detection. Schematic illustration of the production and detection process of rGO/AuNPs/SPCE. This figure illustrates the step-by-step preparation and application of a reduced graphene oxide (rGO) and gold nanoparticles (AuNPs) composite on a screen-printed carbon electrode (SPCE) for the detection of arsenite (As(III)) ions. The steps involved Graphene Oxide Preparation, Reduction to rGO, Composite Formation, Activation and Modification of SPCE, Arsenite Detection, and Electrochemical Measurement. This comprehensive method enhances the sensitivity and selectivity of arsenite ion detection in environmental samples, demonstrating the application of advanced nanocomposite materials in environmental monitoring (adapted with permission [87]) (© 2020 by the authors. Licensee MDPI, Basel, Switzerland. This article is an open access article distributed under the terms and conditions of the Creative Commons Attribution (CC BY) license (http://creativecommons.org/licenses/by/4.0/).).

using anodic stripping voltammetry (ASV) as the detection method. During the analysis, Hg^{2+} was preconcentrated on the Ti3C2TxR surface and reduced to Hg^0 because of the electrostatic attraction of Hg^{2+} and the reducing power of Ti3C2TxR toward metal ions. Consequently, the low LOD, potential linearity range, and good sensitivity of the Ti3C2TxR/CNT/GCE sensor demonstrated the effectiveness of the developed method for the sensitive detection of mercury ions without electrodeposition in real water samples [88].

3.1.2 Inorganic nanosensors

The significant health and environmental dangers associated with HMs pollution have led to the development of highly selective and sensitive electrochemical sensors for heavy-metal ions. Metal nanocomposite such as noble metals (gold (Au), silver (Ag), platinum (Pt)) and metal oxides (titanium dioxide (TiO_2), ferric oxide (Fe_2O_3), nickel oxide (NiO), and copper oxide (CuO)) are good candidates for electrochemical sensor construction owing to their excellent properties such as superior conductivity, catalytic effect, mechanical constancy, and biocompatibility. These characteristics are highly dependent on the shape and size of the NPs, which contribute to the design of sensitive electrochemical sensors [89]. As shown in **Table 2**, various studies have investigated novel sensor designs using diverse materials to achieve excellent detection capabilities. For example, ion-imprinted polymers (IIPs) have demonstrated promising outcomes in sensor design, with a sensitivity five orders of magnitude higher than that of nonimprinted polymeric films [94].

Narouei et al. discovered poly(aniline-co-o-aminophenol) nanofiber decorated with gold nanoparticles to promote sensitivity of mercury ion detection. The sensor has a good detection limit and linear range for the selective determination of Hg^{2+} and monitoring HMs contamination in water and food samples [96]. Additionally, the utilization of innovative nanomaterials such as yttrium porous coordination polymers for lead ion detection (Y-BTC) [90] and heterostructure composite MIL-53 (iron)/silver dichromate/glass carbon electrode (MIL-53(Fe)/Ag_2CrO_4/GCE) has been proven for the simultaneous detection of Pb^{2+} and Cu^{2+} metal ions in tap and river water samples [91].

Naseri et al. developed simultaneous determination of Cd^{2+}, Pb^{2+}, Cu^{2+}, and Hg^{2+} ion using a combination of semiconductor device and transition metal oxides. The design of silver nanostructures on screen-printed carbon electrodes (AgNS/SPCE) has led to the discovery of highly efficient sensors for the concurrent analysis of Cd^{2+}, Pb^{2+}, Cu^{2+}, and Hg^{2+} in water samples [92]. These advancements emphasize the significance of noble metal-based materials for producing efficient electrochemical sensors for the detection of HMIs, paving the way for improved water quality control and environmental monitoring. Similarly, Durai and Badhulika reported a step solid-state process for the production of ANC (Al2NiCoO5) nanoflakes for the simultaneous determination of Cu^{2+}, Pb^{2+}, Hg^{2+}, and Cd^{2+} ions in aqueous and simulated blood serum samples. ASDPV analysis revealed that ANC-functionalized GCE (ANC/GCE) sensors showed a low sensitivity and excellent selectivity for HMIs detection in the existence of various interfering contaminants [93]. The sensor's effectively trace low-level detection of HMIs in various water samples and blood serum samples renders it a potential detection technique for various environmental and bioanalytical applications.

Cadmium is a toxic HMs with cancerogenic effects, and this necessitates sensitive sensors to detect trace amounts. Costa et al. reported a convenient electrochemical sensor for Cd^{2+} determination using an ion-imprinted polymer film (IIPs). Using

Nanomaterial sensor	RE and AE	NPs Synthesis method	Determination technique	HMs	Sample	LOD	Linearity range	Reference
rGO/AuNPs	—	Sonication & Ultrasonic irradiation	CV & DPV	As^{3+}	Blood samples	10 ppb	—	[84]
SiNPs/SPCE	—	Drop-casting method	CV & LSASV	As^{3+}	Real water samples	6.2 ppb	5–30 ppb	[87]
Ti_3C_2TxR/CNT/GCE	—	—	ASV	Hg^{2+}	Real water samples	1.043 ppb	2.0 ppb–1.404 ppm	[88]
Y-BTC	Ag|AgCl & Pt wire	Hydrothermal technique	DPV	Pb^{2+}	Coastal water, Pesticide, and pesticide loaded fruit	0.207 ppb	—	[90]
MIL-53(Fe)/ Ag_2CrO_4 /GCE, (MA 1:1)	Ag|AgCl & Pt wire	Solvothermal method	DPV	Pb^{2+}	River & tap water	15.0 ppb	0.794–39.6 ppm	[91]
				Cu^{2+}		9.6 ppb	0.397–19.9 ppm	
AgNS/SPCE	—	Electrodeposition process AgNS on SPCE	DPASV	Cd^{2+}	Rainwater, Lake water, Tap water	0.4 ppb	5-300 ppb	[92]
				Pb^{2+}		2.5 ppb	5-300 ppb	
				Cu^{2+}		7.3 ppb	50-500 ppb	
				Hg^{2+}		0.7 ppb	5-100 ppb	
Al_2NiCoO_5/GCE (ANC/GCE)	Ag|AgCl & Pt wire	Solid-state synthesis	ASDPV	Cu^{2+}	Simulated blood serum, drinking water, tap water	2.61×10^{-6} ppm	1.0×10^{-5}–1.0 ppm	[93]
				Hg^{2+}		2.32×10^{-6} ppm	1.0×10^{-5}–1.0 ppm	
				Cd^{2+}		1.14×10^{-6} ppm	1.0×10^{-5}–1.0 ppm	
				Pb^{2+}		1.54×10^{-6} ppm	1.0×10^{-5}–1.0 ppm	

Nanomaterial sensor	RE and AE	NPs Synthesis method	Determination technique	HMs	Sample	LOD	Linearity range	Reference
Cd(II)-IIP film/ SPCE	Ag/AgCl & Pt electrode	Electro-polymerization	CV	Cd^{2+}	Marine water, drinking and tap water	0.192 ppb	1.124 ppb– 0.135 ppm	[94]
α-Fe_2O_3/NiO/GCE	Ag/AgCl & Pt electrode	—	SWASV	Pb^{2+}	Water and human blood samples	4.144 ppb	10.36 ppb – 0.186 ppm	[95]

Table 2.
Comparison of different organic and inorganic nano sensor for the detection of toxic HMs.

4-aminophenylacetic acid (4-APA) electropolymerization on screen-printed carbon electrodes, the sensor allowed for selective detection of Cd^{2+} ions in water matrices via CV. As a result, the IIP sensor potentially allows for water quality control and heavy metal monitoring owing to its high sensitivity and selectivity for Cd^{2+} detection [94].

Wei et al. discovered α-Fe_2O_3/NiO nanocomposites to cover glassy carbon electrodes (GCEs) in order to detect lead ions with high sensitivity and selectivity. Compared to other HMIs, the low diffusion energy barrier of Pb atoms at the electrode interface permits Pb(II) to flow to the electrode surface and produce elevated stripping signals. The efficiency of this electrochemical sensor in detecting lead ions is due to its insensitivity to other heavy metal interference and shows strong potential for environmental and biological sample analysis [95].

3.1.3 Metal: organic frameworks/MOFs/based nanosensors

Although GCEs are frequently utilized as WE in electrochemical approaches to measure HMIs, their slow enzymatic activity and poor charge-transfer kinetics limit their specificity and selectivity [97]. Thus, MOFs can increase the conductivity as surface modifiers by offering additional active areas for the interaction between the metal and the electrode. By using hydrogen bonding and electrostatic forces to interact with the target analytes, MOF-based electrochemical sensors can enhance their specificity and selectivity. MOFs have limited conductivity because of their bulk organic linkers and micro-size. To improve the conductivity and signal reproducibility, they are combined with highly conductive materials, such as metal oxides or carbon materials, or MOF functionalization, and the size is controlled to the nano level [82, 97]. The different MOFs nanosensors used for the detection of toxic HMs are summarized in **Table 3**.

3.2 Colorimetric analysis

Selection of appropriate molecules as metal-ion-specific receptors is essential for the production of heavy-metal-ion sensors to ensure selectivity, sensitivity, and reproducibility. Green and biochemically synthesized colorimetric nanosensors work by affinity or electrostatic interactions between sensors and HMIs, bringing a change in absorbance. As shown in **Figure 9**, the block diagram of colorimetric sensors uses aggregation or etching approaches to change the color of the noble metal NPs. Colorimetric nanosensors of the noble metal type have a unique absorption band due to the localized surface plasmon resonance (LSPR) processes, making them useful for the selective and sensitive detection of HMIs. Additionally, nanozyme assisted colorimetric sensors use nanosize materials with catalytic properties (activation or inhibition owing to the binding of HMs to nanozymes) to detect trace levels of HMIs [104, 105]. Thus, several colorimetric sensor types for detection of HMs were examined in this section and are summerized in **Table 4**.

To detect As(III) ions in aqueous solutions using LSPR, Ge et al. discovered a dithiothreitol-capped Au nanorod (DTT-AuNR) colorimetric sensor. The detection of As^{3+} ions by the probe was optimized for influencing parameters, such as DTT concentration, pH, reaction time, and NaCl concentration. Under optimized analytical conditions, the probe showed high sensitivity and successfully determined As^{3+} ions in environmental water samples [106].

MOFs sensor/WE	RE & AE	Detection technique	HMs	Sample	LOD	Linearity range	Reference
Cr-BDC/GCE [a]	Ag/AgCl & Platinum	DPASV [b]	Cd^{2+}	Water	0.021 ppb	0–1.124 ppb	[97]
			Pb^{2+}	Water	0.024 ppb	0–2.072 ppb	
			Hg^{2+}	Water	0.124 nM	0–2.006 ppb	
Ag@MOF-199/GCE [c]	Ag/AgCl & Pt	CV	Ni^{2+}	Water	0.355 ppb	0.5869–5.869 ppb	[98]
AuNP@Gr/MOF-5[d]	Ag/AgCl	SWV [e]	Cd^{2+}	Real water sample	0.562 ppb	1.124 ppb–3.372 ppm	[99]
NH2-MIL-53(Fe)/GCE[f]	—	SWASV [g]	Cd^{2+}	Spiked & real water samples	3.372 ppb	4.946 ppb–0.249 ppm	[100]
Chitosan-PCN222/GCE[h]	Ag/AgCl & Platinum	Amperometric method.	Cu^{2+}	Environmental & bioanalytical fields	3.177 ppb	0.025–0.826 ppm	[101]
Co-TIC$_4$R-I/GCE[i]	Ag/AgCl & Platinum	SWASV	Cd^{2+}	tap water, mineral water, and lake water	0.753 ppb	0.0112–1.911 ppm	[102]
			Pb^{2+}		0.559 ppb	0.0136–3.315 ppm	
			Cu^{2+}		0.406 ppb	0.0032–0.635 ppm	
			Hg^{2+}		0.742 ppb	0.160–3.008 ppm	
GCE/MIL-88B(Fe)-NH2[j]	Ag/AgCl & Platinum	DPV	Cd^{2+}	Aqueous solution	0.0224 ppb	2.810 ppb–0.112 ppm,	[103]
			Pb^{2+}		0.0397 ppm	0.062–2.072 ppm	
			Cu^{2+}		0.0242 ppm	0.0381–0.635 ppm	

[a]*Cr-BDC/GCE—Chromium-benzene dicarboxylate/glassy carbon electrode.*
[b]*DPASV: differential pulse anode stripping voltammetry.*
[c]*Ag@MOF-199/GCE —silver-metal organic framework –199/glass carbon electrode.*
[d]*AuNP@Gr/MOF-5—gold nanoparticle—graphene/metal organic framework.*
[e]*SWV—square wave voltammetry.*
[f]*NH2-MIL-53(Fe)/GCE –amine—Materials of Institut Lavoisier –53 (iron)/glass carbon electrode.*
[g]*SWASV—square wave anode stripping voltammetry.*
[h]*Chitosan-PCN222/GCE—Chitosan-porous coordination network 222/glass carbon electrode.*
[i]*Co-TIC$_4$R-I/GCE—cobalt (II) Thiacalix [4] arene/glass carbon electrode.*
[j]*GCE/MIL-88B(Fe)-NH2—glass carbon electrode/materials of Institut Lavoisier -88B (iron)—amine.*

Table 3.
Comparison of different MOFs nanosensors for the detection of toxic HMs.

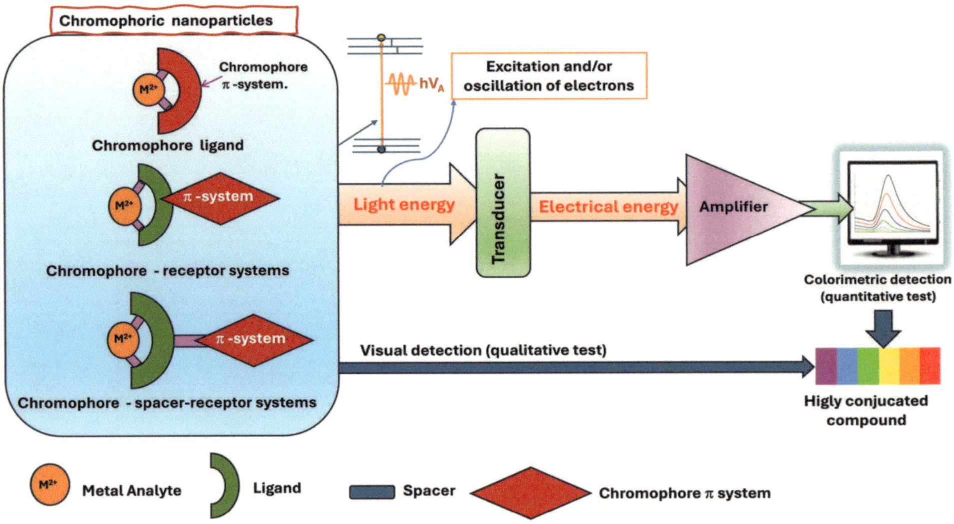

Figure 9.
Schematic representation of chromophoric nanoparticle-based detection systems. This figure illustrates the mechanism of chromophoric nanoparticle-based detection systems. The process highlights the excitation and/or oscillation of electrons in response to light energy, demonstrating the principle behind qualitative and quantitative tests in nanoparticle-based detection technologies. The diagram shows the interaction between metal analytes and various chromophore systems, including chromophore ligand, chromophore-receptor systems, and chromophore-spacer-receptor systems. Upon binding with the ligand, a change in the π-system occurs, which can be detected visually or through a transducer that converts light energy into electrical energy. This electrical signal is then amplified for colorimetric detection of highly conjugated compounds.

Peng et al. develop a colorimetric assay method using gold nanoparticles to detect Pb^{2+} and Cu^{2+} ions in water sample, providing a reference for monitoring metal ions. When Pb^{2+} is added to a solution with a low L-cysteine concentration, it induces Au-S aggregation of nanoparticles due to the interaction of Pb^{2+} and L-cystine. Conversely, Cu^{2+} ions can catalyze the oxidation of L-cysteine, inhibiting the aggregation of AuNPs. Such aggregation and disaggregation of AuNps results in a change in the color of the solution, which can be detected by a colorimetric assay. The UV-vis spectrum of AuNPs is closely related to their aggregation status, allowing the analysis of Cu^{2+} and Pb^{2+} ions in real water samples with good sensitivity and selectivity [107].

Sedghi et al. introduced a colorimetric sensor using Dithizone @titanium dioxide/ poly (tert-Butyl acrylate-acrylic acid (DTZ@TiO2/TBA -AA) nanoparticle for fast detection of Hg^{2+}, Pb^{2+}, and Cd^{2+} in water. Owing to its excellent reproducibility without capacity loss, the sensor provides an economical and effective way to identify heavy metal contamination in water. To create a polymeric nanocomposite, DTZ was included as a chromoionophore in a polymeric membrane containing TBA as a leaving group [108]. This work shows that, under ideal conditions, DTZ in a sensor function as a chromophoric group that enables selective detection of multiple ions of Hg^{2+}, Pb^{2+}, and Cd^{2+} ions in the presence of other interfering metal ions. The colorimetric sensor detection of Hg^{2+}, Pb^{2+}, and Cd^{2+} ions relied on changes in the UV-vis spectrum measurement or detection *via* the color change of the water solution after treatment.

Samuel and Rao investigated the use of cysteine-coated silver nanoparticles (Cys-AgNPs) for individual and combined colorimetric detection of Pb^{2+} and Hg^{2+} ions in water samples. They discovered that the absorption spectrum shifted to shorter wavelengths with decreased intensity when Hg^{2+} ions were complexed with Cys-AgNPs, whereas the absorption spectrum shifted to longer wavelengths with

Colorimetric nano sensors	NPs Synthesis method	Chromophoric ligand	Detection technique	HMs	Sample	LOD	Linearity range	Reference
DTT-AuNRs)	—	DTT	UV-vis absorption band due to LSPR process	As^{3+}	Environmental water samples	2.85 ppb	9.75 ppb–0.75 ppm	[106]
Au NPs	Chemical reduction method	Cystine	Absorbance value at blue color (pink to blue)	Pb^{2+}	Real water sample	10.36 ppb	1.036 ppb–0.518 ppm	[107]
			Absorbance value at pink color (pink to pink)	Cu^{2+}		1.588 ppb	1.588 ppb–19.065 ppb	
DTZ@TiO$_2$/poly (TBA-AA)	Polymerization followed by hydrolysis	Dithizone	Absorbance value at violet color (yellow to violet)	Hg^{2+}	Aqueous solutions	10 ppb	0.01–10 ppm	[108]
			Absorbance value at red color (yellow to red)	Pb^{2+}		20 ppb	20 ppb (visual)	
			Absorbance value at orange color (yellow to orange)	Cd^{2+}		10 ppb	10 ppb (visual)	
Cys-AgNPs	Bio-interfacial sensing approach	Cystine	Absorbance value at white color (yellow to white)	Pb^{2+}	Water samples	10.23 ppb	2.072 ppb –0.518 ppm	[109]
			Absorbance value at dark yellow (yellow to dark yellow)	Hg^{2+}		9.105 ppb	0.20 ppb-0.501 ppm	
PAH–AgNPs	—	PAH	Absorbance value disappearance of yellow color (yellow to color less)	Hg^{2+}	Real water sample	0.2 ppb	0.02 ppm–0.501 ppm	[110]
4-NBT & 4-MBA AgNPs	—	4-NBT & 4-MBA	Absorbance value at purple color (yellow to purple)	Cr^{3+}	Real water & drinking water	0.26 ppb	0.26 ppb-0.104 ppm	[111]
MSA-AuNPs	—	MSA	UV Absorbance value at blue color (red to blue)	Cd^{2+}	Water sample	7.868 ppm	7.868 ppm–22.482 ppm	[112]

Table 4.
Comparison of different colorimetric nanosensors for detection of toxic HMs.

increased absorbance intensity with Pb^{2+} ions [109]. In another study, they developed a colorimetric method for detecting Hg^{2+} ions using poly allylamine hydrochloride (PAH)-modified silver nanoparticles (AgNPs). Following the addition of Hg^{2+} ions, the yellowish-brown color of the synthesized AgNPs changed to a colorless suspension, which was directly proportional to the Hg^{2+} ion concentration. This probe has a low detection limit of 1 nM, making it suitable for real-time analysis [110].

Zhang et al. develop a method for detection of Cr^{3+} using 4-nitrobenzenethiol (4-NBT) and 4-mercaptobenzoic acid (4-MBA) co-functionalized AgNPs. The existence of Cr^{3+} ion brings cooperative metal-ligand interactions to induce AgNP aggregation, developed a purple color from yellow and such color change was controlled using UV-vis spectroscopy or the naked eye. The co-functionalized AgNPs were highly sensitive, and the absorbance ratio was linear with the Cr^{3+} ion concentration. The sensor was tested in drinking water and showed good correlation with ICP-MS results [111].

Chen et al. developed a colorimetric assay for Cd^{2+} ion detection using gold nanoparticles modified with DL-mercapto-succinic acid (MSA-AuNPs). Through aggregation-induced color shift, the probe can detect Cd^{2+} in aqueous solutions with high selectivity [112].

3.3 Fluorescent nanosensors

As shown in **Figure 10**, fluorescence nanosensors can be categorized as fluorescent ligands with identical binding sites and signal sources, fluorescent material conjugated with ligand, fluorescent material combined *via* spacer to ligand systems [113]. The change in intramolecular geometry due to the binding of an analyte (metal) to a

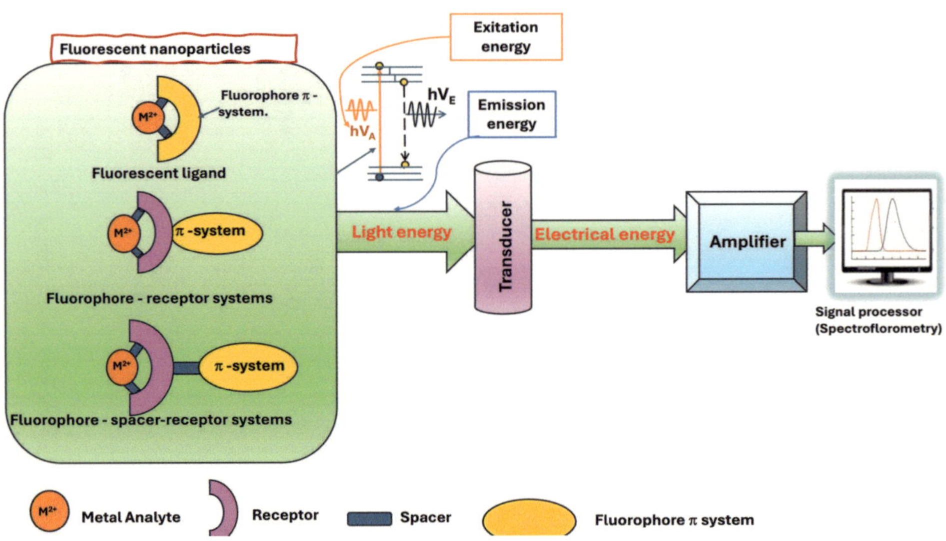

Figure 10.
Schematic representation of fluorescent nanoparticle signal detection system. The figure illustrates the mechanism of fluorescent nanoparticle-based detection systems. The left side of the image depicts three types of fluorescent nanoparticles: fluorophore π-system, fluorophore + receptor systems, and fluorophore + spacer-receptor systems. Each system is represented as a circular entity with labeled components. The right side of the image shows the process flow from excitation energy to emission energy, involving a transducer, an amplifier, and a signal processor (spectrophotometry). This figure highlights the intricate processes involved in fluorescence-based detection and analysis, emphasizing the role of various nanoparticle configurations in enhancing specificity and sensitivity in assays.

Fluorescent nano sensors	NPs synthesis method	Fluorescent ligand	Detection wavelength	HMs	Sample	LOD	Linearity range	Reference
Pyr-NH@SiO2	Amine functionalization followed by peptide coupling reaction	Pyrene	Photoluminescence spectroscopy	Hg^{2+}	Sea water	10 ppb	0-50 ppm	[114]
RA sensor	Schiff base reaction	RA	UV-vis @560 nm & color change (colorless to purple red)	Cu^{2+}	Environmental samples	0.0305 ppb	—	[115]
			Fluorescence detection @580 nm	Hg^{2+}		0.20 ppb	—	
2HA = N-MIL-53(Al) sensor	Schiff base reaction by condensing NH2-Al-MOF and 2-hydroxyacetophenone	NH_2-Al-MOF	Spectrofluorometer@ λ_{EX} = 361 nm & λ_{EM} = 450 nm	AS^{3+}	Real water samples	0.15 ppb	0–0.2 ppm	[116]
APDC-etched CdTe/CdS/SiO2 QDs	Stöber method	APDC	Spectrofluorometer @ λEX: 365 nm and λEM: 622 nm	Cd^{2+}	Seawater, freshwater, and milk samples	0.0387 ppm	0–2 ppm	[117]
N-CDs	Hydrothermal method	l-arginine	Spectrofluorometer @ λEX: 295 nm and λEM: 354 nm	Cd^{2+}	Real food samples	0.0226 ppm,	0–3 ppm,	[118]
				Hg^{2+}		0.0377 ppm	0–10 ppm	
N-SPs	Reaction of TEOS with naphthalene and APTES in ethanol	Naphthalene	Spectrofluorometer	Hg^{2+}	Sea water	1 ppb	0.5 ppb –5.0 ppm	[119]
dsDNA-Fe3O4@ TiO2 NPs	Sol-gel & hydrothermal method	Carboxyl fluorescein labeled DNA	Fluorescent spectrum @525 nm	Pb^{2+}	Polluted water and blood serum samples	10^{-10} ppm	—	[120]

Table 5.
Comparison of different fluorescent nanosensors for the detection of toxic HMs.

receptor results in a significant increase (turn-on) or quenching (turn-off) in fluorescent intensity, indicating the need for an ideal fluorescent nanosensor with strong binding selectivity and signal to avoid environmental interference. The low detection limits and ease of use of fluorescence sensors have drawn considerable attention to their development in recent years [113]. The different fluorescent nanosensors applied for the measuring of toxic HMs are summarized in **Table 5**.

Pyrene-functionalized silica nanoparticles (Pyr-NH@SiO_2 NPs) were reported by Ali et al. as fluorescent chemosensors for measuring Hg^{2+} ions in seawater. After being created and evaluated for Hg^{2+} ion detection, the Pyr-NH@SiO_2 NPs demonstrated excellent selectivity and sensitivity for monitoring environmental pollution. Three steps were involved in the synthesis of Pyr-NH@SiO_2 NPs: First, silica NPs were created by monitoring hydrolysis reaction of tetraethanolorthosilane (TEOS) in ethanol [121]; next, amino-functionalized silica nanoparticles (NH_2@SiO_2 NPs) were produced by condensation with 3-aminopropyltriethoxysilane (APTES); and finally, capped by 1-pyrenecarboxylic acid. The sensor's performance was assessed by evaluating its fluorescence properties in deionized water under ultraviolet light interaction, and its selectivity and sensitivity were determined through photoluminescence spectroscopy [114]. The study highlighted the benefits of using fluorescence-based sensors to detect environmental contaminants, especially mercury ions, due to its quick analysis, high sensitivity, low detection limit, and selectivity. Similarly, Mansha et al. created a highly sensitive mercury detector using a fluorescent chemosensor based on naphthalene-silica nanoparticles in seawater samples. The wide linear detection range (0.5-5 ppb) and sensitivity proved that it accurately detected mercury content in environmental contamination [119].

In another study, Hu et al. discovered fluorometric and colorimetric nanosensors based on rhodamine hydrazide and 9-anthraldehyde (RA) for the effective co-detection of Hg^{2+} and Cu^{2+} ions in environmental samples. A single rhodamine derivative colorimetric sensor was used to assess Cu^{2+} and fluorescently recognize Hg^{2+}, providing a simultaneous detection method for two metal ions. After adding Cu^{2+}, the RA probe exhibited UV-vis absorption at 560 nm, demonstrating a appearance of purple-red color from colorless solution. The fluorescent sensor demonstrated excellent sensitivity to mercury ions, reaching an ultrasensitive level [115].

El-Bindary et al. investigated a fluorescent sensor for the detection of As^{3+} ions in actual water samples by employing a Schiff base reaction and covalent bonding of an amine-functionalized metal (Al) organic framework (NH_2-Al-MOF) and 2-hydroxyacetophenone ligand. The results show that in the presence of interfering species, the fluorescence sensor can quantify As^{3+} ions in a environmetal water sample with excellent LOD and LOQ [116]. In this fluorescent probe's detection of As^{3+} in real sample water, the fluorescence intensity decreased with increasing As^{3+} ion concentration due to fluorescence quenching. Therefore, the hypochromic shift of the fluorescence signal produced was directly proportional to the arsenic concentration, revealing the accurate quantification of arsenic levels in environmental samples.

Chen et al. discovered a unique fluorescent "OFF-ON" quantum dot (QDs) sensor to detect Cd^{2+}ions by etching cadmium telluride/cadmium sulfide/silcon oxide (CdTe/CdS/SiO_2) cores with ammonium pyrrolidine dithiocarbamate (APDC). In the absence of external Cd^{2+} ions, APDC disrupts the Cd-S bond on the QDs and takes Cd to form the APDC-CD complex, which decreases the fluorescence intensity (OFF QDs). APDC-etched CdTe/CdS/SiO2 showed an excellent sensor response, with increased fluorescence intensity associated with increased Cd^{2+} ions amount (ON QDs) due to the reversal of the disrupted Cd-S bond by forming the APDS-Cd complex from the sample Cd ion concentration. Fluorescent QDs detect Cd^{2+} ion concentrations in seawater, freshwater, and milk samples [117].

Tan et al. reported a luminescence method for detecting cadmium and mercury ions in food samples using carbon dots (CDs) as fluorescent detector probes. The probe was synthesized using a hydrothermal method to synthesize nitrogen-doped CDs (N-CDs) for measurement of Cd^{2+} and Hg^{2+}ions. Sodium diphosphate and phytic acid were used as coating agents to detect each ion selectively by avoiding d interfering ions, allowing the sensor to be slaked with Cd^{2+} and Hg^{2+} ions within a short time. Owing to its high selectivity and sensitivity, this sensor has been successfully applied for Cd^{2+} and Hg^{2+} ion detection in fruits and vegetables, providing an innovative assay method for heavy metal ions in foods [118].

Li et al. investigate the fluorescence biosensor, comprised of magnetite/titanium dioxide (Fe_3O_4@TiO_2) core-shell nanosensor modified with a carboxyl fluorescein labeled DNA for effective detection of lead ion (Pb^{2+}) in polluted water and biological samples. Fluorescence biosensors are a potential novel class of biocompatible sensors for measuring HMIs in biological and environmental samples because of their outstanding linearity range, sensitivity, accuracy, recovery, and high adsorbent capacity [119].

4. Approaches to remediation of heavy metal pollution

HMs pollution poses significant environmental and public health risks due to its persistence, bioaccumulation, and toxicity. Effective remediation techniques require innovative, multidisciplinary technologies that combine environmental chemistry and toxicology. As industrialization has increased with the release of HMs, various remediation strategies have been developed, including chemical and biological approaches and subcategories, such as adsorption, coagulation, electrodialysis, oxidation, biological treatment, membrane filtration, ion exchange, photocatalysis, and nanotechnology [122].

Heavy metal contamination is a chemical issue, making conventional physicochemical treatment methods expensive, environmentally damaging, and generating hazardous waste [123]. Biological approaches, including phytoremediation and bioremediation, have emerged as sustainable and eco-friendly solutions. Phytoremediation involves various mechanisms, such as phytoaccumulation, phytostabilization, phytodegradation, phytovolatilization, and hydraulic control, which use the inherent capabilities of plants to mitigate, immobilize, transform, or remove contaminants from the environment. This "green" technology is cost-effective, preserves the local ecosystem, and does not produce large volumes of hazardous waste. The efficiency of phytoremediation can be enhanced using plant growth-promoting rhizobacteria (PGPR), which convert HMs into more soluble and bioavailable forms, making them more accessible for plant uptake [124–126]. Bioremediation approaches that use microorganisms to degrade or transform HMs are preferred to chemical treatments because of their environmental sustainability. Nanomaterials, which are efficient and versatile, have emerged as promising solutions to these problems, making them promising alternatives to conventional methods [126].

4.1 Advanced nanotechnology approach for heavy metal treatment and removal

Researchers have developed nanomaterials with unique physicochemical properties and ease of surface functionalization for heavy metal remediation with the aim of removing or reducing environmental pollution to WHO-accepted levels. These nanomaterials effectively eliminate pollutants like HMs, dyes, and chlorinated organic compounds [19, 127–130].

Research has shown the effectiveness of various nanomaterials, dendrimers, and nanocomposites for heavy metal remediation in wastewater, soil, and other environments [131, 132]. Nanomaterials have high adsorption capabilities and can be easily tailored to target specific contaminants [132]. Magnetite-chitosan composite nanomaterials have shown excellent adsorption for cadmium and copper ion removal [133]. Metal-organic frameworks (MOFs) have also been studied for their ability to remove HMs from aqueous environments [134]. Nanomaterials' high-energy adsorption sites and strong binding ability make them efficient adsorbents. They can also be separated and reused as raw materials, contributing to sustainability and affordability [131].

Nano-adsorbents have shown significant efficiency in heavy metal remediation, with adsorption trends determined by factors such as the ionic radius, mass, and electronegativity [135]. Various methods including chemical precipitation, adsorption, ion exchange, and membrane filtration have been used to remove HMs from aqueous environments. These methods utilize the inherent properties of nano-adsorbents to target and eliminate toxic metal ions, thereby ensuring water safety and quality. The demand for efficient and cost-effective heavy-metal removal solutions is increasing, with research potentially leading to the development of innovative nanomaterials with improved selectivity, regeneration, and scalability, ultimately reducing heavy metal pollution.

4.2 Types of nanomaterials for heavy metal remediation

Nanomaterials are versatile tools for HMs removal, with various categories including organic, polymeric, metal-based, biological, and hybrid materials. Organic nanomaterials offer a high surface area and reactivity, while polymeric nanomaterials provide flexible structures for specific pollutants. Metal-based nanomaterials are known for their catalytic and adsorptive properties. Biological nanomaterials are biodegradable and environment-friendly. Hybrid nanomaterials combine the advantages of different materials to enhance their performance and reduce environmental impact [136]. Combining nanoremediation with transgenic biotechnological approaches can reduce heavy metal contamination in vegetables and the food chain. Nanomaterials, including oxides of various metals, exhibit high selectivity and removal efficacy for heavy-metal pollutants [137, 138].

4.2.1 Carbon-based nanomaterials

Carbon-based nanomaterials such as CNTs and graphene have become effective for removing HMs from the environment because of their large surface area, ease of chemical modification, and strong adsorption capabilities [139]. Surface functionalization enhances their performance; however, potential toxicity and environmental impact must be considered for safe and sustainable use. Activated carbon (AC), produced from coal, wood, and agricultural residues, has a high surface area and mechanical strength, making it suitable for various applications, including catalyst supports, capacitors, electrodes, and adsorbents for removing metal ions, organic pollutants, and gases from water [140].

4.2.1.1 Carbon nanotubes

Carbon nanotubes (CNTs) are cylindrical nanostructures made of graphene sheets, which are known for their tensile strength, electrical conductivity, and chemical stability. They have been used to remove HMs, such as Pb^{2+}, Hg^{2+}, Cd^{2+}, and As^{3+} ions from aqueous solutions. CNTs' large surface area of CNTs provides numerous adsorption sites, making them suitable for environmental remediation, particularly for water treatment. Multiwall carbon nanotubes (MWCNTs) have also emerged with exceptional properties, including a high surface area and conductivity. They effectively adsorb heavy metal ions, such as chromium, lead, and cadmium, through chemical interactions with their functional groups. Plasma-oxidized MWCNTs are particularly effective because of their higher number of oxygenated functional groups. MWCNT composites with compounds such as iron oxide and manganese oxide exhibited enhanced adsorption performance. Functionalized MWCNTs are up to 20 times more effective for metal ion adsorption than unoxidized MWCNTs, making them highly promising for heavy metal remediation in water [139].

4.2.1.2 Graphene nanomaterials

Graphene, a two-dimensional honeycomb lattice with a single layer of carbon atoms, is a powerful material for the removal of HMs from contaminated water. Its derivatives, such as GO and rGO, adsorb HMs *via* surface complexation, electrostatic attraction, and ion exchange processes. The oxygen-containing functional groups on graphene oxide enhance its adsorption capability. Graphene-based materials are effective in removing HMs such as Cr^{6+}, Pb^{2+}, and Hg^{2+} from polluted water and can be engineered to increase selectivity and adsorption capacity [122, 141, 142]. Research has shown that graphene oxide can remove up to 98% of lead ions from water within a short contact time. Reduced graphene oxide (rGO) has high adsorption capacities for silver and chromium ions, making it a promising material for heavy metal remediation. The removal mechanism involves surface complexation, electrostatic attraction, and hydrogen bonding, which facilitate the removal of HMs from contaminated environments [122, 141–143].

4.2.2 Inorganic nanoparticles

Inorganic nanoparticles, such as metal oxides and zero-valent metals, are widely used in heavy metal remediation because of their high reactivity and specificity toward heavy metal ions [144]. Metal oxide nanoparticles, such as TiO_3, ZnO, and Fe_3O_4, have significant adsorption capacities and can form stable complexes with HMs, facilitating their removal from contaminated environments [145, 146]. These nanoparticles operate through surface complexation and redox reactions to adsorb and immobilize HMs such as As^{3+}, Pb^{2+}, Cd^{2+}, and Cr^{6+}. Iron oxide nanoparticles, particularly magnetite, have shown exceptional efficacy in removing arsenic from groundwater, with removal efficiencies exceeding 90% [147–149]. Zero-valent iron nanoparticles (NZVI) are distinguished by their high reactivity, which enables them to reduce and immobilize a broad spectrum of HMs. They have a large surface area and can reduce metal ions to elemental or less toxic forms, which is important for the remediation of contaminated groundwater and soil. NZVI particles have been particularly successful in addressing hexavalent chromium contamination, reducing it to Cr^{3+}, and precipitating as $Cr(OH)_3$ [144, 150–154].

4.3 Factors influencing nanomaterials' effectiveness in heavy metal removal

The efficiency of nanomaterials in removing HMs from contaminated environments is influenced by various factors, which can be optimized to improve the remediation performance [138].

4.3.1 pH of the solution environment

PH impacts the nanomaterial surface charge and heavy metal ionization. Low pH increases the number of metal cations and enhances adsorption onto negatively charged surfaces. Iron oxide nanoparticles adsorb optimally at pH levels where the metal hydroxides are less soluble [138]. The adsorption mechanisms of metal ions on functionalized nano-adsorbents depend on pH of the solution environment. For example, in acidic pH there is competition between metal ions with hydronium ions (H_3O^+) for binding sites on the carboxylate groups ($COO-$). On protonated functional groups ($COOH^+$) at low pH conditions, which reduces the availability of binding sites for metal ions. At a high pH, hydroxylated complexes of functional groups form enhancing the adsorption capacity for metal ions owing to the increased availability of negatively charged sites [138]. Adjusting the pH to promote favorable interactions can improve the adsorption efficiency.

4.3.2 Temperature

Temperature significantly impacts the adsorption and desorption processes, with higher temperatures potentially increasing metal ion mobility and improving adsorption rates. It is crucial to identify the optimal temperature range for the maximum adsorption capacity. For example, lead ion adsorption on CNTs increases with increasing temperature, while endothermic processes, such as Hg^{2+}adsorption on chitosan-alginate nanoparticles (CANPs), show a peak at an optimal temperature (30°C) before declining because of increased metal ion mobility and weakened electrostatic interactions. Endothermic processes, such as Pb(II) adsorption on Fe_3O_4 nano-adsorbents, show improved efficiency with increasing temperature, owing to greater ionic mobility. Understanding these temperature effects is essential for optimizing the adsorption conditions and enhancing the process efficiency [155].

4.3.3 Contact time

Equilibrium adsorption is crucial for efficient removal of HMs from nanomaterials. The rate of adsorption varies depending on the type of nanomaterial and heavy metal involved. The optimal contact time ensures the maximum removal efficiency. Studies have shown that cadmium adsorption on graphene oxide occurs quickly, with the efficiency increasing during the initial phase. The adsorption kinetics of Pb^{2+} and Cd^{2+} ions showed that within the first 20 min, the process achieved completion rates of 100 and 91% for Pb^{2+} and Cd^{2+} ions, respectively. Beyond this initial period, the adsorption rate plateaus indicate the importance of optimizing contact time for effective adsorption [156].

4.3.4 Dosage of nano-adsorbent

The amount of nano-adsorbent in a solution affects its capacity for metal-ion removal. A higher dosage provides more active sites but may also lead to aggregation, reducing the surface area. Balancing the dosage is crucial for efficient remediation and cost-effectiveness. For example, increasing the dosage of nanoscale zero-valent iron (NZVI) particles enhances arsenic removal, but requires careful management to prevent particle agglomeration. The adsorbent dose plays a critical role in the adsorption processes, with an initial increase in enhancing efficiency owing to the greater number of active sites. However, an excessive dosage can lead to nanoparticle agglomeration, reducing the effective surface area and number of accessible active sites, and diminishing the adsorption efficiency. Empirical studies have shown that an optimal dosage threshold exists, and beyond this threshold, the efficiency diminishes owing to nanoparticle agglomeration. Therefore, optimizing the adsorbent dosage is essential for maximizing the adsorption efficiency [157, 158].

4.3.5 Regeneration of adsorbents

Remediation of contaminants involves the regeneration and reuse of nano-adsorbents, reducing costs and ensuring sustainable use. The main approaches include desorption, thermal treatment, and electrochemical methods. Desorption techniques use acids, bases, or chelating agents to desorb HMs from nano-adsorbents, whereas thermal treatment involves heating the nano-adsorbents to desorb and volatilize HMs. Electrochemical methods induce metal ion desorption from the surface of nanomaterials, which is particularly useful for metal nanoparticle (MNP) regeneration. Stability and reusability are crucial for large-scale remediation projects, as they ensure consistent performance and prolong the lifespan of the adsorbents. Research has shown graphene oxide and NZVI particles can be reused for various adsorption and desorption cycles [159].

However, the lifecycle and disposal challenges of nanomaterial-based adsorbents remain unresolved, limiting their commercial application. Successful regeneration requires the appropriate selection of regeneration agents, which must be safe, cost-effective, and nondamaging to the nanomaterial structure. Eluents selectivity depends on the type of nanomaterial and its adsorption mechanism. Effective regeneration uses acids, salts, alkalis, buffer solutions, chelating agents, and deionized water [138].

The recovery rate of the adsorbent (%Ra) and desorption percentage (%D) were determined using the following equations.

$$\%Ra = \frac{m_i}{m_o} \times 100\% \qquad (4)$$

Where: m_0 = initial mass; m_i = dry weight of the adsorbent.

$$\%D = \frac{C_{des}}{C_o} \times 100\% \qquad (5)$$

where C_{des} is the concentration after desorption and C_o is the concentration before desorption.

5. Future prospectives

In this chapter, we gave a comprehensive overview of the advanced state of nanotechnology-based approaches to heavy metal remediation and the application in heavy metal analysis, but they still face challenges in trace detection of specific target ions in complex matrices. Advanced functionalization of nanomaterials (biofunctionalization) has shown potential for improving specific metal analysis. The development of novel solid supports with high stability, and specific adsorption requires optimization of synthesis scheme, surface properties, and geometric arrangement. Hybridization of different types of nanomaterials is a good approach. Future nanomaterial-based metal analysis may integrate *in situ* detection and miniaturization like microfluidics device requiring interdisciplinary research fields are needed to advance new nanomaterial for both laboratory and commercial applications. Besides, nanotechnology holds great promise but requires considerations like scalability, cost, environmental impact, and regulatory approval. Integrating nanotechnology into environmental monitoring and remediation strategies is a significant step toward addressing heavy metal toxicity. Continued research and development are needed to refine effective, cheap, reusable, safe, and sustainable approaches.

6. Conclusion

Heavy metal pollution represents a profound environmental and public health challenge, exacerbated by industrialization and human activities. Monitoring heavy metal concentrations is crucial for environmental preservation and human health protection, necessitating the development of selective and sensitive detection methods.

The urgency of addressing this issue has driven advancements in both detection and remediation technologies and among them nanotechnology is being explored as a new strategy in the fight against heavy metal contamination, offering innovative solutions that surpass traditional methods in sensitivity, selectivity, and efficiency. Nanomaterials have high surface-area-to-volume ratios, catalytic activities, and affinities for HMs. They are also comprised of nanosensors and nano-removers with excellent sensitivity, selectivity, and reactivity.

In summary, researchers have made significant progress in designing nanomaterials capable of detecting heavy metal pollutants in real-world samples. This advanced nanotechnology-based HMI analysis and remediation hold great promise for routine environmental and agricultural sample analysis. Hence, the chapter review serves as a foundation for policy makers, researchers, chemists, and material scientists for designing next-generation nanomaterial sensor for various environmental applications.

Author details

Yimer Seid Ali[1,2†], Ibrahim Shaw[1†], Yang Liu[1] and Chuanpin Chen[1*]

1 Xiangya School of Pharmaceutical Sciences, Central South University, Changsha, Hunan, People's Republic of China

2 Department of Pharmacy, Wollo University College of Medicine and Health Science, Wollo Tertiary Care and Teaching Hospital, Dessie, Ethiopia

*Address all correspondence to: ccpin2000@hotmail.com

†Contributed equally to this book chapter (co-first author).

References

[1] Kinuthia GK et al. Levels of heavy metals in wastewater and soil samples from open drainage channels in Nairobi, Kenya: Community health implication. Scientific Reports. 2020;**10**(1):8434

[2] Dagdag O et al. An overview of heavy metal pollution and control. In: Heavy Metals in the Environment: Management Strategies for Global Pollution. Washington DC: American Chemical Society; 2023. pp. 3-24

[3] Panqing Y et al. Human health-risk assessment of heavy metal–contaminated soil based on Monte Carlo simulation. Scientific Reports. 2023;**13**(1):7033

[4] Dökmeci AH. Environmental impacts of heavy metals and their bioremediation. Heavy Metals-Their Environmental Impacts and Mitigation. IntechOpen; 2020. pp. 1-16

[5] Agency for Toxic Substances and Disease Registry: Substance priority list. 2022. pp. 1-12

[6] Witkowska D, Słowik J, Chilicka K. Heavy metals and human health: Possible exposure pathways and the competition for protein binding sites. Molecules. 2021;**26**(19):1-16

[7] Glicklich D, Shin CT, Frishman WH. Heavy metal toxicity in chronic renal failure and cardiovascular disease: Possible role for chelation therapy. Cardiology in Review. 2020;**28**(6):312-318

[8] Renu K et al. Molecular mechanism of heavy metals (Lead, chromium, arsenic, mercury, nickel and cadmium) - induced hepatotoxicity – A review. Chemosphere. 2021;**271**:129735

[9] Lin X et al. Combined toxicity of heavy metal mixtures in liver cells. Journal of Applied Toxicology. 2016;**36**(9):1163-1172

[10] Ijomone OM et al. The aging brain: Impact of heavy metal neurotoxicity. Critical Reviews in Toxicology. 2020;**50**(9):801-814

[11] Kim HS, Kim YJ, Seo YR. An overview of carcinogenic heavy metal: Molecular toxicity mechanism and prevention. Journal of Cancer Prevention. 2015;**20**(4):232-240

[12] Yuan Y et al. In vitro toxicity evaluation of heavy metals in urban air particulate matter on human lung epithelial cells. Science of the Total Environment. 2019;**678**:301-308

[13] Nemery B. Metal toxicity and the respiratory tract. European Respiratory Journal. 1990;**3**(2):202-219

[14] Bhardwaj JK, Paliwal A, Saraf P. Effects of heavy metals on reproduction owing to infertility. Journal of Biochemical and Molecular Toxicology. 2021;**35**(8):e22823

[15] Alissa EM, Ferns GA. Heavy metal poisoning and cardiovascular disease. Journal of Toxicology. 2011;**2011**:870125

[16] Engwa GA et al. Mechanism and health effects of heavy metal toxicity in humans. Poisoning in the Modern World-New Tricks for An Old Dog. IntechOpen. 2019;**10**:70-90

[17] Bello S et al. Carcinogenic and non-carcinogenic health risk assessment of heavy metals exposure from Shanono and Bagwai artisanal gold mines, Kano state, Nigeria. Scientific African. 2019;**6**:e00197

[18] Inobeme A et al. Recent advances in nanotechnology for remediation of heavy metals. Environmental Monitoring and Assessment. 2023;**195**(1):111

[19] Gong Z et al. Application of nanotechnology in analysis and removal of heavy metals in food and water resources. Nanomaterials. 2021;**11**(7):1792

[20] Briffa J, Sinagra E, Blundell R. Heavy metal pollution in the environment and their toxicological effects on humans. Heliyon. 2020;**6**(9):e04691

[21] Pan L et al. A review of heavy metal pollution levels and health risk assessment of urban soils in Chinese cities. Environmental Science and Pollution Research. 2018;**25**(2):1055-1069

[22] Arora D et al. Unleashing the feasibility of nanotechnology in phytoremediation of heavy metal–contaminated soil: A critical review towards sustainable approach. Water, Air, & Soil Pollution. 2024;**235**(1):57

[23] Yu R et al. Heavy metal pollution and health risk in China. Global Health Journal. 2017;**1**(1):47-55

[24] Abouian Jahromi M, Jamshidi-Zanjani A, Khodadadi Darban A. Heavy metal pollution and human health risk assessment for exposure to surface soil of mining area: A comprehensive study. Environmental Earth Sciences. 2020;**79**(14):365

[25] Waheed A, Mansha M, Ullah N. Nanomaterials-based electrochemical detection of heavy metals in water: Current status, challenges and future direction. TrAC Trends in Analytical Chemistry. 2018;**105**:37-51

[26] Ariño C et al. Voltammetric determination of metal ions beyond mercury electrodes. A review. Analytica Chimica Acta. 2017;**990**:11-53

[27] Tchounwou PB et al. Heavy metal toxicity and the environment. Molecular, clinical and environmental toxicology. Environmental Toxicology. 2012;**3**:133-164

[28] Kumar M, Puri A. A review of permissible limits of drinking water. Indian Journal of Occupational and Environmental Medicine. 2012;**16**(1):40-44

[29] Fizal ANS et al. Assessment of the chemical hazard awareness of petrol tanker driver: A case study. Heliyon. 2019;**5**(8)

[30] Martin S, Griswold W. Human health effects of heavy metals. Environmental Science and Technology Briefs for Citizens. 2009;**15**(5):1-6

[31] Oyem HH, Oyem IM, Usese AI. Iron, manganese, cadmium, chromium, zinc and arsenic groundwater contents of Agbor and Owa communities of Nigeria. Springerplus. 2015;**4**:104

[32] WHO. Nickel in Drinking Water: Background Document for Development of WHO Guidelines for Drinking-Water Quality. Geneva: World Health Organization; 2021. pp. 1-36

[33] Cui Y et al. Assessment of heavy metal contamination levels and health risks in environmental media in the northeast region. Sustainable Cities and Society. 2022;**80**:103796

[34] Balali-Mood M et al. Toxic mechanisms of five heavy metals: Mercury, lead, chromium, cadmium, and arsenic. Frontiers in Pharmacology. 2021;**12**:643972

[35] Jaishankar M et al. Toxicity, mechanism and health effects of

some heavy metals. Interdisciplinary Toxicology. 2014;7(2):60-72

[36] Fairbrother A et al. Framework for metals risk assessment. Ecotoxicology and Environmental Safety. 2007;**68**(2):145-227

[37] Antoine JMR, Fung LAH, Grant CN. Assessment of the potential health risks associated with the aluminium, arsenic, cadmium and lead content in selected fruits and vegetables grown in Jamaica. Toxicology Reports. 2017;**4**:181-187

[38] Huang Z et al. Characteristics of heavy metal accumulation and risk assessment in understory Panax notoginseng planting system. Environmental Geochemistry and Health. 2023;**45**(12):9029-9040

[39] Bayda S et al. The history of nanoscience and nanotechnology: From chemical-physical applications to nanomedicine. Molecules. 2019;**25**(1):112

[40] Wang L et al. A holistic and state-of-the-art review of nanotechnology in solar cells. Sustainable Energy Technologies and Assessments. 2022;**54**:102864

[41] Ekrami E et al. Nanotechnology: A sustainable solution for heavy metals remediation. Environmental Nanotechnology, Monitoring & Management. 2022;**18**:100718

[42] Salek, Maghsoudi A et al. Recent advances in nanotechnology-based biosensors development for detection of arsenic, lead, mercury, and cadmium. International Journal of Nanomedicine. 2021;**16**:803-832

[43] Mekuye B, Abera B. Nanomaterials: An overview of synthesis, classification, characterization, and applications. Nano Select. 2023;**4**(8):486-501

[44] Jose A et al. Porous inorganic nanomaterials: Their evolution towards hierarchical porous nanostructures. Micro. 2024;**4**(2):229-280

[45] Debnath SK, Srivastava R. Drug delivery with carbon-based nanomaterials as versatile nanocarriers: Progress and prospects. Frontiers in Nanotechnology. 2021;**3**:644564

[46] Benjamin SR, Miranda Ribeiro EJ Junior. Graphene-based electrochemical sensors for detection of environmental pollutants. Current Opinion in Environmental Science & Health. 2022;**29**:100381

[47] Anas NAA et al. Development of graphene quantum dots-based optical sensor for toxic metal ion detection. Sensors (Basel). 2019;**19**(18):3850

[48] Kolahdouz M et al. Carbon-related materials: Graphene and carbon nanotubes in semiconductor applications and design. Micromachines (Basel). 2022;**13**(8):1257

[49] Alhalili Z. Metal oxides nanoparticles: General structural description, chemical, physical, and biological synthesis methods, role in pesticides and heavy metal removal through wastewater treatment. Molecules. 2023;**28**(7):3086

[50] Altammar KA. A review on nanoparticles: Characteristics, synthesis, applications, and challenges. Frontiers in Microbiology. 2023;**14**:1155622

[51] Iravani S. Nanomaterials and nanotechnology for water treatment: Recent advances. Inorganic and Nano-Metal Chemistry. 2021;**51**(12):1615-1645

[52] Karunakaran G et al. Biosynthesis of nanoparticles from various biological

sources and its biomedical applications. Molecules. 2023;**28**(11):4527

[53] Sipuka DS et al. Dendrimer–gold nanocomposite-based electrochemical aptasensor for the detection of dopamine. ACS Omega. 2023;**8**(37):33403-33411

[54] Baig N, Kammakakam I, Falath W. Nanomaterials: A review of synthesis methods, properties, recent progress, and challenges. Materials Advances. 2021;**2**(6):1821-1871

[55] Singh J et al. 'Green' synthesis of metals and their oxide nanoparticles: Applications for environmental remediation. Journal of Nanobiotechnology. 2018;**16**:1-24

[56] Yaqoob AA et al. Recent advances in metal decorated nanomaterials and their various biological applications: A review. Frontiers in Chemistry. 2020;**8**:341

[57] Okafor SN et al. Quantitative determination of selected heavy metals and micronutrients in branded dairy products sold in Nigeria using atomic absorption spectroscopy (AAS). African Journal of Pharmaceutical Research and Development. 2024;**16**(1):33-38

[58] Das DD, Sharma N, Chawla PA. Neutron activation analysis: An excellent nondestructive analytical technique for trace metal analysis. Critical Reviews in Analytical Chemistry. 2023:1-17

[59] Povarov VG et al. Quantitative determination of trace heavy metals and selected rock-forming elements in porous carbon materials by the X-ray fluorescence method. ACS Omega. 2021;**6**(38):24595-24601

[60] He X et al. Analysis of four toxic metals in a single rice seed by matrix solid phase dispersion -inductively coupled plasma mass spectrometry. Scientific Reports. 2016;**6**(1):38472

[61] Massadeh AM, El-Rjoob A-WO, Gharaibeh SA. Analysis of selected heavy metals in tap water by inductively coupled plasma-optical emission spectrometry after pre-concentration using Chelex-100 ion exchange resin. Water, Air, & Soil Pollution. 2020;**231**:1-14

[62] Ullah N et al. Nanomaterial-based optical chemical sensors for the detection of heavy metals in water: Recent advances and challenges. TrAC Trends in Analytical Chemistry. 2018;**100**:155-166

[63] Aragay G, Pons J, Merkoçi A. Recent trends in macro-, micro-, and nanomaterial-based tools and strategies for heavy-metal detection. Chemical Reviews. 2011;**111**(5):3433-3458

[64] Zhang B et al. Facile and green fabrication of size-controlled AuNPs/CNFs hybrids for the highly sensitive simultaneous detection of heavy metal ions. Electrochimica Acta. 2016;**196**:422-430

[65] Agoro MA et al. Heavy metals in wastewater and sewage sludge from selected municipal treatment plants in Eastern Cape Province, South Africa. Water. 2020;**12**(10):2746

[66] Qasem NA, Mohammed RH, Lawal DU. Removal of heavy metal ions from wastewater: A comprehensive and critical review. npj Clean Water. 2021;**4**(1):1-15

[67] Ariya P et al. General applications of nanomaterials. Futuristic Trends in Chemical Material Sciences & Nano Technology. 2024;**3**(13):115-134

[68] Shetty SS et al. Environmental pollutants and their effects on human health. Heliyon. 2023;**9**(9):e19496

[69] Yang J et al. Nanomaterials for the removal of heavy metals from wastewater. Nanomaterials. 2019;**9**(3):424

[70] Alshammari BH et al. Organic and inorganic nanomaterials: Fabrication, properties and applications. RSC Advances. 2023;**13**(20):13735-13785

[71] Khan I, Saeed K, Khan I. Nanoparticles: Properties, applications and toxicities. Arabian Journal of Chemistry. 2019;**12**(7):908-931

[72] Malik S et al. Nanomaterials-based biosensor and their applications: A review. Heliyon. 2023;**9**(9):e19929

[73] Pikna Lu, Heželová M, Kováčová Z. Optimization of simultaneous electrochemical determination of Cd (II), Pb (II), Cu (II) and Hg (II) at carbon nanotube-modified graphite electrodes. Journal of Environmental Science and Health, Part A. 2015;**50**(8):874-881

[74] Li Y-K et al. Recent advances in nanomaterials for analysis of trace heavy metals. Critical Reviews in Analytical Chemistry. 2021;**51**(4):353-372

[75] Li M et al. Nanostructured sensors for detection of heavy metals: A review. ACS Sustainable Chemistry & Engineering. 2013;**1**(7):713-723

[76] March G, Nguyen TD, Piro B. Modified electrodes used for electrochemical detection of metal ions in environmental analysis. Biosensors. 2015;**5**(2):241-275

[77] Song H et al. Carbon nanomaterials-based electrochemical sensors for heavy metal detection. Critical Reviews in Analytical Chemistry. 2022:1-20

[78] Pandey P et al. Application of thiolated gold nanoparticles for the enhancement of glucose oxidase activity. Langmuir. 2007;**23**(6):3333-3337

[79] Pérez-López B, Merkoçi A. Nanoparticles for the development of improved (bio) sensing systems. Analytical and Bioanalytical Chemistry. 2011;**399**:1577-1590

[80] Sulthana SF et al. Electrochemical sensors for heavy metal ion detection in aqueous medium: A systematic review. ACS Omega. 2024;**9**(24):25493-25512

[81] Vasudevan M et al. A comprehensive review on biopolymer mediated nanomaterial composites and their applications in electrochemical sensors. Critical Reviews in Analytical Chemistry. 2022:1-24

[82] Hasan Alzaimoor EF, Khan E. Metal-organic frameworks (MOFs)-based sensors for the detection of heavy metals: A review. Critical Reviews in Analytical Chemistry. 2023:1-22

[83] WHO. WHO guidelines approved by the guidelines review committee. In: Guidelines for Drinking-Water Quality: Fourth Edition Incorporating the First and Second Addenda. Geneva: World Health Organization; 2022

[84] Gautam V et al. Trace level electrochemical analysis of arsenite in human serum utilising rGO/AuNPs based sensor platform. International Journal of Environmental Analytical Chemistry. 2024;**104**(8):1881-1895

[85] Gautam V et al. Silicon nanowires/reduced graphene oxide nanocomposite based novel sensor platform for detection of cyclohexane and formaldehyde.

Materials Science in Semiconductor Processing. 2021;**123**:105571

[86] Kimling J et al. Turkevich method for gold nanoparticle synthesis revisited. The Journal of Physical Chemistry B. 2006;**110**(32):15700-15707

[87] Ismail S et al. Electrochemical detection of Arsenite using a silica nanoparticles-modified screen-printed carbon electrode. Materials (Basel). 2020;**13**(14):3168

[88] Xia Y et al. Innovative strategy based on novel Ti3C2Tx MXenes nanoribbons/carbon nanotubes hybrids for anodic stripping voltammetry sensing of mercury ion. Sensors and Actuators B: Chemical. 2022;**355**:131247

[89] Ghalkhani M et al. Application of nanomaterials in development of electrochemical sensors and drug delivery systems for anticancer drugs and cancer biomarkers. Critical Reviews in Analytical Chemistry. 2022;**52**(3):481-503

[90] Patil SS et al. Ultrasensitive and selective electrochemical sensor based on yttrium benzenetricarboxylate porous coordination polymer (Y-BTC) for detection of Pb2+ from bio-analytes. ECS Journal of Solid State Science and Technology. 2023;**12**(5):057002

[91] Guo X et al. Sensitive, selective and simultaneous monitor of multiple heavy metals in environment using a low-cost MIL-53(Fe)/Ag2CrO4 modified GCE sensor. Journal of the Electrochemical Society. 2022;**169**(9):097508

[92] Naseri M et al. A robust electrochemical sensor based on butterfly-shaped silver nanostructure for concurrent quantification of heavy metals in water samples. Electroanalysis. 2023;**35**(2):e202200114

[93] Durai L, Badhulika S. Stripping voltammetry and chemometrics assisted ultra-selective, simultaneous detection of trace amounts of heavy metal ions in aqua and blood serum samples. Sensors and Actuators Reports. 2022;**4**:100097

[94] Costa M et al. Disposable electrochemical sensor based on ion imprinted polymeric receptor for Cd(II) ion monitoring in waters. Sensors and Actuators B: Chemical. 2023;**383**:133559

[95] Wei J et al. Highly sensitive and selective electrochemical detection of Pb (II) in serum via an α-Fe2O3/NiO heterostructure: Evidence from theoretical calculations and adsorption investigation. Sensors and Actuators B: Chemical. 2021;**344**:130295

[96] Narouei FH et al. Highly sensitive mercury detection using electroactive gold-decorated polymer nanofibers. Sensors and Actuators B: Chemical. 2021;**329**:129267

[97] Deore KB et al. Chromium-benzenedicarboxylates metal organic framework for supersensitive and selective electrochemical sensor of toxic Cd2+, Pb2+, and Hg2+ metal ions: Study of their interactive mechanism. Journal of the Electrochemical Society. 2023;**170**(4):046505

[98] Bodkhe GA et al. Ag@ MOF-199 metal organic framework for selective detection of nickel ions in aqueous media. Ceramics International. 2023;**49**(4):6772-6779

[99] Ganash AA, Alotaibi MM. An electrochemical sensor for cadmium ion detection that is based on an AuNP-functionalized metal–organic framework-graphene nanocomposite. Journal of Materials Research. 2022;**37**(23):4368-4380

[100] Garg N, Deep A, Sharma AL. Sensitive detection of cadmium using amine-functionalized Fe-MOF by anodic stripping voltammetry. Industrial & Engineering Chemistry Research. 2023;**62**(25):9735-9746

[101] Xu Z et al. Selective sensing of copper ions by mesoporous porphyrinic metal–organic framework Nanoovals. Analytical Chemistry. 2020;**92**(2):2201-2206

[102] Guo T-T et al. Sulfur-bridged Co(II)-thiacalix[4]arene metal-organic framework as an electrochemical sensor for the determination of toxic heavy metals. Inorganic Chemistry. 2023;**62**(11):4485-4494

[103] Tran LT et al. MIL-88B (Fe)-NH 2: An amine-functionalized metal–organic framework for application in a sensitive electrochemical sensor for Cd 2+, Pb 2+, and Cu 2+ ion detection. RSC Advances. 2023;**13**(32):21861-21872

[104] Fakayode SO et al. Electrochemical and colorimetric nanosensors for detection of heavy metal ions: A review. Sensors. 2023;**23**(22):9080

[105] Xu X et al. Nanomaterial-based sensors and strategies for heavy metal ion detection. Green Analytical Chemistry. 2022;**2**:100020

[106] Ge K et al. A colorimetric probe based on functionalized gold nanorods for sensitive and selective detection of as (III) ions. Sensors. 2018;**18**(7):2372

[107] Peng LJ et al. Determination of copper and lead ions using gold nanoparticles as a colorimetric probe with the aid of cysteine. Separation Science Plus. 2022;**5**(9):529-538

[108] Sedghi R et al. Design and synthesis of colorimetric sensor based on dithizone@TiO2/poly (tert-butyl acrylate-acrylic acid) nanocomposite for fast visual detection of mercury, lead and cadmium ions in aqueous media. Environmental Nanotechnology, Monitoring & Management. 2022;**18**:100670

[109] Samuel VR, Rao KJ. A rapid colorimetric dual sensor for the detection of mercury and lead ions in water using cysteine capped silver nanoparticles. Chemical Physics Impact. 2023;**6**:100161

[110] Samuel VR, Rao KJ. A colorimetric sensor for the stable and selective detection of mercury ions using PAH-capped silver nanoparticles in an aqueous medium. Applied Nanoscience. 2024;**14**(1):33-42

[111] Zhang Z et al. Colorimetric detection of Cr 3+ in aqueous solution based on cofunctionalized silver nanoparticles modified with 4-nitrobenzenethiol and 4-mercaptobenzoic acid. Nano. 2015;**10**(07):1550095

[112] Chen N et al. Colorimetric detection of cadmium ions using DL-mercaptosuccinic acid-modified gold nanoparticles. Journal of Nanoscience and Nanotechnology. 2016;**16**(1):840-843

[113] Kowser Z et al. A brief review on novel pyrene based fluorometric and colorimetric chemosensors for the detection of Cu 2+. Materials Chemistry Frontiers. 2021;**5**(5):2173-2200

[114] Ali S et al. Cost-effective and selective fluorescent chemosensor (Pyr-NH@ SiO2 NPs) for mercury detection in seawater. Nanomaterials. 2022;**12**(8):1249

[115] Hu JP et al. Fluorescence and colorimetric-based sensing of Hg2+

and Cu2+ using rhodamine derivative. Applied Organometallic Chemistry. 2023;**37**(2):e6938

[116] El-Bindary MA et al. Synthesis of novel fluorescent sensor based on a modified amino Al-MOF for rapid, sensitive, and selective detection of arsenic in aqueous solution. Applied Organometallic Chemistry. 2023;**37**(6):e7102

[117] Chen J et al. An "off-on" fluorescent nanosensor for the detection of cadmium ions based on APDC-etched CdTe/CdS/SiO(2) quantum dots. Heliyon. 2024;**10**(5):e26980

[118] Tan Q et al. One-step synthesis of highly fluorescent carbon dots as fluorescence sensors for the parallel detection of cadmium and mercury ions. Frontiers in Chemistry. 2022;**10**:1005231

[119] Mansha M et al. Naphthalene-based silica nanoparticles as a highly sensitive fluorescent chemosensor for mercury detection in real seawater. Journal of Molecular Liquids. 2023;**374**:121294

[120] Li J et al. Detecting Pb2+ by a 'turn-on' fluorescence sensor based on DNA functionalized magnetic nanocomposites. Nanotechnology. 2022;**33**(7):075603

[121] Singh LP et al. Preparation of silica nanoparticles and its beneficial role in cementitious materials. Nanomaterials and Nanotechnology. 2011;**1**:9

[122] Ahmad SZN et al. Adsorptive removal of heavy metal ions using graphene-based nanomaterials: Toxicity, roles of functional groups and mechanisms. Chemosphere. 2020;**248**:126008

[123] Roy Chowdhury A et al. Heavy metal pollution and remediation. In: Green Chemistry. Elsevier; 2018. pp. 359-373

[124] Ashraf MA et al. Advances in microbe-assisted reclamation of heavy metal contaminated soils over the last decade: A review. Journal of Environmental Management. 2017;**198**:132-143

[125] Muthusaravanan S et al. Phytoremediation of heavy metals: Mechanisms, methods and enhancements. Environmental Chemistry Letters. 2018;**16**:1339-1359

[126] Akün ME. Heavy metal contamination and remediation of water and soil with case studies from Cyprus. Heavy Metal Toxicity in Public Health. IntechOpen. 2020:3

[127] Guerra FD et al. Nanotechnology for environmental remediation: Materials and applications. Molecules. 2018;**23**(7):1760

[128] Janani R et al. Advancements in heavy metals removal from effluents employing nano-adsorbents: Way towards cleaner production. Environmental Research. 2022;**203**:111815

[129] Ganie AS et al. Nanoremediation technologies for sustainable remediation of contaminated environments: Recent advances and challenges. Chemosphere. 2021;**275**:130065

[130] Jawed A, Saxena V, Pandey LM. Engineered nanomaterials and their surface functionalization for the removal of heavy metals: A review. Journal of Water Process Engineering. 2020;**33**:101009

[131] Chaudhary S et al. Detection and removal of heavy metals from wastewater using nanomaterials.

Pollutants and Water Management: Resources, Strategies and Scarcity. Willy online Library. 2021:241-272

[132] Burakov AE et al. Adsorption of heavy metals on conventional and nanostructured materials for wastewater treatment purposes: A review. Ecotoxicology and Environmental Safety. 2018;**148**:702-712

[133] Hu C et al. Adsorption characterization of Cu (II) and Cd (II) by a magnetite–chitosan composite: Kinetic, thermodynamic and equilibrium studies. Polymers. 2023;**15**(12):2710

[134] Kobielska PA et al. Metal–organic frameworks for heavy metal removal from water. Coordination Chemistry Reviews. 2018;**358**:92-107

[135] Wadhawan S et al. Role of nanomaterials as adsorbents in heavy metal ion removal from waste water: A review. Journal of Water Process Engineering. 2020;**33**:101038

[136] Tahoon MA et al. Effective heavy metals removal from water using nanomaterials: A review. PRO. 2020;**8**(6):645

[137] Kumar S et al. Hazardous heavy metals contamination of vegetables and food chain: Role of sustainable remediation approaches-A review. Environmental Research. 2019;**179**:108792

[138] Singh S et al. Mechanism and kinetics of adsorption and removal of heavy metals from wastewater using nanomaterials. Environmental Chemistry Letters. 2021;**19**(3):2351-2381

[139] Baby R, Saifullah B, Hussein MZ. Carbon nanomaterials for the treatment of heavy metal-contaminated water and environmental remediation. Nanoscale Research Letters. 2019;**14**(1):341

[140] Dong L et al. Preparation, characterization, and application of macroporous activated carbon (MAC) suitable for the BAC water treatment process. Science of the Total Environment. 2019;**647**:1359-1367

[141] Samuel MS et al. Adsorption of Pb (II) from aqueous solution using a magnetic chitosan/graphene oxide composite and its toxicity studies. International Journal of Biological Macromolecules. 2018;**115**:1142-1150

[142] Zhang K et al. Synthesis of reduced graphene oxide/NiO nanocomposites for the removal of Cr (VI) from aqueous water by adsorption. Microporous and Mesoporous Materials. 2018;**255**:7-14

[143] Sahraei R, Ghaemy M. Synthesis of modified gum tragacanth/graphene oxide composite hydrogel for heavy metal ions removal and preparation of silver nanocomposite for antibacterial activity. Carbohydrate Polymers. 2017;**157**:823-833

[144] Fajardo C et al. Heavy metals immobilization capability of two iron-based nanoparticles (nZVI and Fe3O4): Soil and freshwater bioassays to assess ecotoxicological impact. Science of the Total Environment. 2019;**656**:421-432

[145] Cai C et al. Utilization of nanomaterials for in-situ remediation of heavy metal (loid) contaminated sediments: A review. Science of the Total Environment. 2019;**662**:205-217

[146] Hua M et al. Heavy metal removal from water/wastewater by nanosized metal oxides: A review. Journal of Hazardous Materials. 2012;**211**:317-331

[147] Qian W et al. A porous biochar supported nanoscale zero-valent iron material highly efficient for the simultaneous remediation of

cadmium and lead contaminated soil. Journal of Environmental Sciences. 2022;**113**:231-241

[148] Priyadarshni N, Nath P, Chanda N. Sustainable removal of arsenate, arsenite and bacterial contamination from water using biochar stabilized iron and copper oxide nanoparticles and associated mechanism of the remediation process. Journal of Water Process Engineering. 2020;**37**:101495

[149] Wei Y et al. Efficient removal of arsenic from groundwater using iron oxide nanoneedle array-decorated biochar fibers with high Fe utilization and fast adsorption kinetics. Water Research. 2019;**167**:115107

[150] Tarekegn MM, Hiruy AM, Dekebo AH. Nano zero valent iron (nZVI) particles for the removal of heavy metals (Cd 2+, Cu 2+ and Pb 2+) from aqueous solutions. RSC Advances. 2021;**11**(30):18539-18551

[151] Pasinszki T, Krebsz M. Synthesis and application of zero-Valent iron nanoparticles in water treatment, environmental remediation, catalysis, and their biological effects. Nanomaterials (Basel). 2020;**10**(5):917

[152] Fu R et al. The removal of chromium (VI) and lead (II) from groundwater using sepiolite-supported nanoscale zero-valent iron (S-NZVI). Chemosphere. 2015;**138**:726-734

[153] Zhao X et al. An overview of preparation and applications of stabilized zero-valent iron nanoparticles for soil and groundwater remediation. Water Research. 2016;**100**:245-266

[154] Su H et al. Remediation of hexavalent chromium contaminated soil by biochar-supported zero-valent iron nanoparticles. Journal of Hazardous Materials. 2016;**318**:533-540

[155] Dubey R, Bajpai J, Bajpai AK. Chitosan-alginate nanoparticles (CANPs) as potential nanosorbent for removal of Hg (II) ions. Environmental Nanotechnology, Monitoring & Management. 2016;**6**:32-44

[156] Hasanzadeh R et al. Effective removal of toxic metal ions from aqueous solutions: 2-Bifunctional magnetic nanocomposite base on novel reactive PGMA-MAn copolymer@Fe(3)O(4) nanoparticles. Journal of Colloid and Interface Science. 2017;**490**:727-746

[157] Padmavathy KS, Madhu G, Haseena PV. A study on effects of pH, adsorbent dosage, time, initial concentration and adsorption isotherm study for the removal of hexavalent chromium (Cr (VI)) from wastewater by magnetite nanoparticles. Procedia Technology. 2016;**24**:585-594

[158] Garg UK et al. Removal of hexavalent chromium from aqueous solution by agricultural waste biomass. Journal of Hazardous Materials. 2007;**140**(1-2):60-68

[159] Li J et al. Nanoscale zero-valent iron particles supported on reduced graphene oxides by using a plasma technique and their application for removal of heavy-metal ions. Chemistry, An Asian Journal. 2015;**10**(6):1410-1417